AF684264

Copyright © 2024 Kelsie Keir

All rights reserved. No part of this publication may be reproduced, distributed, or transmitted in any form or by any means, including photocopying, recording, or other electronic or mechanical methods, without the prior written permission of the publisher, except in the case of brief quotations embodied in critical reviews and certain other noncommercial uses permitted by copyright law.

ISBN

ISBN: 978-1-964079-00-4

CISSP STUDY GUIDE

Disclaimer

The information provided in this CISSP exam prep guide is intended for general informational and educational purposes only. It does not constitute legal, professional security, or exam preparation advice.

While the authors and publishers strive to ensure the accuracy and up-to-date nature of the information, the cybersecurity field and the CISSP exam content evolve rapidly. We make no guarantees, representations, or warranties of any kind, either express or implied, regarding the completeness, accuracy, timeliness, reliability, suitability, or availability of this guide's content for any particular purpose. You rely on this information at your own risk.

This guide is not a substitute for obtaining qualified professional advice on cybersecurity, risk management, or legal matters. Laws and regulations differ across jurisdictions. You should always consult with appropriate professionals for advice specific to your individual situation.

The authors and publishers of this guide are not affiliated with or endorsed by (ISC)². The CISSP certification is a registered trademark of (ISC)². Use of (ISC)² trademarks does not imply endorsement of this guide or its contents.

Practice exams, questions, and simulations in this guide are designed to aid your understanding of CISSP concepts but should not be considered predictive of the actual CISSP exam. (ISC)² does not release actual exam questions, and attempting to memorize or replicate practice questions is not a substitute for comprehensive understanding of the material.

The authors and publishers of this book shall not be liable under any circumstances or legal theory for any special, incidental, indirect, or consequential damages arising out of or in connection with the use of this guide or its contents.

We reserve the right to update, modify, or amend this disclaimer and the contents of the guide at any time. Your continued use of the guide after any changes constitutes your acceptance of the revised terms.

TABLE OF CONTENTS

Welcome

Picture your ideal career in cybersecurity. Maybe you're protecting sensitive data from relentless hackers, or designing secure systems for a cutting-edge tech company. Perhaps you see yourself as a leader, shaping the cybersecurity strategies that keep organizations safe. The CISSP certification is a major milestone on that journey.

This isn't just about passing another exam. This is about validating your knowledge, proving your determination, and unlocking doors to incredible opportunities. Yes, the CISSP is demanding. But, like anything worthwhile, the rewards match the effort.

Think of this book as your roadmap. We'll break down the exam's core concepts, from risk management to network security. It won't always be easy, but we'll tackle it together, one step at a time. I'll share insights and strategies from those who've aced the exam, plus plenty of real-world scenarios to test your understanding along the way.

I can't guarantee success – that's ultimately up to you. But I can promise you this: if you commit to the journey, put in the hours, and don't give up, you'll transform your theoretical knowledge into practical expertise. With the CISSP credential behind your name, the possibilities become limitless.

Ready to start? Let's dive into the world of cybersecurity!
The CISSP certification isn't just another line on your resume – it's a badge of expertise, recognized worldwide. Organizations of all sizes, from tech giants to government agencies, seek out CISSP holders. Why? They understand that this credential signifies a deep understanding of cybersecurity principles and a commitment to continuous learning in this ever-evolving field.
Let's unpack the value of CISSP, starting from the organization's perspective:

- Proven Competence: When an organization hires a CISSP, they're not taking a gamble on someone's potential. The rigorous exam validates that you possess the skills to design, implement, and manage robust cybersecurity programs aligned with global best practices.
- Reduced Risk: Cyberthreats cost companies millions each year. CISSP professionals are equipped to proactively identify vulnerabilities, mitigate risks, and respond effectively to incidents. That translates to a more secure organization and greater peace of mind.
- Competitive Edge: In many industries, cybersecurity is no longer a "nice to have" – it's a critical factor for winning contracts, complying with regulations, and building client trust. The CISSP credential demonstrates an organization's serious commitment to protecting sensitive data.

Now, what about the benefits for you as an individual?

- Increased Earning Potential: CISSP holders are among the highest-paid cybersecurity professionals. This certification opens doors to more lucrative positions and greater compensation potential.
- Expanded Career Opportunities: The CISSP qualifies you for a range of cybersecurity roles – from technical specialist to leadership positions like Chief Information Security Officer (CISO). It unlocks possibilities across industries and sectors.

- Professional Recognition: Earning your CISSP places you in an elite group of professionals. It's a testament to your knowledge, dedication, and the ability to think strategically about cybersecurity.

The bottom line? The investment you make in preparing for and passing the CISSP exam pays dividends throughout your career. Let's dive deeper into the specific knowledge domains that the CISSP validates... Think of cybersecurity like protecting a castle. High walls alone aren't enough – you need layers of defenses, from archers on the battlements to a well-guarded treasury deep within. This is the essence of the cybersecurity model known as "defense in depth." But before we dive into that, let's lay the groundwork with the fundamental goals of information security.

The CIA Triad

Confidentiality, Integrity, and Availability – these three pillars underpin everything we do in cybersecurity.

Confidentiality: Think "secrecy." We restrict access to sensitive data, like customer records, financial information, or company trade secrets, so only authorized personnel can view it.
Integrity: Simply put, this means ensuring data remains accurate and hasn't been tampered with. Whether accidental or a malicious attack, changes to critical systems or data need to be prevented or detected.
Availability: Systems and the data they hold must be accessible to those who need them, when they need them. Think of a website crashing during a peak shopping season – that's an availability problem.
The Ever-Changing Threat Landscape

Cybersecurity isn't a one-time fix. It's a continuous battle against a relentless and rapidly evolving range of threats:

Hackers: From lone individuals launching ransomware attacks to state-sponsored groups targeting critical infrastructure, the methods and motives vary, but the damage potential is high.
Malware: Viruses, worms, and Trojan horses are just a few examples of malicious software designed to steal data, disrupt systems, or take control for nefarious purposes.
Social Engineering: Tricking people into giving up sensitive information (think phishing emails) is often the easiest way for attackers to gain a foothold in a system.
Zero Trust: Assume Nothing

Traditionally, organizations focused on building strong perimeter defenses, like a firewall. The "zero trust" model flips this inside out. It assumes no user, device, or network traffic should be inherently trusted, even if they're within your organization's walls. It emphasizes continuous authentication and authorization.

Defense in Depth: The Layered Approach

This is where we circle back to our castle analogy. Defense in depth means having multiple, overlapping security measures. Think firewalls, encryption, intrusion detection systems, strict access controls, and employee awareness training. Even if one defense fails, others remain in place.

The CISSP exam will test your deep understanding of these core concepts and how to apply them in real-world situations. Moving forward, we'll unpack each area in greater detail...

Let's start by understanding the organization behind the CISSP – the International Information System Security Certification Consortium, otherwise known as (ISC)². This non-profit organization sets the standards, administers the exams, and oversees professional development for the respected CISSP credential.

Now, let's dig into the nuts and bolts of the CISSP exam itself:

- Format: The CISSP exam is delivered in a Computerized Adaptive Testing (CAT) format. This means the questions get progressively harder or easier based on how you're performing. It's more efficient than a traditional fixed-question exam and ensures your score accurately reflects your knowledge base.
- Length and Questions: The exam has a maximum of 150 questions (though you might get fewer if you reach the required competency level early). You'll have three hours to complete it.
- Question Types: Expect primarily multiple-choice questions, with some scenario-based formats demanding you analyze a situation and choose the best course of action.
- Scoring: The CISSP isn't graded on a percentage basis. You need a scaled score of 700 out of 1000 to pass. (ISC)² doesn't release details of the exact scoring algorithm, and there's no penalty for incorrect answers – so answer every question, even if you're unsure!

Keeping Your CISSP Current: Recertification

Earning the CISSP is a major accomplishment, but it doesn't end there. To maintain your certification, you'll need to engage in continuing education and earn CPEs (Continuing Professional Education) credits. Here's the breakdown:

- CPE Requirements: You need to earn 40 CPEs each year and a minimum of 120 CPEs over a three-year cycle. These credits can be gained through activities like attending conferences, taking courses, teaching cybersecurity, or publishing research.
- Annual Maintenance Fees (AMF): In addition to CPEs, there's an annual fee to keep your certification active.

Think of recertification as proof of your ongoing commitment to staying current in the ever-evolving field of cybersecurity.

Moving forward, we'll dive into the meat and potatoes of the exam – the eight domains of the CISSP Common Body of Knowledge (CBK)...

Think of these as the foundational pillars supporting your entire cybersecurity knowledge base. Here's a quick overview to get us started:

Domain 1: Security and Risk Management We'll cover concepts like governance, compliance, risk assessment methodologies, and how to develop security policies that actually make sense for the business.

Domain 2: Asset Security How do you classify and protect data? What's the deal with data ownership? This section focuses on the full lifecycle of sensitive assets, from creation to disposal.

Domain 3: Security Architecture and Engineering Time to get technical! We'll explore cryptography, design principles for secure systems, and how to select appropriate security controls for different IT environments.

Domain 4: Communication and Network Security From network topologies and protocols to firewalls and intrusion detection systems, this domain is all about protecting the flow of information across your network.

Domain 5: Identity and Access Management (IAM) Controlling who can access what, and how! Here we'll cover authentication, authorization, access control models, and those pesky privileged accounts that can be both powerful and risky.

Domain 6: Security Assessment and Testing Vulnerability scanning, penetration testing, audits...this domain is about knowing how to find the chinks in your organization's cyber armor and then fixing them.

Domain 7: Security Operations This is where the rubber meets the road. Incident response, continuous monitoring, disaster recovery, and all the day-to-day activities that keep your organization secure will be our focus here.

Domain 8: Software Development Security Security needs to be baked in from the start! We'll discuss the secure software development lifecycle (SDLC), secure coding practices, and how to address vulnerabilities in existing applications.
Ready to start our deep dive? Let's tackle Domain 1 – Security and Risk Management.

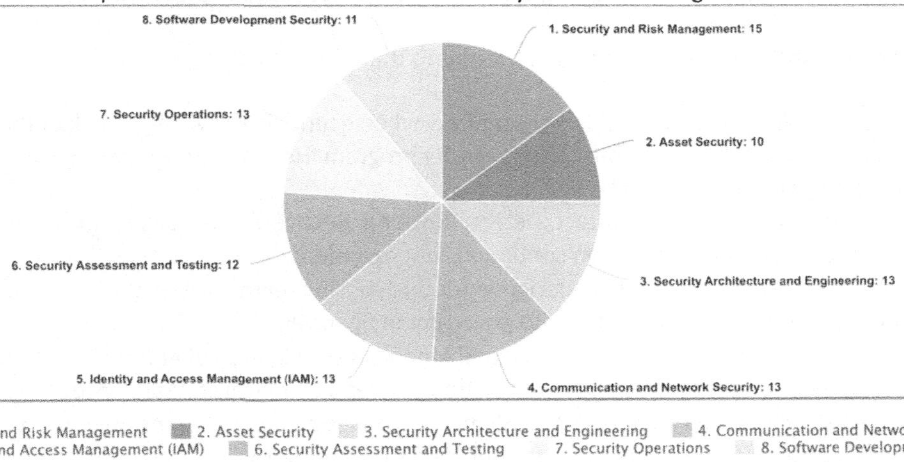

I. Security and Risk Management

Picture walking a tightrope. With each step, you carefully balance immediate risks against the rewards of reaching your goal. Security and risk management is the cybersecurity version of that tightrope walk. It's about finding that delicate balance between protecting your organization's assets while still enabling it to function and innovate.

This section of the CISSP exam deep dives into the heart of making smart cybersecurity decisions. It's not just about technical solutions. Think of it like this: understanding threats and vulnerabilities is like identifying the strong winds buffeting your tightrope. Risk analysis is about figuring out how likely those winds are to knock you off balance, and what the consequences might be if you fall.

We'll cover key concepts like risk assessments, business continuity planning, and how to develop security policies that align with your organization's goals. It might seem a bit dry at first, but trust me, this is where you gain the strategic understanding that sets apart a true CISSP professional.

Ready to start making calculated moves? Let's dive into the principles of security and risk management.

Let's break down some of the most widely recognized cybersecurity frameworks! Think of them as different blueprints for building a robust cybersecurity program. Here's your comparative analysis:

NIST Cybersecurity Framework (CSF)

- Core Functions: "Identify, Protect, Detect, Respond, Recover." The CSF provides a flexible, risk-based approach with a focus on continuous improvement.
- Industry Agnostic: The CSF isn't designed for one specific sector. This makes it broadly adaptable, from healthcare to manufacturing to government agencies.
- Strength: Its straightforward language and emphasis on aligning cybersecurity with business goals make it a great tool for communication between technical staff and leadership.
- Weakness: The broad nature of the CSF means it relies on organizations mapping it to more specific standards and regulations for detailed implementation.

ISO 27001/2

- Core Functions: ISO 27001 provides detailed requirements for establishing an Information Security Management System (ISMS). ISO 27002 offers further guidance with best practice security controls.
- Global Recognition: Achieving ISO 27001 certification demonstrates a commitment to international security standards, often attractive for companies operating in multiple countries.
- Strength: The emphasis on a management system, with processes for risk assessment, auditing, and continuous improvement brings a structured approach to cybersecurity.
- Weakness: Implementation can be resource-intensive and the certification process itself can feel rigid or bureaucratic for some organizations.

PCI DSS

- Core Functions: Focused specifically on protecting cardholder data and the systems that handle it. Provides requirements for secure payment processing.

- Mandatory for Merchants: If your organization processes credit or debit cards, PCI DSS compliance isn't optional, it's a requirement.
- Strength: The detailed technical and operational controls in PCI DSS provide a clear roadmap for protecting sensitive payment information.
- Weakness: Its narrow focus on cardholder data means it won't fully address all your organization's cybersecurity risks.

Which One is Right?

The best framework depends on your organization's specific needs, industry, and regulatory requirements. Here are some additional points to consider:

- Critical Infrastructure: The NIST CSF is often favored due to its emphasis on operational resilience and alignment with other NIST standards.
- International Operations: ISO 27001 certification may provide a competitive advantage with global clients.
- Businesses Handling Payments: PCI DSS compliance will be non-negotiable.
- Smaller Organizations: The NIST CSF offers a good starting point due to its scalability.

Can you use multiple frameworks?

Absolutely! Organizations frequently combine elements of different frameworks. For example, you could use the NIST CSF for your overarching cybersecurity strategy, ISO 27001 for a structured ISMS, and layer in PCI DSS compliance if you process card payments.

Navigating the world of compliance feels a bit like traversing a tangled labyrinth. HIPAA, PCI DSS, GDPR, CCPA... these acronyms alone can give you a headache! But understanding where they overlap and differ is critical for any cybersecurity professional and a significant focus of the CISSP exam. Let's untangle these threads:

Common Ground

At their core, these regulations share a few fundamental goals:

- Protecting Sensitive Data: Whether it's health information, credit card numbers, or someone's online browsing habits, a core concern is preventing the unauthorized access, use, or disclosure of such data.
- Risk-Based Approach: Regulations increasingly emphasize the need to identify assets that contain sensitive data, gauge vulnerabilities, and prioritize security measures based on the level of risk. No one-size-fits-all solution exists!
- Accountability: Implementing technical safeguards is essential, but so is having documented policies, incident response procedures, and employee training around data protection.

Drilling Down into Distinctions

- Industry and Scope: HIPAA is specific to healthcare and related industries. PCI DSS, as the name implies, focuses on payment card data. GDPR (the EU's General Data Protection Regulation) and CCPA (California Consumer Privacy Act) have a broader reach, with a focus on protecting personal data.
- Data Subject Rights: GDPR and CCPA grant individuals significant rights over their data, like the right to access or request deletion. HIPAA has similar provisions, but with more nuance depending on the specifics of the healthcare scenario.
- Enforcement and Penalties: Fines for non-compliance can be hefty, but they vary by regulation. Understanding the regulatory landscape in the geographic regions where your organization operates is crucial.

Strategies for the CISSP

- Mapping Controls: Don't reinvent the wheel! Examine where security controls demanded by one regulation also contribute to meeting requirements in others. For instance, strong encryption satisfies elements of both HIPAA and PCI DSS.
- Consider the Highest Bar: Often, the strictest regulation becomes your guiding star. If you're GDPR compliant, you'll likely cover most of the bases for CCPA, although some California-specific details will still need attention.
- Don't Neglect the Basics: Across regulations, you'll find recurring themes like access controls, vulnerability management, and incident response. A strong security foundation benefits you regardless of the specific compliance checklist.

Keep in mind, compliance is a moving target! Regulations get amended, and new ones pop up. This is why the CISSP emphasizes a principle-based approach.

Let's be honest, sometimes cybersecurity compliance can feel like a soul-crushing box-ticking exercise. But aligning security with true business needs isn't just a "nice to have," it's what sets a CISSP apart as a strategic thinker. Here's how to break out of the compliance rut:

Think Like an Attacker

NIST CSF or ISO 27001 provide a solid foundation, but they aren't a crystal ball. Perform threat modeling to brainstorm realistic ways attackers could target your organization. Ask questions like:

What are your crown jewels? (Customer data, intellectual property, trade secrets – the things a breach would devastate your business)
Who's most likely to target those assets and what are their tactics? (Nation-state actors, cybercriminals, disgruntled employees all pose different risks)
Mapping Vulnerabilities: Where are you exposed? Think technical flaws (unpatched systems), procedural gaps (lax password policies), and even human factors (susceptibility to social engineering).
Prioritize Based on Impact

Not all risks are created equal. A vulnerability in your guest WiFi network is less urgent than a critical flaw in your payment processing system. Consider:

Likelihood: How probable is an attack scenario?
Potential Damage: What's the worst-case scenario if a breach occurs? Look at financial loss, reputational harm, business disruption, or even safety risks.
Choosing Controls: It's Not One-Size-Fits-All

Mapping compliance requirements to controls is where your risk analysis pays dividends. Consider these examples:

Technical: Encryption and access controls fall under many regulations and protect data directly.
Administrative: Your compliance-mandated incident response plan becomes far more valuable if aligned with realistic, well-practiced scenarios specific to your business.
Physical: If your business involves sensitive materials or operations, controls like access control to facilities and secure disposal of documents might be prioritized.
Resilience: Beyond the Bad Day

True security isn't just preventing breaches, it's about bouncing back if the worst happens. Consider:

Disaster Recovery: How robust and well-tested are your backups? Aligning this with business continuity priorities can literally save your company in a crisis.
Cyber Insurance: While controversial, exploring cyber insurance forces you to analyze risks in financial terms, which can be eye-opening for business leaders.
The Power of Storytelling

Boardrooms don't get excited by compliance lists, but they understand loss. Frame your security recommendations in terms of business impact. Rather than just stating a regulation, paint a vivid picture of what could happen if a specific risk isn't addressed.

The CISSP exam will test your ability to bridge the gap between technical details and business value.

Imagine your organization is building a new state-of-the-art manufacturing facility. You need strong safety protocols, detailed blueprints for the construction, and a plan for keeping everything running smoothly long-term. This is a perfect analogy for how we can use different frameworks together! Let's break it down:
Scenario: A mid-sized company with international clientele manufactures custom medical devices and holds sensitive patient data. They're scaling rapidly and need to mature their cybersecurity posture.

- NIST CSF: The Big Picture – We start with the 'Identify, Protect, Detect, Respond, Recover' functions of the CSF. This helps set overarching goals and align cybersecurity with business objectives like protecting trade secrets, complying with healthcare regulations, and safeguarding customer trust.
- ISO 27001/2: The Detailed Blueprint – ISO 27001's structured ISMS (Information Security Management System) approach provides granular requirements for risk assessments, policy development, and auditing. ISO 27002 gives us a catalog of practical security controls to implement (think access controls, encryption, incident response procedures).
- COBIT: Where IT Meets Governance – COBIT helps ensure our cybersecurity program supports the company's broader IT goals. Do our security processes facilitate efficient data handling? Are we getting the right metrics on cyber risk to support decision-making at the leadership level?

Why this combo works well:

- Adaptability: The NIST CSF lets us tailor our approach to the company's specific risks and industry needs.
- International Credibility: ISO 27001 certification may be attractive for landing contracts with global healthcare partners.
- "Whole Company" View: COBIT keeps us from operating in a security silo, ensuring our cybersecurity investments contribute to the smooth operation of the business.

Things to Consider:

- Budget & Resources: ISO 27001 certification can be a significant undertaking. We may start with NIST CSF to build a strong foundation and move towards ISO certification in the future.
- Vendor Management: Third-party suppliers may introduce risk. COBIT principles can help us implement security requirements for vendor contracts and assessments.

The CISSP exam will test your ability to integrate the strengths of different frameworks for real-world scenarios just like this.

Picture this: You're the newly hired CISO at a fast-growing healthcare startup. The company just landed a significant contract to develop a software platform for managing patient records. Excitement is high, but you know this also brings greater security risks. Let's analyze this using both qualitative and quantitative risk assessment approaches:

Scenario: Potential Data Breach Impact

- **Possible Attack Vectors:** Unpatched web application vulnerabilities, insider leaks of data, phishing attacks exploiting staff who are rushing to meet deadlines.
- **Assets at Stake:** Sensitive patient data, intellectual property in the software code, reputation of both your company and its new, high-profile client.

Qualitative Analysis: The Story of Impact

- **Reputational Harm:** A breach could erode trust, jeopardizing future contracts in the heavily regulated healthcare sector.
- **Regulatory Penalties:** Potential HIPAA violations resulting in hefty fines.
- **Operational Disruption:** An incident could cripple development, delaying the project significantly and angering the client.

Qualitative analysis paints a high-level picture, emphasizing risks that are difficult to quantify with a dollar amount. This helps communicate implications to non-technical stakeholders.

Quantitative Analysis: Putting Numbers to It

- **Annualized Loss Expectancy (ALE):** We estimate the likelihood of a breach based on industry data and our current security posture. Multiply this by the potential financial loss (fines, lost revenue, incident response costs), resulting in a specific dollar figure for the risk over a year.
- **Cost-Benefit Analysis:** We research various security solutions (e.g., web application firewall, employee awareness training). Can we justify the cost of the control by comparing it to the potential ALE it reduces?

Quantitative analysis is crucial when making budget requests or choosing between security measures with varying price tags.

When to Choose One Over the Other

- **Early Stages or Limited Resources:** Qualitative assessments can get you started quickly, guiding initial focus areas even with incomplete data.
- **Getting Executive Buy-In:** Stories of reputational damage often resonate more than spreadsheet projections in your initial pitch for increased security investment.
- **Quantifiable Impact:** Calculating specific ALEs for risks like a ransomware attack, where lost business revenue is easier to estimate, can be incredibly persuasive.

CISSP Takeaway: Don't get bogged down in complex formulas if data isn't readily available. A skilled CISO uses both qualitative and quantitative analysis strategically, knowing the strengths and limitations of each for the given situation!

Now lets get into the nuances of risk treatment, using your example of a potential data breach due to malware infection.

Risk Treatment Options – Quick Reminder

Accept: Acknowledge the risk and its potential consequences, choosing to take no action (usually with small, manageable risks).

Mitigate: Reduce the likelihood or impact of the risk through security controls (ex: firewalls, vulnerability patching, strong password policies).

Avoid: Change or cease the activity that creates the risk (extreme option, might mean abandoning a project or technology).
Transfer: Shift some of the risk to a third party (like cyber insurance or outsourcing a risky process).
Scenario: Data Breach due to Malware

Let's consider different circumstances, where each risk treatment option might be the most sensible:

Situation 1: Small Non-Profit with Tight Budget

Impact: Even a minor breach could be devastating financially and reputationally.
Risk Tolerance: Very low.
Best Strategy: Likely a combination of mitigation (strict security hygiene, robust backups) and transfer (cyber insurance may be their best bet to cover potential losses).
Situation 2: Tech Company Handling Sensitive IP

Impact: Leak of trade secrets could cripple the company.
Risk Tolerance: Extremely low.
Best Strategy: Heavy emphasis on mitigation (layered defenses, access controls, employee training). Might consider avoidance if malware risks on certain platforms seem unmanageable.
Situation 3: Healthcare Provider Bound by HIPAA

Impact: Breach carries serious regulatory penalties, beyond just financial loss.
Risk Tolerance: Low, as non-compliance is non-negotiable.
Best Strategy: Prioritize mitigation that aligns with HIPAA requirements. Acceptance is rarely an option due to regulatory obligations.

There's no one-size-fits all approach to risk treatment. The best choice depends heavily on your organization's unique circumstances.
Risk appetite isn't static – it might shift over time, triggering a re-evaluation of previous risk decisions.
Understanding the business impact of a breach allows us to set priorities, even if certain risks can only be accepted due to constraints.
The CISSP exam expects you to think strategically, considering these business and regulatory factors alongside the technical analysis of the risk itself. Let's explore a different cyber threat, or would you like to delve into one of the risk treatment strategies in more depth?

Understanding risk appetite and tolerance is like trying to navigate with a map and a compass. The map is your technical risk analysis, the compass is your organization's unique risk mindset. Let's unpack how that mindset is shaped:

Factors Influencing Risk Appetite & Tolerance

- **Industry:** A highly regulated healthcare provider handling sensitive data will naturally have a lower risk tolerance than a scrappy startup pushing out experimental apps.
- **Regulatory Environment:** Strict compliance mandates like HIPAA or PCI DSS will dictate that certain risks simply cannot be accepted.
- **Financial Position:** A cash-strapped organization might be forced to accept more risk in the short term, even though longer-term they'd prefer stronger controls.

- **Corporate Culture:** A risk-averse company with a 'security first' attitude will make different decisions compared to one that prioritizes rapid innovation and sees some risk as a cost of doing business.

Tailoring Your Communication for the C-Suite

Risk analysis is useless if the executives don't understand or act on it. Here's how to bridge the gap:

- **Skip the Jargon:** Avoid getting bogged down in CVSS scores or overly technical lingo. Focus on business impact.
- **Frame It Positively:** Don't just present problems. Link risk mitigation to achieving goals: "By investing in X now, we reduce the likelihood of a breach that would disrupt our product launch by Y%."
- **Know Your Audience:** Is your CEO financially driven? Quantify risk in terms of potential losses. Do they value reputation above all? Emphasize how a security incident could damage the company's brand.

It's a Conversation, Not a Lecture:

Invite questions. Actively listen to their concerns. This helps you 'calibrate' your future risk assessments and communication strategies.

The exam will test your ability to analyze both the technical aspects of risk and the softer organizational factors influencing how that risk is perceived. It's this combination that makes a true CISSP an effective advisor and not just a technician.

Imagine your risk assessment is a snapshot in time. But the world of cybersecurity moves faster than a photographer can say "cheese"! A truly effective CISSP knows that risk management is an ongoing, dynamic process. Let's explore how to make sure your assessments stay relevant:

Why Constant Vigilance is Key

- Threat Landscape Evolution: New exploits, zero-days, and changes in attacker tactics emerge constantly. Yesterday's strong defense could be tomorrow's glaring vulnerability.
- Internal Shifts: Did your company launch a new product line, enter a foreign market, or undergo a merger? Each brings unique risks and regulatory considerations.
- The Human Factor: Employee turnover, lapses in security discipline, or outsourcing parts of your operation all impact your risk profile.

Methods for Continuous Monitoring and Reassessment

- Vulnerability Scans: Don't just run these after a patch. Scheduled recurring scans help identify new flaws as they pop up in systems and applications.
- Threat Intelligence: Subscribing to industry feeds or joining ISACs (Information Sharing and Analysis Centers) keeps you informed of attacks targeting similar organizations.
- Log Analysis: SIEMs and log review can flag anomalous activity potentially signaling a compromise early on, minimizing damage.
- Trend Spotting: Keep an eye on security news and vendor advisories to proactively identify if your systems are likely targets of emerging threats.
- Regularly Scheduled Reviews: Embed checkpoints into your calendar (quarterly, annually) to formally revisit risk assumptions built into your management plans.

Adapting Your Mitigation Strategies

Reassessment might reveal the need to:

- Patch Quickly: A new, critical vulnerability demands a faster patching timeline than your standard maintenance window.
- Upgrade Controls: Maybe your antivirus is sufficient for everyday threats but struggles with advanced malware variants. It's time to invest in a more robust solution.

- Rethink Training: If phishing remains a success vector for attackers, it's time for a fresh approach to employee awareness.
- Third-Party Scrutiny: Changes in your supply chain or increased reliance on cloud services warrant revisiting your vendor risk assessments.

CISSP Mindset: Embrace the fact that in cybersecurity, the finish line is always moving! The ability to continuously adapt and iterate on your risk management processes is what sets you apart from those who treat it as a checkbox exercise.

Think of security policies as the rulebook for your cybersecurity team's game. A clear, well-crafted policy sets everyone up for success. A poorly written one leads to confusion, corner-cutting, and a whole lot of security headaches.

Key Ingredients of a Strong Security Policy

Scope: Clearly define what the policy covers (e.g., acceptable use of company devices, data encryption standards, password complexity)
Purpose: Briefly explain why the policy matters. Tie it back to protecting the business and its assets.
Roles and Responsibilities: Outline who owns the policy (usually CISO or CIO), who enforces it (IT, HR, etc.), and what everyday employees are expected to do.
Enforcement: State the consequences of violations. Be specific, but ensure they match the severity of the violation.
Let's Dissect a Bad Example

Vague Policy: "Employees must use secure passwords."

Why It Stinks: What defines secure? How often do passwords change? Are password managers allowed? This leaves too much to interpretation.

Transforming Vague to Actionable

Scope: Password Security Policy for Company Systems
Purpose: Protect sensitive data by preventing unauthorized access through weak or compromised passwords.
Roles & Responsibilities:
IT Dept: Sets password complexity rules, enforces expiration, provides tools (password manager)
Employees: Create strong passwords, never share them, report suspected compromises immediately.
Enforcement:
1st Violation: Warning + mandatory reset
Multiple Violations: Disciplinary action up to termination (for serious or intentional cases)
Tips for the CISSP

Clarity is King: Avoid overly technical jargon that non-IT staff won't grasp.
Involve Stakeholders: Getting input during the drafting process improves buy-in.
Regular Review: Policies aren't meant to gather dust. Revisit them as technology and your organization change.
Remember, a policy is useless if no one knows it exists or bothers to follow it!

Imagine you're building a house and your contractor hands you a vague sketch instead of detailed blueprints. Chances are you'll end up with wonky doorways and mismatched windows! Standards in cybersecurity serve as those blueprints, ensuring everyone is working towards the same solid, secure structure.

The Role of Standards

Think of standards as technical rulebooks specifying:

- **Configuration Settings:** "Passwords must be at least 12 characters, contain a mix of symbols, expire every 90 days..." This removes guesswork and makes security consistent across the environment.
- **System Hardening:** Guidance to disable unnecessary services, remove default accounts, or configure firewalls according to best practices.
- **Secure Coding Practices:** How developers should write software to minimize vulnerabilities, such as input validation to thwart injection attacks.

Where do Standards Come From?

- **Industry-Wide Benchmarks:** The Center for Internet Security (CIS) Benchmarks are a goldmine. They provide configuration guidelines for a wide range of operating systems, applications, and network devices.
- **Vendor-Specific:** Companies like Microsoft and Cisco often release hardening guides tailored to their products, aligning security with their best practices.
- **Internal to your Organization:** As your company matures, you might customize standards to reflect unique systems or compliance requirements.

Using Standards: A Balancing Act

It's tempting to embrace standards as the ultimate answer – just follow them and you're secure, right? Not quite! Keep in mind:

- **Risk Assessment Still Matters:** A standard is a good starting point, but you'll need to tailor it based on your specific threat landscape and business needs.
- **Usability Trade-offs:** Ultra-strict password rules might lead to employees writing them on sticky notes stuck to their monitor!
- **Regular Updates:** Standards evolve as threats and technologies change. Make sure you're using the most recent versions.

The CISSP Edge

The exam expects you to know both the importance of standards and when some flexibility or customization is necessary for a truly effective security posture.
Let's explore how standards fit into a broader context!

Imagine procedures as the bridge between high-level policies (your destination) and the nitty-gritty of day-to-day security operations (how you actually get there). Poor procedures are like a rickety bridge, leaving employees feeling lost and more likely to take dangerous shortcuts.

Let's dissect why truly effective procedures are more than just a checklist:

They Reinforce the "Why"
A step like "Run vulnerability scans weekly" may be followed blindly. But explaining that this helps proactively find weaknesses before attackers exploit them encourages diligence, even when staff are busy.

Connecting the Dots to Policies
Including a reference to the relevant policy (e.g., "As outlined in our Acceptable Use Policy...") strengthens the connection between everyday actions and the overarching security goals of the organization.

Addressing the Rules: Compliance Check!
If a procedure is tied to a HIPAA requirement, for instance, clearly state this. Employees are more invested in following it knowing it's not just an arbitrary IT task, but key to avoiding fines or reputational harm.

Clarity = Adherence
Procedures riddled with jargon or assumptions only cause frustration. Use plain language, diagrams when helpful, and define any technical terms.

The Room for "What if?"
Anticipate questions and build some decision points into the procedure. Example: "If vulnerability scan flags a critical flaw in a production system, escalate to team lead BEFORE patching." This empowers employees to act correctly even in non-standard situations.

Practical Example
Here's a weak vs. strong excerpt from an incident response procedure:

Weak: "Submit a ticket in the IT helpdesk system."

Strong: "Submit a ticket in the IT helpdesk system (marked 'Security Incident'). This ensures proper prioritization, as per our Incident Response Policy, and helps us meet the 72-hour breach notification requirement under [relevant regulation]."
The exam wants you to see beyond the technical steps. Great procedures promote a culture of security. They help employees grasp not just what to do, but why it matters, building both competence and a sense of shared responsibility.
Picture security like building good health habits. It's not just about occasional checkups, but about making the right choices part of your daily routine. This is what we aim for when we bake security into the very fabric of how an organization operates. Let's break it down:

Embedding Security into the SDLC (Software Development Lifecycle)

- **Shift Left Philosophy:** Finding a vulnerability after release is like discovering an unhealthy ingredient in your meal after you've already eaten it! 'Shift left' means security testing occurs early and often during development.
- **Tools and Automation:** Integrating code scanning tools and security-focused test cases streamlines the process for developers.
- **Training is Key:** Empower developers to write secure code from the start to minimize costly rewrites later.

Change Management: Security as a Gatekeeper

- **Risk Assessment as Standard:** Before any change is approved (new software, system updates), explicitly assess its potential impact on security posture.
- **Documentation and Accountability:** Clear records of security review within your change approval process promote transparency.

- **Change + Vulnerability Management = BFFs:** Did a change introduce known vulnerabilities? Have a process to quickly patch or rollback if needed.

Incident Response: Lessons That Stick

- **Postmortems That Matter:** After an incident, go beyond fixing the immediate issue. Analyze how it could have been prevented or detected sooner.
- **Updating Processes and Controls:** Are your playbooks and security tools failing to keep up with attacker techniques? Use incidents to drive improvement.
- **Communication is Key:** Were non-technical staff confused about their role during the response? Lessons learned can shape better training and awareness programs.

Why This Approach Strengthens Cybersecurity

Embedding security this way creates a ripple effect:

- **Efficiency:** It's cheaper to fix a bug early than a production breach.
- **Proactive, Not Reactive:** You're constantly improving your defenses, staying ahead of threats rather than playing catch-up.
- **Shared Ownership:** Security isn't just an IT burden; everyone understands their role in protecting the organization.

The exam will test your ability to see the big picture – how isolated security activities fit into the broader health and success of the business.

The (ISC)² Code of Ethics is a great starting point, but when those black-and-white rules get tested in the messy real-world, things get complicated fast. Let's explore some other ways to think about ethical dilemmas:

Deontological Ethics: Duty and Rules

Emphasis: Focuses on the inherent rightness or wrongness of an action, regardless of consequences. Think of it like following your moral compass, even if it leads to a difficult outcome.
Example: Lying is always wrong, even if it could achieve a 'good' outcome.
Utilitarianism: The Greatest Good

Emphasis: Aims to make decisions resulting in the greatest good for the greatest number of people. It involves a cost-benefit analysis with a focus on overall outcome.
Example: Weighing potential harm caused by a decision against the potential benefits to a wider group of stakeholders.
Scenario: To Patch or Not to Patch?

Imagine a hospital's critical patient monitoring systems run on an outdated OS with a known, severe vulnerability.

Patching Dilemma: Applying the patch risks temporarily disrupting the system, potentially jeopardizing care. Doing nothing leaves patients exposed to a cyberattack, which could have even worse consequences.
How Frameworks Diverge:

Deontological View: A security professional might argue that honesty about the risk is paramount, even if it causes fear. Patient safety is a non-negotiable duty.

Utilitarian View: Prioritizing the well-being of the majority of patients might justify delaying the patch until a carefully planned maintenance window is available, minimizing the disruption risk.
The exam wants you to understand that there's rarely a single 'right' answer in cybersecurity ethics.
Consider these points:

The (ISC)² Code is your Foundation:** It provides overarching principles like protecting society and acting honorably.
Context is King:** The severity of the threat, impact on vulnerable populations, and urgency of the situation all matter.
Document Your Reasoning:** If forced to choose a less-than-ideal option, clearly articulate why you believe it aligns best with your ethical obligations given the circumstances.
Let's craft a scenario that brings those ethical gray areas into sharp focus. Imagine you're the head of security at a social media company. Here's what you're grappling with:

- **The Government Request:** Law enforcement agencies reach out with a subpoena demanding user data related to an ongoing investigation. The scope of the request seems excessive, and you worry about the privacy implications for your users.
- **Internal Pressure:** Your company's legal team urges compliance. Angering the authorities could impact the company's reputation and even regulatory standing.
- **Your Conscience is Nagging:** Is the investigation justified? Are your users being unfairly targeted? Does turning over this much data cross a line, even if technically 'legal'?

Strategies for the Ethical Tightrope

1. **Don't React, Examine:** Knee-jerk decisions made under pressure rarely end well. Take a step back and thoroughly analyze the situation:
    - Understand the Law: What are your precise legal obligations? Are there grounds to challenge any part of the request?
    - Company Policies & Precedent: Does your organization have internal policies on handling such requests? Has this happened before?
    - (ISC)² Code of Ethics: How do principles like "act honorably, justly, responsibly, and legally" apply in this specific situation?
2. **Seek Counsel (Carefully):**
    - Internal: Involve your company's legal team, but remember, their focus is protecting the company, not necessarily individual rights.
    - External: If allowed, consulting an outside lawyer specializing in privacy law could be invaluable for understanding your options.
    - Professional Mentors: If you have trusted mentors in the cybersecurity field, their experience with similar dilemmas might offer guidance.
3. **Negotiation, Not Capitulation:**
    - Push Back: Can you narrow the scope of the request and protect most user data, while still partially complying?
    - Document Everything: Maintain detailed records of the request, your concerns, actions taken, and justifications for your choices.

Important Considerations

- Your Company Culture: Is there space for ethical dissent? Or will raising concerns make you a target? This sadly impacts your strategy.
- Risk Tolerance: How far are you personally willing to go? Losing your job might be a price you're willing to pay on principle.

- Whistleblowing: Is this a 'blow the whistle to the press' scenario? Proceed with extreme caution – the stakes here are incredibly high.

The exam wants to see that you recognize your duty isn't blind obedience. It's about finding the most ethical path within complicated, often imperfect, constraints.

Data can have incredible power – to cure diseases, improve our lives, or cause significant harm in the wrong hands. Understanding data ethics is essential for any cybersecurity professional and heavily emphasized on the CISSP exam. Let's delve into this:

Security Research: A Double-Edged Sword

The Goal: Security researchers often analyze large datasets of real-world cyberattacks or user behavior to identify patterns and vulnerabilities. This helps strengthen our defenses.

The Risks: Even with the best intentions, this data might contain...

Personally Identifiable Information (PII): Names, addresses, health records... a breach of this in itself would be a disaster!

Implicit Bias: Data might reflect societal biases. If an algorithm is trained on this, it risks perpetuating unfairness or discrimination.

Data Ethics in Action

Anonymization Isn't Foolproof: Removing names doesn't guarantee privacy. Enough data points combined can often re-identify an individual. Consider using techniques like differential privacy, which adds 'noise' to mask individual contributions.

Informed Consent: When possible, obtain clear consent for using data. Transparency builds trust, even if full technical understanding may be unlikely for the average user.

Purpose Limitation: Data collected for research shouldn't end up sold to marketers! Define and enforce strict limits on how the data can be used.

Think Like an Attacker: Can you piece together an individual's identity even from seemingly innocuous data? Adopt an adversarial mindset to anticipate unintended consequences.

Real-World Complexity

Often, completely 'clean' datasets don't exist. Consider these dilemmas:

Security Flaw Disclosure: Do you reveal a zero-day if the vendor is unresponsive, even if it means some users get attacked while awaiting a patch?

Analyzing Breached Data: Is it ethical to study leaked data from criminals, even if it could help prevent future attacks?

Data ethics isn't about perfect answers; it's about asking hard questions. The exam probes your ability to weigh privacy risks, potential benefits, and your obligations to both protect individuals and strengthen the security of the larger community.

Think of emerging technologies like powerful but unpredictable fire. They have the potential to solve incredible problems...or burn everything to the ground. Our ethics as cybersecurity professionals play a pivotal role in determining which outcome prevails. Let's tackle this challenge area by area:

Artificial Intelligence (AI)

- AI Bias: If datasets used for training AI models reflect existing prejudices (for example, in hiring or criminal justice), the resulting algorithms will perpetuate those biases at a much larger scale.

- Lack of Explainability: Complex AI systems can produce results that even their creators don't fully grasp. This makes accountability difficult when things go wrong.
- Autonomous Weapons: The concept of AI systems making life-or-death decisions on the battlefield raises deeply troubling questions.

Internet of Things (IoT)

- Pervasive Surveillance: Smart devices in our homes and cities collect a staggering amount of data. Who owns this data, how can it be used, and what safeguards prevent abuse?
- Insecure by Design: Many IoT devices prioritize convenience over security, creating vast networks of easily exploitable entry points for attackers.
- Botnet Fuel: Compromised IoT devices often become unwilling participants in DDoS attacks, harming others.

Surveillance Technologies

- Function Creep: Tools initially designed for a specific purpose (like facial recognition to catch terrorists) can quickly be expanded for less justifiable mass surveillance.
- Disproportionate Impact: These technologies often worsen existing societal bias, unfairly targeting marginalized groups.
- Chilling Effect: Knowing you're constantly monitored crushes freedom of expression, dissent, and association.

Guidelines for Responsible Tech

- Privacy by Design: Make strong privacy protection a cornerstone, not an afterthought, during the development process.
- Algorithmic Auditing: Regularly scrutinize AI models for bias, especially in high-stakes applications.
- Transparency: To the extent possible, be open about how these technologies work and the decisions they are making.
- Human Oversight: Especially for critical decisions, maintain a 'human in the loop' to prevent AI systems from causing unchecked harm.
- Clear Regulation: While technology outpaces law, we desperately need stronger frameworks balancing innovation with ethical guardrails.

The exam will test your ability to see beyond the technical coolness factor. Consider:

- Duty to Warn: If you foresee the potential for your work to be misused, you have an ethical obligation to speak up.
- Think Like the Adversary: How could these technologies be weaponized by malicious actors?
- Be Part of the Solution: Engage in discussions about ethical tech policy and advocate for responsible practices within your organization.

Let's choose a specific technology, like AI-powered facial recognition.

II. Asset Security

Picture your organization's critical assets – sensitive client data, intellectual property, the very systems keeping your business running. Asset security is about far more than just firewalls and antivirus. It's about understanding, classifying, and safeguarding the lifeblood of your organization.

In this section, we'll dive into the nitty-gritty of securing assets throughout their entire lifecycle. Think of it like this:

Know Your Treasure: You can't protect what you don't understand. We'll cover data classification, ownership models, and why not all assets are created equal.
Lifecycle Approach: From the moment an asset is created to its eventual disposal, security is paramount. We'll discuss secure storage, handling practices, and ensuring data doesn't linger after it's no longer needed.
It's a Team Effort: Asset security isn't just an IT problem. We'll explore how HR, legal, and even facilities personnel play crucial roles in protecting your organization's crown jewels.

Let's get practical! Imagine you've just been hired as the head of information security at a mid-sized healthcare software company. Patient data, upcoming product designs... there's a lot at stake. Right now, their data protection is an ad-hoc mess, and your first order of business is building a rock-solid classification schema.

Types of Data & Their Value

- Patient Records: The most critical asset. Compliance with HIPAA isn't optional, and a breach here could devastate the company.
- Intellectual Property (IP): Source code and proprietary algorithms represent future revenue streams. Competitors would love to get their hands on this!
- Financial Data: Typical stuff like budgets, and transaction records... less sensitive, but still needs protection.
- Employee PII: Personal data like social security numbers, home addresses, etc. While not core to operations, a breach here damages trust and opens the door to identity theft.

Designing Our Schema: Keep it Simple

We don't want a system so complex no one uses it. Let's propose:

- **Restricted:**
  - Definition: Patient data, source code, key design documents.
  - Access: Strict need-to-know basis. Requires approvals at a senior level.
  - Controls: Encryption at rest and in transit, strong physical access controls for storage areas.
- **Confidential:**
  - Definition: Financial data, employee PII
  - Access: Employees whose jobs require access. Logging and auditing in place.
  - Controls: Encryption for highly sensitive files, regular security awareness training for staff.

- **Public:**
  - Definition: Marketing materials, approved external content, etc.
  - Access: Openly accessible to staff or the public as intended.

Why This Works

- Aligns with Business Needs: Protects highly sensitive assets without hindering day-to-day work.
- Room for Nuance: Within each category, we can further specify access levels as the need arises.
- Foundation for More: This creates a framework for data handling procedures, storage guidelines, and employee training.

The exam emphasizes practical design, not just theory. Things to consider:

- Data Retention: Do we have deletion policies? Keeping stale data increases our risk surface area.
- Over-classification Trap: When too much is labeled "restricted," people find workarounds, defeating the purpose.
- Communicate and Iterate: Rolling out a classification schema isn't a one-and-done activity. Get buy-in, then refine based on real-world use!

Let's explore how our classification schema influences those tool choices and other security measures:

Impact on Data Protection Tools

- Restricted:
  - Encryption: Strong mandatory encryption for data both at rest (think full-disk encryption on storage devices) and in transit (enforced use of secure protocols like HTTPS).
  - Access Controls: Role-based access controls (RBAC) become essential. Not every developer needs access to patient records! Consider multi-factor authentication as an additional layer here.
  - Physical Security: Servers or storage areas holding restricted data might require enhanced measures like access logs, cameras, and limited entry points.
- Confidential:
  - Encryption: Use selectively for highly sensitive files or consider email encryption solutions.
  - Access Control: Still important, but emphasis shifts to logging and auditing to catch anomalous behavior early.
  - Data Loss Prevention (DLP): Tools for monitoring what confidential data is being copied, emailed, etc., can help detect potential leaks.
- Public:
  - Focus on Integrity: Ensure these systems are patched, have anti-malware, and are monitored for any unauthorized modification.

Beyond Tools: How the Schema Shapes Our Approach

- Training: Tailored to roles! Staff working with Restricted data get rigorous training on HIPAA compliance and incident response, while those mainly handling Confidential get focused sessions on phishing dangers and social engineering tactics.
- Incident Response: Our plan needs to prioritize stopping the leak of Restricted data. Incident responders might need specialized training and clearance to handle these cases.
- Vulnerability Management: Systems containing Restricted data likely get prioritized for patching and more frequent scanning to reduce risk.

Notice how technical decisions flow directly from our understanding of the business and its data. This alignment is what the exam is looking for. It's also why merely memorizing buzzwords won't cut it!

A few more things to consider:

- Cloud Storage: If we use it, where does each classification tier fit within our cloud provider's security offerings?
- Vendors: Contracts for third parties who handle our data should reflect our classification schema and compliance requirements.

Data labeling is like putting the right stickers on all your boxes before a big move. It sounds straightforward, but in the chaotic sprawl of most organizations, it gets messy fast! Let's break down the methods and the hurdles:

Labeling Techniques

Manual: Humans directly apply labels. This could be as simple as adding tags in document properties or more structured, with employees selecting from a dropdown aligned with your classification schema.
Automated: Tools can analyze content using keywords, patterns, or file location. Might be suitable for initial labeling but often needs human fine-tuning to be accurate.
Metadata-based: Labels are embedded in the file itself. This travels with the data and can be actioned by security tools automatically (like DLP blocking transmission of a "Restricted" file).
Challenges Abound

Legacy Data: What about that mountain of old files no one's touched in years? Massive retroactive labeling projects are costly and resource-intensive.
User Resistance: If it's confusing or adds tedious steps, employees will find ways to circumvent the system.
Accuracy: Humans make mistakes; algorithms miss context. Incorrect labels are worse than no labels at all, creating a false sense of security.
Usability vs. Security: Strict labeling that blocks actions users see as 'normal work' will breed resentment, not compliance.
Maintenance: As our business changes, so should our schema. Keeping labels updated is an ongoing process, not a one-time fix.
Strategies for Success

Phase It In: Start with new data creation. Mandate clear labeling from the get-go for all newly generated or received files.
Prioritization is Key: Focus on your crown jewels – those high-risk data stores. A strategic, incremental approach is more realistic.
Tool Choice Matters: User-friendly interfaces and automation, where feasible, go a long way in reducing friction.
Incentives over Punishment: Can proper labeling be tied to performance reviews or gamified? Making it a positive can increase adoption.
The exam wants you to understand that perfect solutions rarely exist. Consider:

Risk-Based Approach: Where would mislabeling be most disastrous? Aim for strong protection there, even if other areas remain less strict.
Communication is Everything: Explain why this matters to the business, not just in a generic "security is important" way.

Remember that data classification scheme we designed earlier? Now's where we put it to work! Imagine the classification levels as keys that unlock increasingly secure rooms within our digital fortress. Let's see how this plays out in practice:

Access Controls

- Restricted: Need-to-know takes on real meaning here. Access Control Lists (ACLs) get granular. Think individual users or small teams, not broad departments, being granted permissions. Multi-factor authentication might be mandated.
- Confidential: Access is generally role-based. HR staff can access the bulk of employee PII, marketing can't. Logging is key to spot anomalies early.
- Public: Few restrictions, unless modifications to these files could endanger systems.

Encryption

- Restricted: Mandatory, both in transit and at rest. Look into hardware-based encryption solutions for the most sensitive repositories.
- Confidential: Encryption at rest for designated high-risk datasets. Enforced, secure transmission protocols for when it's in motion.
- Public: Not a priority, but integrity monitoring still matters.

Storage Procedures

- Restricted: Physical access controls get layered. Think locked server rooms, or designated, highly secure areas within a datacenter. Backup procedures are meticulously documented and tested.
- Confidential: Encryption and strong access control likely suffice, but separation from Restricted data stores is a good practice.
- Public: Redundant storage and focus on availability over secrecy.

Other Impacts of Classification

- Incident Response: Breaches involving Restricted data trigger stricter reporting timelines, might require specialized responders.
- Data Retention: Do we need to keep Confidential data as long as Restricted? Clear deletion schedules reduce our attack surface.
- Cloud Adoption: Our schema influences vendor selection. Can their solution provide the segmentation and fine-grained controls we need for Restricted data?

The exam tests your ability to translate business risk into actionable controls. Notice:

- No One-Size-Fits-All: Solutions are tailored to the sensitivity level.
- Justification Matters: A CISSP can explain to stakeholders why those expensive hardware encryption modules are non-negotiable for some data.
- Continuous Refinement: As the threat landscape and business needs evolve, so should our control mapping.

Understanding data roles is like figuring out who's in charge of your luggage during a trip. The owner decides what's packed and its final destination, but the custodian (airline, hotel bellhop, etc.) is responsible for it not getting lost or damaged along the way. Let's unpack this:

Data Owner

- The Decision Maker: Usually a senior business stakeholder (ex: Head of Product for customer data, Chief Medical Officer for patient records).
- Sets the Rules: They define acceptable use, retention periods, and overall risk tolerance for the data they 'own.'
- Accountability: Ultimately answerable if a breach or misuse of the data occurs. This responsibility doesn't disappear, even though they delegate tasks!

Data Custodian

- The Doer: Typically IT staff, database admins, etc. They implement the security controls defined by the owner.
- Day-to-Day Guardians: Responsible for backups, access controls, incident response, and ensuring the data's integrity is maintained.
- Limited Authority: Custodians shouldn't be unilaterally changing access permissions or making independent decisions about data use.

Where Things Get Tricky

- Fuzzy Ownership: In many organizations, it's unclear who truly 'owns' certain datasets, especially when data is used for multiple purposes.
- Urgent Requests: A data custodian receives an urgent request seeming to come from the owner...but was it spoofed? What's their process for verifying?
- Conflicting Priorities: Owner wants data accessible for analytics, custodian worries about making it too easy to exfiltrate.
- The Buck Stops... Where? When something goes wrong, finger-pointing between owner and custodian is unproductive.

Strategies for Success

- Formal Definitions: Clarify roles in security policies. Don't assume everyone has the same understanding of these terms!
- Shared Responsibility: Frameworks like RACI (Responsible, Accountable, Consulted, Informed) can map out ownership at a finer level.
- Communication Channels: Regular meetings between owners and custodians proactively address questions and tensions.

The exam will test your ability to see how breakdowns in these roles lead to security risks. Think through:

- Audit Implications: How can we demonstrate to auditors that both owners and custodians are fulfilling their obligations?
- Third-Party Risk: If a vendor is the custodian, can we simply hand off all accountability to them? (Spoiler alert: No.)

Let's tackle a classic conflict: marketing vs. security with customer data in the crossfire.

The Scenario

The marketing team wants access to a wider range of customer data stored in the CRM system. Their goal is to create hyper-targeted campaigns and improve personalization. They argue this directly supports the company's revenue targets. Currently, they only see basic contact information and purchase history. They want additional fields like browsing behavior and support ticket interactions.

Data Ownership Considerations

- Who's the Owner? This is where things get fuzzy. CMO might see it 'marketing's data,' while the privacy officer might disagree.
- Customer Consent: Was the data originally collected under a clear privacy policy? Using it beyond that scope could violate compliance and erode trust.
- Risk Assessment: Does marketing have a legitimate need, or is this a nice-to-have? Expanding the 'attack surface' always needs careful justification.

Data Custodian Concerns

- Security Headaches: More access points, more potential for leaks, accidental or otherwise. Does IT have the ability to monitor this new access effectively?
- Technical Feasibility: Segregating this data for marketing's use might be complex and time-consuming to implement.

- Playing Gatekeeper: Security teams often end up as the 'bad guys' denying requests, damaging relationships across the company.

Finding a Middle Ground (Maybe)

- Temporary & Limited: Could granting marketing temporary access for a specific project, with enhanced auditing, be a compromise?
- Anonymization/Masking: Can the data be partially anonymized to fulfill marketing's needs while protecting the most privacy-sensitive parts?
- Training as Solution: Perhaps better training for marketing on secure data handling and privacy regulations would ease concerns on both sides.

This scenario has no easy answer – which is the point! The exam wants you to:

- Weigh competing interests: Increasing revenue vs. protecting customer privacy and upholding compliance.
- Propose Solutions: Technical, procedural, and policy-based options should all be on the table.
- Communicate Risk Effectively: IT needs to translate their concerns into a language the CMO and other business leaders understand.

Let's get into the complex world of healthcare records, where compliance and ethics are as important as the data itself.

Data Lifecycle Mapping: Patient Records

1. **Creation**

- Sources: Patient intake forms, physician notes, lab results, medical device readings, imaging scans, etc.
- Input Controls: Emphasis on accuracy and integrity. Dropdowns for common entries instead of freeform text reduces errors.
- Access: Strictly limited to authorized healthcare providers directly involved in care.

2. **Storage**

- System: Electronic Health Record (EHR) system, which must be HIPAA-compliant.
- Encryption: Mandatory both at rest and in transit.
- Backups & Redundancy: Disaster recovery plans in place. Off-site backups protect against physical incidents.

3. **Access**

- Role-Based Controls: Fine-grained permissions based on provider role, relationship to patient, and reason for access.
- Logging and Auditing: Every access attempt is logged to deter misuse and aid in investigations if needed.

4. **Use**

- Primary: Direct patient treatment by authorized personnel.
- Secondary: De-identified data pooled for research (strict protocols if individual consent isn't obtained).

5. **Sharing**

- Patient Transfers: Secure mechanisms, ensuring the full record is transmitted only to the new care provider.
- External Specialists: May require temporary, limited access with clear expiration.
- Regulatory Reporting: Some reporting to public health authorities may be mandated.

6. **Archiving**

- Retention Periods: Dictated by HIPAA and state laws (can be surprisingly long!)

- Secure Storage: Might migrate to lower-tier storage systems, but encryption and access controls shouldn't weaken.

7. **Destruction**

- Certified Methods: Not just deletion! Files overwritten, media physically destroyed according to NIST standards.
- Documentation: Chain of custody and proof of destruction for compliance purposes.

CISSP Considerations

- Incident Response: Breaches involving patient data have heavier consequences and urgency. Specialized procedures are needed.
- Evolving Technology: AI-powered diagnostic tools, patient-wearable data... how do these feed into the lifecycle and our security controls?
- Patient Rights: HIPAA gives patients rights regarding their data access and correction. How does our system facilitate this?

Important Notes

- This is simplified! A real-world EHR system would require far more detail at each step.
- Laws change, so must our lifecycle map. This is a continuous process, not a checkbox activity.

Let's talk trade-offs: Stricter access controls to improve security might slightly slow down care delivery in an emergency situation.

Let's take a deep dive into security controls tailored to each stage of our patient record lifecycle. Remember, the goal isn't merely to have controls but to choose the right ones for the specific risks we face.

1. Creation

Input Validation: Preventing malformed or intentionally malicious data entry. Reduces the chance of system crashes or corruption.

Digital Signatures: Where feasible, especially for high-stakes entries, this ensures non-repudiation and helps establish chain of custody.

Timestamps: Accurate timestamps are crucial for both medical and potential legal purposes.

2. Storage

Encryption: Strong, industry-approved algorithms. Consider both 'field-level' encryption for sensitive data and full disk encryption.

Access Control Lists (ACLs): Fine-grained permissions. Not just at the system level, but potentially down to specific record types.

Backups: Include these in your disaster recovery plan. Test restores regularly!

3. Access

Role-Based Access Control (RBAC): Job function drives access, not individual whims.

Audit Logging: Every interaction (read, modify, delete) leaves an unalterable trail. Proactive monitoring seeks out anomalies.

Break-the-Glass Procedures: For true emergencies, bypass normal controls, but with heavy logging and justification after the fact.

4. Use

Data Masking: For research, obscure details like name and address while preserving medically relevant information.

Watermarking: Embedded identifiers can help track down the source of leaks of de-identified data.

DLP: Tools to prevent patient records from being emailed outside, copied to USB drives, etc.

5. Sharing

Secure Transmission: HTTPS is the bare minimum. Consider client-side certificates as an additional layer of authentication for transfers.
Consent Management: How does our system track a patient's choices regarding what can be shared and with whom?
Time & Scope Limits: Sharing for specialist consultations should be automatically revoked when no longer needed.
6. Archiving
Integrity Checks: Before archiving, ensure data hasn't been corrupted or tampered with.
Access Controls Remain: "Archived" doesn't mean open season. Access here is rarer but still needs strong controls.
7. Destruction
NIST 800-88 Guidelines: Overwriting, degaussing, or physical destruction methods depending on the media.
Certificates of Destruction: Especially when using a third-party vendor, this creates accountability.

The exam emphasizes practical defenses. Consider:
Threat Modeling: For each stage, what are realistic attacks, and do our controls counter them?
Usability Impact: If doctors resort to sticky notes because the EHR is too cumbersome, security fails.
Evolving Landscape: New regulations, zero-day vulnerabilities...all demand adaptability in our controls.
Regulations like HIPAA and GDPR aren't just a box-ticking exercise; they significantly shape how we must handle data throughout its entire lifecycle. Let's dissect this, focusing on patient records to get specific:
Influence on Data Lifecycle

- Creation:
  - HIPAA: Enforces the minimum necessary principle. We can't collect more data than strictly needed for treatment or legally mandated purposes.
  - GDPR: Emphasizes transparency and purpose limitation. Patients must clearly understand what data is collected and how it will be used.
- Storage:
  - HIPAA: Requires safeguards around encryption, access controls, and integrity checks, often with more stringency than for other types of data.
  - GDPR: Places restrictions on how long personal data can be stored and strengthens the "right to be forgotten" (erasure upon request).
- Access:
  - HIPAA: Strict controls on who can access data, mandates audit logging to deter snooping.
  - GDPR: Emphasizes data subject access rights – patients must have a way to easily obtain a copy of their own data or request corrections.
- Use:
  - HIPAA: Explicit rules govern use for research or disclosure to third parties, often requiring patient consent.
  - GDPR: Data minimization – even within your organization, patient data shouldn't be more widely accessible than necessary for specific tasks.
- Sharing:
  - HIPAA: Notice of Privacy Practices must outline with whom data might be shared (insurers, labs, etc.). Patient authorization is frequently required.

- GDPR: Cross-border data transfers get extra scrutiny. Sending patient data outside the EU has stricter requirements.
- Archiving & Destruction:
  - HIPAA & State Laws: Dictate surprisingly long retention periods (sometimes decades!). Secure archiving is just as important as for active data.
  - GDPR: "Right to be forgotten" adds a layer of complexity. Deletion requests must be honored, with potential exceptions.

Key Compliance Requirements

- Retention Policies: Must align with legal minimums. Develop a clear deletion schedule defensible to auditors.
- Secure Disposal: Not just hitting 'delete'! HIPAA mandates procedures preventing data recovery, especially for sensitive health records.
- Audit Trails: Indispensable for investigations. Log not just access but changes, system events...enough to reconstruct 'what happened' for both breach analysis and potential legal proceedings.

The exam expects you to understand the why behind the controls. Think about these:

- Conflicting Laws: What if HIPAA requires 7-year storage, but GDPR limits it to 5? Legal consultation is probably needed!
- Risk-Based Approach: Even within HIPAA, patient records have more stringent requirements than general HR data.
- Impact on Incident Response: Breach notification timelines under HIPAA are notoriously short – your IR plan must reflect this.

Let's add complexity: think about how a patient data lifecycle and regulatory requirements might differ for a US-based healthcare provider vs. one operating in the EU.

The data lifecycle used to be neatly contained within our organization's firewalls. Emerging technologies blow those boundaries wide open, presenting us with a double-edged sword of opportunity and risk. Let's dissect this:

Cloud Storage

- Opportunities: Scalability, reduced hardware costs, potential for improved disaster recovery with geo-redundant cloud backups.
- Challenges:
  - Loss of Direct Control: Are the cloud provider's security practices up to snuff? Contract negotiation becomes crucial.
  - Regulatory Hurdles: Can we meet HIPAA compliance when storing patient data in a cloud environment? It demands deeper scrutiny.
  - Vendor Lock-in: Migrating data out of a cloud service can prove surprisingly difficult or costly.

Big Data Analytics

- Opportunities: Unlocking insights from vast pools of data could lead to medical breakthroughs or improved operational efficiency.
- Challenges:
  - Anonymization Pitfalls: Is our data truly anonymized, or could individuals be re-identified with enough effort?
  - Bias Amplification: If our datasets are biased, the resulting analytics will further perpetuate unfairness.

- Function Creep: Data collected for research might end up feeding marketing algorithms instead...yikes!

Collaboration Tools

- Opportunities: Real-time sharing speeds up work, especially for distributed medical teams consulting on a complex case.
- Challenges:
  - Shadow IT: Staff adopt slick but unsanctioned tools. How do we ensure data in those tools stays secure?
  - Data Sprawl: Is patient data now scattered across multiple platforms, increasing the risk of accidental leaks?
  - Version Control Chaos: Conflicting copies of sensitive reports compromise integrity and might have legal ramifications.

The exam wants you to see beyond the shiny tech. Consider:

- Data Classification is Key: What's truly sensitive enough to warrant staying on-premises, even if it's less convenient?
- Contractual Safeguards: If using cloud vendors, security clauses and SLAs aren't just legal boilerplate, but central to our risk posture.
- Continuous Monitoring: The ease of spinning up new tools means IT might be blindsided. Proactive vulnerability scanning is essential.

Let's put our encryption knowledge to the test with a real-world challenge.

Scenario

Acme Corp, a manufacturing firm, regularly needs to exchange highly sensitive design specifications with a key supplier, Beta Innovations. Currently, they're using email with password-protected zip files... and everyone knows this is less than ideal. Your task is to recommend a more robust encryption solution.

Requirements

Confidentiality: Only Acme and Beta should be able to access the design data.
Integrity: Ensure the files haven't been tampered with in transit.
Non-Repudiation: A way to prove which party sent a specific file, if disputes arise.
Usability: The solution shouldn't introduce too much friction for the non-technical users involved in the exchange.
Options Analysis

Let's consider a few approaches:

Option 1: AES Only

Pros: Fast, strong encryption for data at rest and in transit.
Cons: Key exchange is a mess! Sharing the AES key insecurely voids everything. No way to achieve non-repudiation.
Option 2: RSA Only

Pros: Solves key exchange somewhat – Acme encrypts with Beta's public key, etc. Provides digital signatures.
Cons: Slower, especially for large design files. Key management overhead for each party involved.
Option 3: Hybrid Approach

Pros: Leverages the strengths of both! Use AES for bulk data encryption and RSA to securely exchange the AES key and create digital signatures.
Cons: The most complex to implement and for users to understand.
Trade-offs & Recommendation

The hybrid approach offers the best security, but let's be realistic:

User Training: Technical complexity introduces room for error and can breed resentment.
Is It Overkill?: How sensitive is the data truly? Strong AES and a secure key exchange protocol (not email!) might suffice, with less overhead.
CISSP Considerations

Risk Tolerance: Is Acme in a highly targeted industry where advanced attacks are likely?
Regulatory Landscape: Do they have to comply with standards that might mandate specific encryption methods?
Long Game: The "good enough" solution now might need upgrading later as threats evolve.
Exam Nugget: There's often not one right answer! The CISSP values justification – explain your choices based on the specific situation.

Think of Data Loss Prevention (DLP) like a vigilant airport security guard for your sensitive data. It aims to catch risky behavior before it becomes a headline-grabbing breach. But, just like long TSA lines can frustrate travelers, overly strict DLP can strangle productivity. Let's find the right balance.
DLP Implementation: A Step-by-Step Approach

1. **It Starts with Classification:** Without knowing your crown jewels, DLP is just blindly flailing. Our earlier data classification schema becomes the foundation upon which DLP rules are built.
2. **Know Your Enemy (Or Careless Users):** What are your primary data leakage risks? Malicious employees? Accidental emails to the wrong address? Lost laptops? This shapes your DLP strategy.
3. **Deployment Points:**
    - Network-Based: Monitors traffic exiting your perimeter. Catches broad data uploads, sensitive data heading to odd destinations, etc.
    - Endpoint-Based: Agents on user devices enforce finer-grained control (block USB writes, clipboard actions, etc.). Can prevent both intentional and accidental leaks.
4. **Detection Techniques:**
    - Exact Matches: Blocks known Social Security Numbers, credit card formats, etc. Prone to false positives if not precisely designed.
    - Contextual Analysis: Understands 'projects' or 'departments.' A patient record sent from HR to an external address is a huge red flag!
    - Behavioral Baselines: Learns what's "normal" for a user. Sudden mass file copying is an alert, even if the content isn't inherently sensitive.
5. **Action, Not Just Alerts:**
    - Blocking: The heavy hand. Might be necessary for your most critical data.

- Quarantine: Holds data for review, allowing less disruptive intervention.
- Encryption: Can be triggered automatically (sensitive data tries to email as plaintext? Force encryption!)

Impact on Users

- False Positives Erode Trust: If good work is constantly blocked, users find workarounds...defeating the system.
- Performance Overhead: Especially with endpoint DLP, there might be a noticeable impact on machines.
- Privacy vs. Protection: Heavy monitoring can feel intrusive, damaging morale. Transparent communication is key.

The exam wants you to design practical DLP, not recite textbook features. Consider:

- Phased Rollout: Start with monitoring mode, identify pain points, then gradually tighten controls.
- Tailored to Risk: Maybe marketing doesn't need draconian DLP on par with the R&D department.
- DLP Alone Isn't Enough: Employee training and strong device management are complementary to DLP.

Let's analyze a DLP use case: preventing employees from accidentally uploading sensitive documents to personal cloud storage solutions. How would endpoint vs. network DLP differ in this scenario?

Now get into the world of Digital Rights Management (DRM), where it goes beyond just making sure you paid for that movie download. DRM is all about enforcing control, even after your data has left your cozy network perimeter.

Traditional Access Controls vs. DRM

- Traditional: Think bouncers at your data nightclub. File permissions, ACLs... they guard the door. Once someone's 'in,' they can party (read, copy, modify) as their permissions allow.
- DRM: Like a magic tag attached to the data itself. Even if someone smuggles a copy out, it might "expire," refuse to print, or prevent itself from being edited.

When DRM Supplements Traditional Controls

- Sensitive Design Files: Leaving your network doesn't mean your engineering IP should be openly editable by a contractor. DRM can enforce view-only restrictions, even on the other side of the world.
- Leaked Data: Zero-day breaches happen. DRM isn't preventative, but if leaked documents are watermarked or self-destruct after a time, it limits the damage.
- Licensing & Monetization: Media companies obviously love DRM (though consumers often hate it!). It lets them enforce pay-per-view models and complex usage rights.
- BYOD Headache: Data on personal devices is hard to control. DRM can step in to allow viewing while restricting copying or forwarding sensitive content.

DRM in the Cloud: Challenges Abound

- It Gets Complex Fast: DRM systems need licensing servers, specialized client software... SaaS vendors may or may not play nicely.
- User Experience Suffers: Making every file open in a proprietary viewer isn't exactly seamless.
- Questionable Effectiveness: Savvy users often find ways to circumvent DRM (screenshots, anyone?). It creates a false sense of security if not balanced with other safeguards.

The exam wants you to see when DRM fills gaps, not replaces other things entirely. Consider:

- Legal, Not Just Technical: DRM is often enforced as much by contract as by technology. "You'll be sued if you violate this" can be a stronger deterrent than the protection itself.
- Is It Worth the Hassle?: Trade-offs between control and usability must be weighed, especially for internal data.

- DRM Isn't Foolproof: Focus on detection and response, assuming some level of DRM circumvention will likely occur.

Let's Get Specific: Imagine a pharmaceutical company collaborating on a new drug with external researchers. What's a reasonable DRM strategy for data exchange in this highly sensitive context?

Let's venture to the exciting frontiers of data security! Here, the rules are still being written, and the potential for solving age-old problems is immense.

Homomorphic Encryption (HE)

The Idea: Perform computations on encrypted data without ever decrypting it! Results are themselves encrypted but would decrypt correctly if you had the key.

Why This Is Mind-Bending: Like performing surgery inside a sealed box you never open.

Potential Use Cases:

Secure Cloud Computing: Outsource processing to the cloud without giving the cloud provider access to your raw data.

Privacy-Preserving Analytics: Aggregate medical research across institutions without sharing individual patient records.

Limitations:

Computationally Expensive: Very slow compared to operations on plaintext.

Limited Operations: HE schemes have varying capabilities, not all math functions are feasible.

Still Nascent: Security and efficiency are actively being researched.

Secure Multi-Party Computation (SMPC)

The Idea: Multiple parties hold pieces of data. They compute a function together without revealing their inputs to each other!

Think of It Like...: A group of people each hold a number, they calculate the average without telling anyone their individual number.

Potential Use Cases:

Secure Auctions: Determine the winning bid without anyone knowing what others bid.

Collaborative Fraud Detection: Banks share suspicious activity patterns without divulging customer details.

Limitations:

Complexity Explosion: Overhead grows dramatically with more parties and complex calculations.

Vulnerability to Collusion: If enough parties maliciously work together, they might reconstruct the inputs.

Blockchain (Beyond Bitcoin)

The Idea: Decentralized tamper-evident ledger. Less about confidentiality, more about integrity and preventing unauthorized changes.

Potential Use Cases:

Supply Chain Tracking: Immutable record of where a product originated, each step of its journey, etc. (Counterfeit prevention).
Secure Voting Systems: Reduce fraud by having an auditable, unalterable record of votes.
Limitations:

Not a Cure-All: Garbage in, garbage out. Data on the blockchain can still be false if the origin was tainted.
Performance Overhead: Some consensus mechanisms (proof-of-work) are energy-intensive and slow on purpose.
Immutability Cuts Both Ways: Reversing errors or deleting data violating compliance rules becomes very tricky.
These are powerful, but not silver bullets! The exam focuses on your ability to:

Separate Hype from Reality: Understand the basic concepts, not just buzzwords.
Match Tool to the Problem: Could traditional encryption suffice, or does the scenario truly demand something like HE?
Consider the Long Game: These technologies are evolving. Solutions built on them need adaptability baked in.
Let's Design: Imagine a healthcare consortium wants a shared database of treatment outcomes, but without compromising patient privacy. Could any of these technologies play a role?

III. Security Architecture and Engineering

Let's embark on the journey of transforming abstract security goals into the steel and concrete of robust systems! In Security Architecture and Engineering, we'll play the role of both visionary and meticulous builder.

Think of it like this:

- **Security Architects:** We design the grand fortresses. Where are the walls, the moats, the watchtowers positioned for maximum protection? How do we balance impenetrable defenses with the need for functional 'gates' for authorized access?
- **Security Engineers:** We turn blueprints into reality. Choosing the right bricks (encryption algorithms, firewalls, access control systems), ensuring they're laid expertly, and rigorously testing the fortress withstands attacks.

This domain is all about:

- **Defense in Depth:** It's not just one magical solution, but layers that frustrate attackers at every turn.
- **Threat Modeling:** If we don't understand our enemies' methods, our defenses will have gaping holes.
- **Standards & Frameworks:** We won't reinvent the wheel. NIST, ISO, and others provide battle-tested guidance.
- **The Human Factor:** Even the strongest technology can be undone by a user clicking a bad link. Security awareness is inseparable from our architecture.

Get ready to dive deep into cryptography, secure network designs, system hardening, and the exciting (and sometimes frustrating) world of security testing. Remember, a cybersecurity professional is never truly done – we must constantly adapt as adversaries evolve!

Let's craft a fortress of least privilege around the deployment of a brand new web server. Our mission is to make an attacker's life incredibly difficult even if they manage to gain a toehold within our system.

Scenario: Public-Facing Web Server

- Purpose: Hosts a company website and a simple customer portal for account information.
- Key Assets: Website code/content, customer database (backend)

Applying Least Privilege

1. **User Accounts:**

- No Admin Rights!: Regular employees interacting with the website (content updates, etc.) get tightly restricted accounts. Can't install software, no changes to system settings.
- Separation of Duties: Developers pushing code updates have a separate account from the content team – limits the impact of compromised credentials.

2. **Service Accounts:**

- One Account Per Service: The web server process itself should NEVER run as 'root' or a high-level admin. Separate accounts for the web service, database, etc.
- Lockdown!: These accounts should be unable to log in interactively, restricted only to the files and system resources their specific service absolutely needs.

3. **Network Access**

- Firewalls Are Your Friend: Strict rules limiting inbound traffic to the web server's essential ports (80, 443 likely). No random connections allowed.
- Outbound Traffic: Does the web server legitimately need to initiate connections out to the internet? Often this can be heavily restricted, blocking avenues for malware.
- Segmentation: This web server shouldn't freely talk to the HR file server... network segmentation limits lateral movement if an attacker does get in.

4. **Data Permissions**

- Web-Writeable: Only specific folders are writable by the web server account (uploads maybe). Prevents malware from defacing your whole site.
- Database Cautions: The web service account has read-only access where possible. SQL Injection gets much less scary if they can't drop tables!
- No Secrets in Code: Connection strings, database passwords... These live in configuration files with the strictest file permissions, not hardcoded.

CISSP Mindset

- It's An Iterative Process: We start with strong restrictions, then loosen them slightly as absolutely needed for operation.
- Monitoring Is Key: Anomalous behavior (a service account suddenly trying to access the finance share) should raise red flags.
- User Frustration: A delicate balance! Explain why these restrictions ultimately protect THEM, too.

Let's complicate it: What if our website needs to accept file uploads from customers? How does this change our least privilege approach?

Picture separation of duties as a complex dance – when done well, it's both secure and graceful. But sometimes, the music of operational efficiency clashes with the rhythm of security best practices. Let's examine one such scenario.

Scenario: Small Business Struggles

You're the de facto IT/security person for a small but growing company. The same person handles:

- Network Administration: Firewall changes, new user setup, etc.
- Helpdesk: Resolving day-to-day user issues, password resets, software installs.
- Development: They built the in-house inventory management system, and maintain it... because no one else can.

Risks Abound

- Segregation of Duties? What's That?: One person has the keys to the whole kingdom. They could abuse access, install malware, or just make mistakes with disastrous impact.
- No Checks & Balances: Risky changes can be pushed through with nobody providing a second, critical look.
- Burnout & Resentment: This person is overworked and likely feels underappreciated... a recipe for disgruntled insiders, a major threat.

Mitigation Strategies (Not Ideal, But Realistic)

- Dual Control: For sensitive changes, mandate that another person (a manager, even if less technical) reviews and co-approves.
- Job Rotation: If feasible, bring in another partially trained employee to share the load. Reduces the 'single point of failure' risk.
- Transaction Logs & Audits: Meticulous logging of admin actions. Regular review becomes non-negotiable, knowing someone IS watching.

- Privileged Account Management (PAM): Investing in a tool might be worthwhile. It can vault passwords, enforce session monitoring, and sometimes alert on anomalous behavior.

The exam wants you to see beyond the textbook ideal, to the messy reality companies face. Consider:

- Risk Acceptance: This company might consciously accept the risk, knowing full segregation is impossible with their budget. It's a bad, but informed decision.
- Compensating Controls: Auditing becomes paramount if they can't prevent risky combinations of duty.
- Temporary Workarounds: During a crisis, sometimes you break the rules out of necessity, then restore them after. Document the heck out of it!

Let's explore a different situation – a large financial institution with strict separation of duties. What new tensions might emerge in that environment?

Fail-safe defaults are like having your security system's alarm permanently blaring. Sure, they're incredibly safe...but also maddening for everyone involved. Let's find a way to turn down the volume without leaving the doors wide open.

The Challenge of "Deny-All"

Let's picture a brand new firewall, configured with an iron-clad fail-safe: every single packet that isn't explicitly allowed gets dropped. This seems like great security, until...

- Nothing Works!: Employees can't access websites, email, or essential internal systems. Business grinds to a halt.
- Frantic Helpdesk: Flooded with calls from users who mistakenly assume their computer is broken.
- Shadow IT Rises: Frustrated departments circumvent IT entirely, using unsanctioned tools that open new security holes.

It's Not Just Firewalls...

- New Software Installations: System lockdown policies block users from installing even approved productivity tools.
- Strict Password Policies: Excessively complex password rules lead to sticky notes under keyboards and a never-ending stream of resets.
- Restrictive File Permissions: A necessary collaboration suddenly fails because 'nobody' has write access anymore.

Finding a Balance

The CISSP exam wants you to design for the real world, not a theoretical one. Consider:

- Gradual Rollout: Start restrictive in a test environment, identify pain points, then loosen iteratively with careful monitoring.
- User Education: Don't assume they know why things aren't working! Proactive communication and training ease the transition.
- Context-Aware Defaults: Can the firewall tell an internal workstation from a public-facing web server? Deeper context means less 'one-size-fits-all' blocking.
- Exceptions Workflow: There WILL be legitimate needs breaking the rules. Have a fast, clear process for approvals, not an opaque black hole.
- Justify Your Decisions: The more disruptive the default, the stronger your threat model and risk assessment better be.
- Security That's Invisible: Is often the best kind! Can we achieve the security goals in a way users barely notice?

- Usable != Insecure: These don't need to be enemies. Creative solutions often exist to satisfy both.

Let's make it concrete: A marketing team constantly needs to share large design files. Strict file permission defaults make this a nightmare. How can we improve security without hindering them?

Forget those generic diagrams with concentric rings labeled "defense in depth!" Let's make this tangible by building a fortress layer by layer, specifically for protecting those juicy marketing design files while enabling sharing.

Scenario: Secure File Exchange

Goal: External collaborators can upload and download files, marketing has controlled access, sensitive data stays protected.

1. The Perimeter (DMZ)

- Not Our House: Web server hosting the upload portal lives in a DMZ, isolated from the internal network. Compromise here is contained.
- Firewall Rules: Strict! Only ports 80/443 open, specifically to the web server's IP. Source IPs might even be whitelisted (if the partners are known).
- Intrusion Detection System (IDS): Watches DMZ traffic for anomalies, raising an early alert if attackers start probing.

2. Application Layer

- Input Validation: The portal won't accept just any file. Strict rules on name, size, format. This blocks obvious malware masquerading as a design.
- Authentication & Fine-Grained Authorization: Not just "logged in," but roles like "Collaborator" with upload-only permission to a specific project folder.
- Web Application Firewall (WAF): Looking for attacks like SQL injection attempts, cross-site scripting – common ways to hijack web apps.

3. The Treasure Chest (Database)

- Encryption at Rest: Even if the database is breached, raw design files aren't immediately exposed.
- Principle of Least Privilege: The database account used by the web application can only access the specific tables and data it needs.
- Logging and Auditing: Every file access, modification, or deletion is meticulously logged, providing a trail for investigations if needed.

How It All Works Together

- Malware Delivery: Blocked at the file upload stage (application layer)
- Zero-day Exploit: Web server compromise is bad, but contained in the DMZ. Strict database permissions limit the attacker's movement.
- Leaked Password: Fine-grained roles mean they access a limited dataset, not the whole archive.
- Insider Threat: Anomalous database access patterns (odd hours, excessive data) might be spotted by the IDS or during log reviews.

The exam wants you to see the synergy, not just a list of technologies. Think like an attacker:

- "If I break through X, what's my next move?": This mindset drives your layered defenses.
- No Single Point of Failure: Even if one control fails, others are there to slow or stop the attack.
- Monitoring is Key: All these defenses are useless if nobody looks at the alerts and logs!

Let's add a wrinkle: We need to allow real-time preview of these design files within the web portal itself. How does this change our risk landscape? What new controls are needed?

Enabling real-time previews adds a whole new dimension to our security challenge! Think of it like letting potential buyers peek inside the treasure chest before they've fully committed. Here's how it shakes things up:

New Attack Surfaces

- File Parsing Vulnerabilities: The image previewer itself becomes a juicy target. A zero-day in that software could let attackers break out of the sandbox.
- Cross-Site Scripting (XSS): If attackers can embed malicious code in the filenames or metadata of design files, it might execute when previewed.
- Performance & Scalability: Preview generation can be resource-intensive. A denial of service attack could now be launched by simply uploading tons of files.

Adapting Our Defenses

1. **Isolation is Key:**
- Sandbox the Preview Process: It runs in a heavily restricted environment, cut off from the main web application and the database. Limits the blast radius.
- Virtualization/Containers: Quick preview instances spin up to handle a file, then self-destruct, preventing lingering compromise.

2. **Input Sanitization (Even More So!)**
- Stricter Filtering: We can't just assume design files are safe. Fuzzing tools might be used to deliberately generate malformed files to test our previewer's resilience.
- Metadata Scrubbing: Remove extraneous metadata from the previews, reducing the attack surface for sneaky embedded code (XSS, etc.).

3. **Monitoring & Anomaly Detection**
- File Upload Patterns: A sudden spike in uploads might be an early sign of a DoS attack attempt. Rate limiting could be implemented.
- Preview Resource Hogging: If the preview process hangs or starts consuming excessive resources, it should be automatically terminated.
- Behavioral Baselines: What's "normal" preview activity for legitimate users? Deviations from this might indicate malicious intent.

Additional Considerations

- Watermarking Previews: Overlays a subtle "Draft" or user-identifying watermark on the preview, discouraging unauthorized reuse.
- Tiered Access: Maybe low-res previews are open to all; full-quality ones require a higher authorization level.
- Client-Side Previews: If feasible, pushing some of the previewing to the user's browser offloads risk, but has UX trade-offs.

CISSP Mindset

- Evolving Threat Landscape: Adding functionality means being vigilant about new attack vectors it might introduce.
- Usability Impacts: Heavy sandboxing might make previews too slow, frustrating legitimate users. Security is always a balancing act.
- "Previewable" Doesn't Equal "Downloadable": Our controls need to ensure the underlying high-res design file remains protected.

Let's dispel the myth that there's always one 'best' encryption algorithm. The choice is all about trade-offs and finding the right tool for the job.

AES-256 vs. AES-128: When Bigger Is Better

- Key Difference: Key length (256 bits vs. 128 bits). More bits = exponentially harder to brute-force.
- Scenarios Favoring AES-256:
  - Top Secret Data: Where confidentiality is paramount for the long term. Today's 'unbreakable' might be a problem a decade from now with faster computers.
  - Regulatory Mandates: Some compliance standards (handling certain types of financial or health data) might specify AES-256 as the minimum.
- But It Comes at a Cost: AES-256 is slightly slower than AES-128, especially noticeable on older or resource-constrained devices.

RSA vs. Elliptic Curve Cryptography (ECC)

- Functionally Different: RSA is older, used for key exchange and digital signatures. ECC offers similar security with smaller key sizes, gaining popularity.
- Scenarios Favoring ECC:
  - Low-Power Devices: Think IoT sensors, smartwatches, etc. ECC is less computationally demanding = better battery life.
  - High-Speed & Scalability: ECC shines in environments with many simultaneous connections (busy web servers)
- But...Compatibility Catches: Not every device and software library supports all flavors of ECC. RSA remains the 'safe' choice for legacy interoperability.

CISSP Mindset

The exam emphasizes practical choices, not just reciting textbook trivia. Consider:

- Risk Appetite: "Unbreakable for 50 years" might be overkill for a newsletter signup form.
- Performance Isn't Everything: A slightly slower, but more secure algorithm might be justified for highly sensitive data.
- Change is the Only Constant: Just because AES-256 is top dog today, doesn't mean a vulnerability won't emerge in the future. Crypto agility is key!

Let's analyze this: You're designing a secure messaging app for smartphones. Space to store keys is limited and you want snappy performance. Would ECC be a good fit?

ECC looks incredibly promising for our secure messaging app scenario! Let's break down why it aligns well with our constraints:

ECC's Strengths Meet Our Needs

Tiny Keys, Big Security: ECC offers equivalent security to RSA with much smaller key sizes. This is precious when storage is limited on mobile devices.

Performance Champ: ECC operations are generally faster than RSA, especially the key generation process. This contributes to a responsive, snappy feel for the user.

Power Miser: Less computationally intensive cryptography means our app won't be a major contributor to battery drain, a big win for users.

Potential Pitfalls (Important to Consider)

Implementation Matters: Not all ECC curves are created equal. Some have known vulnerabilities, others are patent-encumbered. Careful curve selection and using a well-regarded crypto library are essential.

Limited Compatibility: While ECC adoption is growing, you may still run into edge case devices or legacy systems that don't fully support it. Having a fallback to RSA might be necessary.

Side-Channel Susceptibility: Like many algorithms, ECC isn't immune to side-channel attacks (timing analysis, power monitoring). Meticulous implementation is needed to mitigate these.

Let's justify our potential decision to use ECC:
Threat Model: Assuming users are exchanging sensitive conversations, strong encryption is a must. ECC lets us have that without sacrificing the user experience.
Risk Tolerance: The compatibility and side-channel concerns are manageable with careful development and testing. The advantages outweigh the risks in this context.
Future-Proofing: As ECC becomes even more ubiquitous, we've already positioned our app to be at the forefront of security.
Let's put our encryption toolbox to work and understand when each tool is the perfect fit for the task at hand.

Symmetric Encryption (Data at Rest)

- Think: Locking a Safe
- Best For: Bulk data storage – files on a hard drive, database entries, backups shipped offsite, etc.
- Security Goals:
  - Confidentiality: Only those with the shared key can 'open' the safe to view the contents.
  - (Some) Integrity: Modern symmetric modes often detect tampering, not just unauthorized reading.
- Why Symmetric Is Key: Speed! AES is blazingly fast compared to asymmetric methods.
- Example: A patient records database at a hospital. AES keeps the data unreadable even if the hard drive is stolen.

Asymmetric Encryption (Key Exchange)

- Think: A Secure Handshake
- Best For: Establishing the shared secret needed for symmetric encryption, and sometimes digital signatures.
- Security Goals:
  - Secure Key Exchange: Alice can encrypt something ONLY Bob can decrypt, and vice-versa. Basis for things like TLS (https).
  - Authentication: Digital signatures built on RSA prove who 'sent' a document in a way that's hard to forge.
- Why Not Just Asymmetric for Everything?: It's SLOW, especially on large datasets.
- Example: Your browser connects to a website. RSA facilitates an exchange of a unique AES key to be used for the duration of that browsing session.

Hashing (Integrity Verification)

- Think: Unique Fingerprint
- Best For: Password storage, verifying file hasn't been tampered with, ensuring messages haven't been changed in transit.
- Security Goal:
  - Integrity, Not Confidentiality: You CAN'T get the original data back from a well-designed hash. It's a one-way street.
- Why This Matters:
  - Passwords: We store the hash, never the password itself. When a user logs in, we hash their attempt and compare the hashes.
  - Software Downloads: Website provides the SHA256 hash of the file. You download, compute the hash yourself, and they should match.
- Example: You change your bank account settings online. A hash is used to guarantee no attacker modified your changes behind-the-scenes.

These often work in concert:

1. Asymmetric encryption (RSA) securely sends you a symmetric (AES) key.
2. AES encrypts your large dataset for storage.
3. Hashing verifies the dataset wasn't corrupted on disk.

The exam wants you to see this 'dance' of cryptography, not just isolated concepts!

Imagine PKI as the infrastructure that turns "digital yelling across a crowded room" into private, trusted conversations. Let's see how this plays out with secure email using S/MIME.

How PKI Empowers Secure Email

- Certificates: The Backbone. Alice has a certificate issued by a trusted Certificate Authority (CA). It includes her public key and states, essentially, "This CA vouches that this key really belongs to Alice."
- Sending a Secure Message: Bob wants to email Alice confidentially.
    - He gets Alice's certificate (maybe from a public directory).
    - He uses Alice's public key to encrypt his message. Only Alice with her corresponding private key can open it.
- Digital Signatures: Going Further. Alice can 'sign' her reply using her private key. Bob uses her public key to verify the signature. This offers:
    - Authentication: It really WAS Alice who sent it
    - Non-Repudiation: She can't later claim she didn't write it
    - Integrity: The message content hasn't been altered.

Trust Hierarchies & CAs

- Root of Trust: Web browsers and email clients come with a pre-trusted list of root CAs. Certificates issued by these root CAs (and those in the 'chain' beneath them) are inherently trusted.
- This Sounds Fragile!: It kind of is. If a CA gets compromised, all the certificates they've issued become suspect. That's where revocation comes in...

Certificate Revocation Lists (CRLs) & OCSP

- "That Cert's No Good Anymore!": Maybe Alice's private key was stolen. Her certificate needs to be revoked to prevent imposters.
- CRLs: CAs publish a list of revoked certificates' serial numbers. Your email software checks this list before trusting a cert.
- OCSP: A more real-time check, asking the CA directly about the status of a specific certificate.

CISSP Considerations

- PKI is Complex: Managing certificates at scale in a large organization gets messy. Employees need training on how to use S/MIME properly!
- Usability Frictions: Exchanging certificates in a user-friendly way is a hurdle. Not all email clients play nicely.
- Attackers Target CAs: If the root of trust is compromised, the whole system crumbles.
- Doesn't Solve Everything: PKI can prevent a third party from reading Alice's email, but not Bob himself from doing something bad with the data once he's decrypted it.

Let's explore a scenario: Your company wants to start using S/MIME for all internal communication. What factors should you weigh before rolling this out?

Picture this: you build an impenetrable fortress, but then tape the key to the front door. That's the tragicomedy of amazing encryption rendered useless by poor key management. Let's dive into best practices, and how neglecting them blows everything up.

Key Generation: Randomness Rules

Predictable is Poison: Using weak passwords or easily guessable seeds to derive keys is like building your fortress walls out of cardboard.
Cryptographically Secure Randomness: Special tools and sources of (true) randomness are essential. Hardware random number generators, leveraging sources like atmospheric noise, fill this need.
Storage: HSMs and Split Knowledge

Keys Are the Crown Jewels: Treat them as such!
Hardware Security Modules (HSMs): Specialized, tamper-resistant appliances. Keys live here, and perform cryptographic operations without the key itself ever leaving.
Key Escrow: For those "break glass in case of emergency" scenarios. Can involve splitting the key among multiple custodians requiring all to be present to reconstruct it.
Avoid Plaintext Storage: If you absolutely must store keys in software, robust encryption should be the bare minimum.
Rotation: Proactive Defense

Keys Don't Age Well: The longer a key is in use, the more data is encrypted with it, the more likely it is to be compromised someday.
Regular Rotation: Reduces the impact of a single key leak. Policy dictates how often (sensitive data = more frequent rotation).
Automate & Track: Manual key rotation is error-prone. Tools simplify the process and provide an audit trail.
Destruction: Zeroes, Not Deletes

Deletion Ain't Enough: Data remnants often linger on drives. "Deleting" a key file isn't truly wiping it.
Secure Erasure: Specialized tools overwrite the area with random data multiple times, making recovery near impossible.
Cryptographic Key Shredding: Some systems logically 'destroy' a key without wiping its storage, making it unrecoverable even if the media is accessed.
When Things Go Wrong: Impact

Perfect Encryption, Nullified: An attacker gains access to the key? They unlock all the data it previously protected.
Long-Term Compromise: If old keys aren't securely destroyed, decrypting archived data (backups, etc.) becomes possible for an adversary long after the fact.
Loss of Trust: Sloppy key management can shatter reputation, especially when dealing with compliance mandates or legal fallout.
Think risk-based trade-offs:

Storing the key protecting your company picnic signup form requires less rigor than the key to your main customer database.
Procedures Matter: Perfect tech is useless if users write keys on sticky notes!

Weakest Link: A top-tier HSM won't help if an attacker phishes the credentials to access it. Security is always holistic.

Data centers aren't just about keeping servers cool; it's about maintaining the precise environment they thrive in. Let's dive into factors beyond the obvious air conditioning unit.

Humidity: The Hidden Enemy

- Too Dry: Static Electricity! Zap! Those tiny sparks might feel harmless to you, but can fry sensitive electronics components.
- Too Humid: Condensation forms on gear. Picture a cold drink sweating on a hot day... only that 'drink' is your expensive server. Not ideal.
- The Goldilocks Zone: Data center humidity is kept within a specific range (often 45-55% relative humidity), ensuring both equipment safety and human comfort.

Power Problems & Solutions

- Spikes and Sags: Power fluctuates more than you might think! This stresses power supplies and contributes to hardware failures.
- UPS Saves the Day: (Uninterruptible Power Supply). Big batteries + power conditioning keep things running smoothly for short outages. Bridges the gap until the generator kicks in...
- Generator Power: For longer outages, diesel generators take over. Routine testing and fuel contracts are vital – they do no good if they fail to start!
- Redundant Power Feeds: Ideally, the data center draws power from two separate grids, mitigating impact if one feed has problems.

Impact of Neglect

- Hardware Death Spiral: Outside its comfy zone, equipment fails faster. This might not be immediate, but shaves time off component lifespans.
- Unexplained Instability: Random crashes, odd errors... could be traced back to flaky power or electrostatic discharge frying a memory chip.
- Outages = Downtime = $$$ Lost: That critical app goes down, business grinds to a halt, and the frantic scramble to recover is costly.

The exam emphasizes the 'why' behind the 'what':

- Risk Assessment: How much downtime can your business tolerate? That drives the expense of your power safeguards.
- It's Not Just Servers: Humidifiers, dehumidifiers, power monitoring systems – these supporting bits matter just as much!
- Fire Suppression: Gotta consider this too! Water-based systems and a data center are a bad mix... Gas suppression has its own hazards

Let's add complexity: Many companies use colocation facilities (renting space in a data center). How does this change their control over environmental factors?

Let's face the unfortunate reality: Even with top-notch prevention, fires can happen. In a data center, the fire suppression system itself could be as damaging as the flames if you choose the wrong type.

Water-Based Sprinklers

The Classic: Effective at actually extinguishing most fires, and readily available.

Data Center Nightmare: If they go off, your servers are probably toast even if the fire itself was small.

Usage: Might be used in non-data center areas of the facility (offices, etc.) with careful design to contain any water damage.

Gaseous Suppression Systems

These work by displacing oxygen, starving the fire, or chemically interrupting combustion. Several types exist:

Inert Gases (IG-55, Argonite, etc): Safe for humans at breathable concentrations. Floods the room, reducing oxygen levels enough to extinguish the fire.
Chemical Agents (FM-200, etc): Work faster, need lower concentrations. Some have environmental concerns, others break down fairly quickly.
Pros: No water damage! Electronics may survive if the fire is caught early.
Cons: Can be fatal to humans if oxygen levels drop too low. Evacuation and strict safety procedures are mandatory.
Choosing Wisely

Life Safety is Paramount: If the data center is normally occupied, a system safe for humans at the necessary concentrations is a must!
Type of Fires: Electrical fires respond differently than, say, a burning pile of cardboard boxes. Your suppression system needs to match the likely fuel sources.
Speed vs. Damage: In a server room full of irreplaceable data, maybe preventing water damage is worth a slightly higher risk of the fire spreading a tiny bit more before the gas kicks in.
Additional Considerations

Pre-Action Systems: These have a detection stage before releasing the agent, reducing accidental discharge. Often coupled with loud alarms so people can get out!
Localized Suppression: Protecting individual cabinets is possible, containing the problem without shutting down the whole data center.
Environmental Impact: Some older chemical agents are terrible for the ozone layer. Newer facilities likely use more eco-friendly options.
The exam wants you to assess trade-offs, not just regurgitate facts about suppression systems. Consider:

Business Continuity: How quickly can you recover if fire DOES damage equipment? This influences how aggressively you prioritize equipment safety in suppression choice.
The Human Factor: Complex systems reliant on manual activation during a crisis are failure-prone. Automation is good, with robust safeguards!

Imagine trying to break into a bank vault. You'd face layer upon layer of security, not just one obstacle. Physical access controls aim for the same approach to safeguard your most critical areas.

The Layers

- Badges: The Starting Point. Identifies who a person IS at a glance. Often used with electronic card readers to track entry and exit.
- Biometrics: More Certainty. Fingerprint, iris scan, etc. Harder to fake than a stolen badge, proves the body in front of the door is truly the authorized one.
- Mantraps: Controlling Flow. Interlocking doors, only one opens at a time. Prevents tailgating, lets guards carefully check each person.
- Human Guards: The X-Factor. Trained to spot suspicious behavior tech might miss. Can intervene as a deterrent or in a crisis situation.

Two-Factor Authentication (Something You Have + Something You Are)

- High-Security Must-Have: For that server room holding the keys to the kingdom, we need more than a swipe card that could be lost. Badge + Biometric gets us closer.
- It's Not Perfect: Even biometrics can be fooled with enough effort. Defense in depth means even if this combo is bypassed, there are further hurdles.
- Where 2FA Gets Tricky: Areas of high throughput (employee entrance during rush hour). Balancing strong security with preventing a bottleneck is an ongoing challenge.

Situations Demanding 2FA

- Data Centers: Where the business's lifeblood (its data) is stored.
- R&D Labs: Intellectual property central, prevent unauthorized access keeping competitors at bay.
- SCIFs (Sensitive Compartmented Information Facilities): Government/military have strict rules where 2FA is the bare minimum.

The exam tests your judgment, not just memorization of tech. Think about:

- Tailgating: All the tech in the world is defeated by the person holding the door open for a buddy. Employee awareness matters!
- Cost Considerations: Mantraps are expensive! Justify them based on the level of threat and value of the asset being protected.
- Failure Points: What happens if the biometric reader glitches? Backup procedures are needed, without creating new vulnerabilities.

Let's explore this: An executive insists on being exempt from "pesky" 2FA, citing their importance. How do you approach this from a risk-based perspective?

That tricky scenario of the executive demanding special treatment is, sadly, a classic security headache! Let's craft a response that balances security needs with the realities of organizational politics.

Emphasize Risk, Not Rules

- Frame It Their Way: Execs care about the big picture. "This policy protects our competitive advantage/customer trust/shareholder value" gets more attention than "because I said so."
- Targeted Threats: Executives ARE more likely to be targeted by sophisticated spear-phishing or social engineering. Their access makes them a high-value target.
- Data Breach = Their Problem: Draw a clear line between a compromised exec account and the fallout they'll personally face: investigations, potential fines, reputational damage...

Mitigating, Not Necessarily Eliminating, Risk

- Compromise Isn't Impossible: Honesty is key. 2FA dramatically lowers the risk, but a determined attacker might succeed. Our job is making that extremely difficult.
- "Executive-Friendly" Options: Maybe a less intrusive biometric like voice recognition? A physical token instead of an app on their phone? Show you're working with them for usability, not against them.
- Tiered Access: Does the exec truly need access to the source code repository? Fine-grained access control reduces their attack surface.

Additional Considerations

- Written Exception (CYA): If they absolutely refuse, get it signed – by them, ideally with legal counsel present. Makes it clear they assume the risk.
- Increased Monitoring: Extra scrutiny on their account activity. Anomalies are investigated faster, limiting potential damage.
- Training as Key: Execs are often victims because they're taught to trust, not be suspicious. Tailored security awareness helps them be part of the solution.The exam wants you to be the pragmatic bridge between rigid security ideals and business needs. Consider:

- Influence, Not Just Authority: How can you persuade the exec, even if you can't force compliance?
- What if...?: Run tabletop exercises. "Your account was breached, here's the likely impact..." can be more persuasive than abstract warnings.

Often, the easiest way to breach a fortress isn't by smashing through the gate, but by sweetly smiling your way in. Let's explore how attackers exploit the edges of physical security, and the countermeasures that are often overlooked.

Surveillance: It's Not Just About Recording

Spotting the Pre-Attack: Attackers may case the facility. Cameras trained on entrances help spot people who linger far too long, take unusual interest in security measures, etc.
Behavioral Analysis: Modern video analytics can detect suspicious patterns. Someone repeatedly trying doors, or backtracking through access-controlled zones might raise a flag.
The Blind Spots Matter: Cameras deter...but also create predictable areas an attacker can exploit. Are those blind spots truly unavoidable?
Anti-Tailgating: Culture + Tech

The Human Weakness: Even the fanciest mantraps are useless if employees politely hold doors open. Awareness campaigns ("See Something, Say Something") are vital.
'Buddy System' for High-Security: In critical areas, two-person authorization (neither door opens unless two valid badges are used) mitigates casual tailgating.
Turnstiles & Physical Barriers: Make tailgating physically difficult. This forces attackers to engage in more overt behavior, increasing their risk.
The Visitor Conundrum

Escorting Necessary? Does every visitor truly need an escort, or does it depend on where they're going? Risk-based policies are key.
Visitor Badges: Clearly identifiable, often with time limitations or restricted access. Allows for social accountability – an out-of-place visitor stands out.
Sign-In/Sign-Out: Seems basic, but vital! Knowing who is on-site during an emergency is crucial, and logs provide investigative trails.
Social Engineering: The Lockpick of the Mind

Impersonation: The fake repairman, the lost delivery person... Staff need to be trained to politely but firmly challenge anyone not visibly authorized.
Pretexting: That urgent call from "IT" needing the employee to confirm their badge number 'due to a glitch'? Policies on information release are as important as any firewall.
Urgency & Authority: Attackers create scenarios that override good sense. Employees need empowerment to question things that seem 'off' without fear of reprisal.
Think in Scenarios: The exam loves "Which of these is the biggest risk?" Analyze a facility, and its purpose, to determine the likely attack vectors.
Defense in Depth, AGAIN: Tech helps, but if employees leave the backdoor wide open, your perimeter is useless.
The Cost of Inconvenience: Security vs. usability is real. A researcher blocked from entering their lab for 10 minutes due to a tailgating false alarm isn't going to be your security champion.

Let's make it tangible: You're securing a facility with regular employees and frequent contractor visits. What's your priority focus – surveillance, anti-tailgating tech, or a visitor management overhaul?

IV. Communication and Network Security

Welcome to the exciting domain of Communication and Network Security! Think of it like this:

We're building an intricate global highway system designed to move precious cargo (data) at blazing speeds while keeping it safe from thieves and highway bandits.
Robust networks are the foundation upon which everything else in cybersecurity rests. If our roads themselves are compromised, our strongest city walls mean little.
In this domain, we'll dive into:

Network Architecture: How do we design networks that are not only efficient but compartmentalized to limit the spread of attacks? Think of it like a city with well-planned checkpoints and segmented districts.
Secure Protocols: The traffic rules of the internet. We'll look at things like TCP/IP, SSL/TLS, and IPsec, understanding how they provide confidentiality, integrity, and authentication of data in motion.
Firewalls & Intrusion Detection Systems (IDS)/Intrusion Prevention Systems (IPS): Our network's vigilant guards. Controlling what traffic can flow, and spotting intruders trying to sneak past.
Wireless Security: The unique challenges protecting data traveling over invisible airwaves, easily intercepted by the right adversary.
Remote Access & VPNs: Secure tunnels into our network, crucial in a world of remote work, but also ripe for exploitation if configured poorly.
Get ready to think like both network engineers AND skilled attackers, because the only way to design a truly resilient system is to understand how adversaries will try to break it!
The OSI and TCP/IP models are like two maps of the same treacherous networking terrain. One is super detailed, good for theorizing; the other is pragmatic, reflecting reality's tendency to smudge the lines a bit. Let's break it down!

OSI Model: Layers of Abstraction

1. Physical: Raw bits on copper wire, fiber optics, etc. The 'trucks' carrying our data.
2. Data Link: Error correction, turning raw bits into frames. MAC addresses live here.
3. Network: Routing! IP addresses, packets finding their way across interconnected networks.
4. Transport: Enter TCP and UDP. Reliable delivery, ports, keeping track of data flows.
5. Session: Managing long-lived connections between apps. Often gets rolled into upper layers in practice.
6. Presentation: Data formatting (think ASCII/Unicode), sometimes compression or encryption.
7. Application: The protocols users directly interact with: HTTP for websites, SMTP for email, etc.

TCP/IP: What Really Runs the Internet

- Application: Combines OSI's concept of Application, Presentation, and Session into one pragmatic blob.
- Transport: Mostly the same as OSI (TCP, UDP).
- Internet: Equivalent to OSI's Network layer (IP).
- Network Access: Lumps together OSI's Physical and Data Link. Focus is on the tech to get bits onto a specific medium (Ethernet, WiFi, etc.)

Where Things Get Interesting

- HTTP (Hypertext Transfer Protocol): Clearly Application layer in both models.
- TLS (Secure Sockets Layer): OSI would say Presentation (it encrypts data), TCP/IP treats it more like Transport (securing the connection itself).
- BGP (Border Gateway Protocol): The glue holding the internet's routing together. Network layer through and through.
- IPsec: Secures IP packets. Can tunnel traffic. Sits in a fuzzy spot – OSI would see it squarely in the Network layer, TCP/IP treats it more broadly.

Exams will test your understanding of where things conceptually should be, even if the real world is messier. Focus on:

- Encapsulation: As data moves down the layers, headers are added, like boxes within boxes. Understanding this is KEY.
- "Close Enough" Mapping: Could you make an argument for TLS being either Presentation or Transport layer? Yes, and that's the kind of nuanced thinking the CISSP wants.
- Troubleshooting Mindset: A problem with HTTPS is unlikely a physical cabling fault. The models guide where you start your investigation.

Let's journey into the mind of the networking devices that make information flow. Picture them as diligent postal workers, each with a unique way of figuring out where to send your precious data packets.

Layer 3 Routers: IP-Driven Decisions

Address Fanatics: Routers are obsessed with IP addresses (specifically the destination IP in the packet header).
Routing Table: Their guidebook. Lists known network destinations, and which "next hop" router gets the packet closer to its goal.
Finding the Way: Imagine each router is a crossroads. The routing table is their map, and forwarding the packet is like deciding to send it north, south, east, or west to reach its final destination.
Layer 2 Switches: MAC-Based Simplicity

All About Those Addresses: Switches live in the land of MAC addresses, those unique identifiers burned into every network card.
MAC Address Table: They learn which MAC is connected to each of their ports. Imagine a list: Port 1 has Alice's computer, Port 2 has Bob's, etc.
Efficient Delivery: A packet arrives, the switch peeks at the destination MAC, checks its table, and sends it out the specific port where that device lives, and ONLY that port.
Stateful Firewalls: The Traffic Police

Layers 3 & 4 Scrutiny: These firewalls analyze IP addresses and port numbers, used by protocols like TCP and UDP.
The Rulebook: Firewall policies dictate what's allowed: "Web traffic from the internet to our servers? Okay. File sharing from HR to the outside? No way!"
Statefulness: The key difference! They track active connections. An outgoing web request automatically gets the response allowed back in, without a poke a hole for every single incoming packet.
CISSP Focus

The exam wants you to visualize how data moves through a network based on these header examinations. Think about:

Broadcasts vs. Unicasts: A switch seeing a broadcast (destination MAC of all Fs) floods it out every port, a router STOPS broadcasts dead in their tracks.
Efficiency: Switches are fast for local traffic, routers are slower but necessary for getting packets to the wider internet.
Attack Filtering: A stateful firewall denying an incoming connection on port 22 (SSH) likely means someone is trying to remotely brute-force their way into a server.
Let's imagine a misconfigured device plugged into the network starts spewing traffic with spoofed IP addresses. At which layer could this potentially be detected and stopped?
Picture your network as a bustling city. VLANs allow us to create distinct neighborhoods, improving security and traffic flow, but like any urban planning, there are trade-offs to consider.

Scenario: Safeguarding the Research Lab

A pharmaceutical company has a research lab where its crown jewel – experimental drug formulas – reside. We'll use VLANs to create a secure 'enclave' for this sensitive data.

- VLAN 100: General Population. Employees browsing the web, checking email, etc.
- VLAN 200: Research Sanctuary. Only authorized devices and users are assigned to this VLAN.
- The 'Moat': Switch configuration ensures traffic from VLAN 100 CANNOT directly reach VLAN 200, and vice-versa.

Security Benefits

- Containment: A malware infection on an HR computer stays in 'its neighborhood', unlikely to spread to the R&D systems.
- Access Control: We tie VLAN membership to user roles. Only research staff get placed on VLAN 200, adding a layer beyond just passwords.
- Monitoring: We focus scrutiny on traffic crossing the VLAN boundary. Any unusual activity stands out more readily.

Operational Considerations

- Trunk Links: Physical cables that carry multiple VLANs tagged, like multi-lane highways with special signs marking which lane is which.
- Inter-VLAN Routing: Let's be realistic, researchers need SOME internet access. This requires a router or layer 3 capable switch, with strict rules on what traffic can cross between VLANs.
- Management Overhead: More VLANs = more configuration to maintain. Finding the balance between segmentation granularity and administrative headache is key.

Think like an attacker, and how the design thwarts them:

- Lateral Movement: Compromise a general workstation? VLANs limit how far the attacker can easily pivot throughout the network.
- Zero-Trust in Action: Even within VLAN 200, we shouldn't blindly trust every device. Further micro-segmentation might make sense depending on the sensitivity of the research.
- The Human Factor: Best technical design is defeated if a researcher plugs a personal, infected device into the lab's network. Policy and training matter!

Let's add complexity: The company starts collaborating with external researchers needing access to SOME data on VLAN 200. How does this change the design?

Inviting external collaborators into our secure VLAN enclave adds a thrilling layer of complexity (and some potential headaches!) Let's explore how we adapt our network design while minimizing risk.

New Challenges
Trust, But Verify: External researchers aren't under our direct control. Their devices might have unknown vulnerabilities, their security practices less stringent.
Controlled Access: They need data on VLAN 200, but do they need it ALL? Fine-grained access is crucial.
Exfiltration Worries: How do we prevent an external party, whether maliciously or accidentally, from walking away with a treasure trove of sensitive data?
Design Adaptations

Option 1: DMZ-Style Isolation
New VLAN (e.g., VLAN 250): Specifically for external collaborators. Contains a 'sanitized' subset of the research data.
Firewalls & Scrutiny: Traffic between VLAN 250 and VLAN 200 is heavily restricted, monitored for anything anomalous.
One-Way Street?: Perhaps data transfer to VLAN 250 is possible, but going back the other direction is blocked or requires manual review.
Option 2: Remote Access with a Twist

No Direct VLAN Placement: Collaborators don't join VLAN 200 at all. Instead, a remote access solution (VPN or tightly controlled jump box) is used.
Access Tied to Identity: Not just a generic "collaborators" login. Individual accounts with auditing for anything they do on our network.
Data Expiration: Files they access might self-destruct or become inaccessible after a set time, preventing indefinite hoarding.
Additional Considerations

Legal Gets Involved: Contracts specifying data handling, liability if a breach happens due to a collaborator's negligence, etc.
Monitoring is KEY: Assume something might go wrong, and focus on detecting it quickly. Anomalous traffic patterns from the collaborator zone warrant investigation.
The Cost of Complexity: More intricate setups = more potential points of failure and administrative burden.
The exam won't hand you a 'right' answer, but rather make you justify your choices based on specific scenarios. Consider:

Risk Appetite: Collaborating on life-saving drug research? Maybe stricter controls outweigh the operational hit, compared to collaborating on a new marketing campaign.
Zero-Trust, Even with 'Partners': Never assume good intentions replace the need for granular controls and vigilance.
This is always a balancing act! Should we also explore potential impacts on the collaboration itself? Does this level of security hinder the speed of research?
Let's think of the DMZ as a heavily fortified border crossing. It's a place where the outside world and our internal network meet cautiously, under strictly controlled conditions.

Why a DMZ Exists

- Sacrificial Zone: Servers we have to expose publicly (website, email) are placed here. If they're breached, the attacker is contained...hopefully far away from the truly sensitive internal systems.

- Layered Defenses: Firewalls on both sides of the DMZ create chokepoints. Traffic rules get restrictive! We filter out the obvious attacks and scrutinize anything that's allowed through.
- Reduces Internal Complexity: Easier to lock down the internal network as a whole when most external traffic only hits the DMZ.

Typical Inhabitants of the DMZ

- Web Servers: Serving your company's public website (note: database storing sensitive content would NOT live here).
- Email Servers: Receiving mail from the outside world. Often has heavy-duty spam and malware filtering.
- Bastion Hosts: Tightly locked down "jump boxes" for remote admin access into the DMZ. Management traffic, not general purpose use.
- DNS Servers: (Sometimes) Authoritative DNS for your domain might reside here.
- File Servers: If you MUST let the public upload files (with caution!), the DMZ is the place for it.

Firewall Rulesets: The Heart of It All

Let's be specific, assuming a basic three-tier setup (Internet <--> DMZ <--> Internal)

- Internet to DMZ:
    - Allow web traffic (ports 80, 443) to the web server ONLY.
    - Allow email traffic (ports 25, etc.) to the email server ONLY.
    - Block literally everything else by default.
- DMZ to Internal:
    - Block all incoming connections initiated from the DMZ by default.
    - Highly restrictive rules permitting only necessary traffic (maybe the DMZ web server needs to pull *some* data from an internal database).
- Internal to DMZ:
    - Allow limited administrative access for management of DMZ systems.

CISSP Focus

The exam emphasizes practical design, not just DMZ theory. Think about:

- Specificity Matters: "Allow web traffic" is too vague. Consider source IPs, destinations, and whether any unusual ports need to be open.
- Monitoring Outbound: Attackers compromise a DMZ system to pivot to the internal network. Unusual traffic leaving the DMZ is a red flag.
- Imperfect Solution: A skilled attacker might breach the DMZ, but it buys you time. Detection and response are key to stop them before they go further.

Let's imagine a scenario: The company needs to let the public securely upload files to a server in the DMZ. How do our firewall rules change, and what other security controls become crucial?

Think of WPA2 and WPA3 as security guards for your Wi-Fi network, each with different strengths and weaknesses. While WPA2 did a decent job for a while, WPA3 arrives with some much-needed upgrades.

WPA2's Weaknesses

- KRACK Attack: Proved that even WPA2 with a strong password could be cracked under the right circumstances, leaving data open to interception.
- Offline Dictionary Attacks: If an attacker captures the handshake process when a device joins the network, they can brute-force guess weak passwords offline (without repeatedly trying against the live network).
- Open Networks (Coffee Shops, etc.): These lack true encryption, making eavesdropping fairly easy for someone nearby.

WPA3's Improvements

1. Enhanced Open to the Rescue

- Opportunistic Encryption: Even on "open" networks, WPA3 encrypts each client's traffic individually. No more snooping on the person's browsing habits next to you at the coffee shop!
- Still Not Ideal: It's encryption, but not authentication. Sophisticated attackers can still perform man-in-the-middle tricks potentially. Better than nothing, though!

2. Simultaneous Authentication of Equals (SAE)

- Dragonfly Handshake: Replaces the vulnerable handshake in WPA2. Resistant to offline dictionary attacks, an attacker must actively try to crack the connection during the handshake itself.
- Forward Secrecy: Even if a past Wi-Fi session is compromised and decrypted, it doesn't give the attacker the keys to decrypt future sessions.

The Hardware Catch

- Clients & APs Must Dance: To get these WPA3 benefits, both your device (laptop, phone) AND your access point (Wi-Fi router) need to support it.
- Backwards Compatibility: WPA3 devices generally play nicely with older WPA2 devices to some extent, but you sacrifice the security enhancements in mixed mode.
- Upgrade Time?: New devices mostly support WPA3 now, but if you have older hardware, it might be the weak link leaving you vulnerable.

CISSP Mindset

The exam prioritizes the 'why' over just the 'what'! Consider:

- Risk vs. Cost: Upgrading an entire corporate Wi-Fi network is a project, even if security benefits are clear. This impacts when adoption gets truly widespread.
- Defense in Depth: Strong Wi-Fi encryption is vital, but don't forget about device-level security. WPA3 won't save you from a malware-riddled laptop.
- Transition Period: We'll be in a mixed WPA2/WPA3 world for a while. How does this affect designs, especially for hotspots with uncontrolled client devices?

EAP in Wi-Fi authentication is like switching from a rusty padlock to a high-tech security system requiring both a keycard AND a fingerprint scan. Let's break down why it's robust, and the weaknesses hiding in those simpler pre-shared keys.

EAP-TLS: Mutual Trust

Certificates All Around: Both the Wi-Fi access point (authenticator) AND the device trying to connect (supplicant) have unique digital certificates.
It's a Two-Way Street: Each side verifies the other's certificate. No more client blindly trusting any network just because it has the right password.
Tunneling Protects: EAP-TLS creates an encrypted tunnel first. The authentication exchange happens inside this, shielding it from prying eyes.
Pre-Shared Keys (PSK) – Easier, But...

One Key to Rule Them All: PSK in WPA2/3 is essentially a big, fancy password shared by all devices on the Wi-Fi network.
Offline Attacks: Capture a device connecting, and you can try cracking its password at your leisure, unlike the harder 'live capture' needed against EAP-TLS.

Management Woes: Change the PSK, now you need to update every single device. Not scalable in large environments.
Rogue Access Points: Attacker sets up a network with the same name and PSK...users may unwittingly connect, letting the attacker snoop.
Why Isn't Everyone Using EAP-TLS?

Complexity Overload: Certificate infrastructure (issuing, managing, revoking certs) is a beast, especially at organizational scale.
Device Support: Not every gadget out there plays nicely with EAP-TLS, whereas nearly anything understands simple WPA2-PSK.
Legacy Hangover: Switching an existing network is hard! Often, PSK stays for basic devices, EAP-TLS gets rolled out for the high-security stuff.
Exams focus on trade-offs, not just 'which is best?'. Think about:

Threat Model: Home Wi-Fi vs. an enterprise protecting sensitive data have different risk levels, thus justifying different levels of complexity.
User Experience Impact: EAP-TLS can be made user-friendly, but PSK usually wins on 'it just works' simplicity for the average person.
Zero-Trust in Action: EAP-TLS aligns better with not blindly trusting any device on the network, even if it knows the password.
Let's make it tangible: You're securing a warehouse with Wi-Fi access points for inventory scanners. Do you go with EAP-TLS or WPA2-PSK, and why?

Picture an attacker as a pirate, cleverly mimicking a legitimate port to lure unsuspecting ships (your devices!) to their doom. Let's explore their tricks, and how to navigate these treacherous waters.

Types of Rogue Access Points

- Accidental Rogue: An employee innocently plugs a home Wi-Fi router into the corporate network...now anyone nearby can hop on, bypassing security.
- Malicious Rogue: Attacker deliberately sets up an access point near the target, hoping to trick people into connecting to it.
- Evil Twin: Impostor! This closely mimics the name (SSID) of a legitimate network, but has far less friendly intentions. Think "Starbucks Wi-Fi" vs. "Starbuks Free Wi-Fi"

The Attacker's Playbook

- Stronger Signal: They blast out their rogue AP, overpowering the real one. Many devices auto-connect to the strongest signal...oops!
- Social Engineering: Names like "Free Airport Wi-Fi" are tempting. Users craving connectivity often don't scrutinize closely.
- Disruption: Attackers might try jamming the real Wi-Fi, forcing users to seek alternatives (like their malicious one nearby).

Detection & Mitigation

- Wireless Scans: Specialized tools detect rogue APs. Think of it like security radar, spotting things your employees' devices don't.
- Known SSID Whitelists: Configure devices to ONLY trust pre-approved Wi-Fi names. Cuts down on accidental connections and some attacks.
- Traffic Analysis: A rogue AP suddenly handling lots of your internal traffic? Red flag! Intrusion detection systems can look for these oddities.

- Client-Side Matters: Devices with up-to-date OS and software are less likely to fall for certain Evil Twin tricks.

The User: Weakest Link or Strongest Defense?

- Awareness is Key: Employees thinking "any free Wi-Fi is good" are a danger. Training on spotting suspicious networks is vital.
- Verify Before Connecting: Teach them to double-check the SSID and ask if unsure. A bit of inconvenience is worth it.
- Corporate VPN: Enforces a layer of encryption even if the underlying Wi-Fi is compromised, protecting data in transit to some degree.

The exam wants you to see the WHOLE picture:

- Policy + Tech: User awareness reduces accidental rogues. Technical detection spots the malicious ones. Both are needed!
- "Home" Isn't Immune: Attackers target employees at home with evil twins, knowing home networks are often poorly secured and might get them a foot in the door.
- Imperfect Solutions: Even the best systems will have gaps. Focus on making it HARD for attackers, and detecting them quickly if they succeed.

Let's add complexity: Imagine the attacker sets up a rogue AP that forces clients to connect via an open portal (like in hotels) asking for personal info before granting access. How dangerous is this?

Let's think of 802.1x as a vigilant bouncer for your network club. It doesn't just check your ID (username/password), but does a full security pat-down before you're allowed into the party.

How It Works: The Key Players

- The Supplicant: Software on the device asking to join the network (your laptop, phone, etc.). It plays by the club's rules.
- The Authenticator: The network switch or wireless access point. Enforces traffic blocking until the bouncer (authentication server) gives the all-clear.
- RADIUS Server: The brains behind the bouncer. It has the policy: "Antivirus up-to-date? Latest OS patches? Okay, you're in! Otherwise...wait outside."

The Dance of Authentication

1. Wanna Get In? Device tries to connect to the network. The authenticator puts up the velvet rope - limited access only.
2. Prove Yourself: The supplicant sends credentials, plus info on its security posture (think of it as showing its anti-virus scan results).
3. RADIUS Check: Authenticator forwards this to the RADIUS server, which consults its policy rulebook.
4. Decision Time:
   - You're Cool: RADIUS tells the authenticator to open the gates! Device gets full network access.
   - Fix Yourself Up: RADIUS tells the authenticator to grant limited access, maybe just enough for the device to download those missing patches.
   - Get Lost: Device fails the check badly. No network for you!

NAC in Action (Beyond Just 802.1x)

- Granular Access: We might give fully compliant devices access to the sensitive server zone, while patched-but-outdated ones get limited to the internet.
- Guest Handling: 802.1x lets us create a 'quarantine' network – guests get basic access, but are kept safely away from our internal systems.

- Change Detection: The supplicant can report if the device's posture changes post-connection (someone disables their antivirus... uh oh!).

The exam wants you to see how these pieces fit in the larger security landscape:

- Zero-Trust in Action: Continuously verifies devices deserve the current level of access, not just a one-time check at login.
- Defense in Depth: Even if malware gets on the device, NAC limits the damage it can do based on network segmentation.
- Complexity Tradeoff: This adds administrative overhead. Is it worth mandating every single IoT sensor have up-to-date antivirus for your 'smart building'? That's a risk-based call.

Let's explore a scenario: An employee's personal device (BYOD) has outdated antivirus. Can it still get some level of access using 802.1x, and how does that impact security?

Picture IDS/IPS as security cameras for your network. The question is, are we looking for known criminals, or unusual behavior? Each approach has its time and place.

Signature-Based (The Mugshot Database)

- Excels at Known Threats: IDS/IPS has rules for specific exploits, malware patterns, etc. If it sees a match – alarm bells!
- Efficient: Like facial recognition at an airport, it's fast. If a known bad guy shows up, we don't need to deliberate.
- False Positives: Overly broad signatures catch legitimate stuff. "Blocks all suspicious Word attachments" might break normal business flow.
- Blind to Zero-Days: If the attacker uses a brand-new technique, signature-based detection is likely to miss it.

Anomaly-Based (Spotting the Shifty Character)

- Catches the Unknown: Baselines normal traffic patterns. A sudden data exfiltration spike at 3 AM? That stands out even if we don't recognize the specific malware.
- Prone to Errors: False positives are a headache. "Normal" traffic shifts over time, requiring constant tuning to avoid pointless alerts.
- Tricky to Define "Bad": Is a user suddenly downloading larger files than usual a breach, or just a project change? Anomaly detection needs human follow-up.
- Good for Insider Threats: Behavioral anomalies can spot a disgruntled employee doing things outside their normal workflow, which signatures would miss.

When to Use Which

- Simple Network, Well-Known Threats: Signature-based might be enough. Think a small office with mostly web browsing activity.
- Defense in Depth: Critical assets often have BOTH. Signatures block the obvious, anomalies raise flags about the stealthier attacks.
- Zero-Day Worries: Protecting a research lab with cutting-edge (and potentially targetable) data? Anomaly-based becomes your crucial early warning system.

The exam emphasizes the 'why' more than the 'what'

- Tuning is Everything: A poorly configured IDS/IPS is either useless (too many false negatives) or a nightmare for IT (constant false positives).
- Alert Overload: Can your team actually respond to every anomaly alert? If not, even the best system fails due to the human factor.
- Never a Silver Bullet: IDS/IPS are amazing at spotting network attacks happening RIGHT NOW, not so great at preventing them to begin with.

Let's make it real: You're defending a financial system processing credit card data. Is signature-based or anomaly-based detection more important, and why?

Think of a SIEM as a master detective piecing together clues from across your network crime scene. Individually, logs might look innocent enough, but when the SIEM connects the dots, a sinister plot emerges!

Scenario 1: The Multi-Stage Break-In

Firewall: Sees unusual connection attempts from a new IP address, but they're BLOCKED (good job, firewall!).
IDS/IPS: Picks up suspicious activity on an internal web server, odd file transfers...not enough to be sure it's an attack.
Server Logs: Show a user logging in at an unusual time, followed by data access on files they don't normally touch.
In Isolation...we might dismiss each as minor. The login was legitimate, the IDS stuff is often noise, etc.

The SIEM's Brilliance

Correlates Timeline: Sees the firewall deny, followed by odd probes, then the unusual login and activity...from the same source IP the firewall fought off.
Raises the Alert: This is no longer just 'weird stuff'. The SIEM's rule engine screams "Potential compromised account! Investigate NOW!"
Scenario 2: The Slow and Stealthy

Endpoint Logs: Over weeks, a workstation has tiny outbound data transfers, easily lost in normal traffic noise.
Anti-Virus: Picks up odd file modifications on that workstation, but nothing definitively malicious...it quarantines, but the user dismisses it.
SIEM Sees the Pattern: Tiny transfers followed by file changes, repeating over time? That's not normal behavior!
Uncovers Data Exfiltration: Even without catching a smoking-gun malware program, the SIEM spots the likely attempt to sneakily steal data over a long period.
CISSP Focus

The exam tests your understanding of WHY a SIEM is powerful, not just rattling off features. Consider:

Attacker Mindset: They try to stay under the radar of any one device's defenses. The SIEM is there to spot them hopping across your network.
Volume & Variety: Modern networks generate HUGE logs. Humans can't sift through it all, the SIEM is built for it.
Customization is Key: Out-of-the-box SIEM rules are okay. Finely tuned ones based on YOUR network's normal baseline are far more effective.
Let's get tricky: Many attacks start with phishing emails. Can a SIEM help us there, and if so, how would we integrate that data?

While a SIEM's traditional strength lies in analyzing network and device logs, integrating email security data takes your threat detection to the next level. Here's how it can help fight phishing and the attacks that often follow:

Connecting the Email Dots

1. Gateway Logs: Your email security gateway keeps its own logs - what it blocked, what it let through, attachment analysis, etc. Feed these logs into your SIEM.
2. Phishing Reports: Have a "Report Phish" button for users? Those reports become incredibly valuable SIEM data.
3. Endpoint Connection Attempts: The SIEM sees an employee click a suspicious link in an email, AND immediately their machine tries to connect to a newly-registered domain with a bad reputation. That's a massive red flag!

Phishing-Driven Attack Scenarios SIEM Excels At

- Identifying Targeted Campaigns: Not just mass blasts, but did multiple employees get emails tailored to your company? The SIEM spots this pattern.
- Zero-Hour Phishing: Fancy new malware evading the gateway? If it triggers device alerts + came from an email, the SIEM highlights the likely source of infection.
- Post-Click Compromise: User fell for the phish, their account then starts doing weird stuff. Correlating across user audit logs and network activity lets you contain the breach quickly.

How to Make it Work

- Email Header Analysis: SIEMs need to 'understand' email headers to trace the message path, sender reputation info, etc.
- Threat Intelligence Feeds: Augment your SIEM with data on known bad domains, malicious attachment types, etc. This enriches correlation.
- User Behavior Analytics (UBA): Some SIEMs look at 'normal' email behavior for employees. Deviations (like suddenly sending tons of attachments) could hint at compromise, even if the content itself looks innocuous at first.

Remember, the exam focuses on the big picture and practical application:

- Defense in Depth: A SIEM doesn't replace good email gateways or endpoint protection, it augments them.
- Data is Only as Good as Your Rules: It takes effort to create SIEM correlation rules that spot phishing attacks effectively.
- Incident Response: SIEM speeds up connecting the dots of how a breach started. This is vital for effective containment and forensics.

V. Identity and Access Management (IAM)

Let's dive into the exciting realm of Identity and Access Management (IAM)! Think of it as the foundation for controlling who gets into your digital fortress, and what they're allowed to do once inside.

Why IAM Matters (A Lot)

The Keys to the Kingdom: IAM protects your precious data and systems. Weak access controls lead to breaches, plain and simple.
It's About People AND Tech: We'll look at the policies, processes, and technologies that work together to make IAM effective.
Beyond Just Logins: IAM isn't simply usernames and passwords. We'll consider identity lifecycle management from new hire to termination and all the steps between.
Business Enabler, Not Just Gatekeeper: Done right, IAM facilitates secure and efficient collaboration. Done poorly, it's a roadblock users try to go around, creating greater risk.
Key Concepts We'll Explore

Authentication: Proving you are who you say you are (passwords, biometrics, tokens).
Authorization: After the 'who', comes the 'what'. Fine-grained permissions limit what you can access, even if you get inside.
Principle of Least Privilege: Giving users the absolute bare minimum they need to function. Reduces the blast radius if an account is compromised.
Identity Governance: Policies and procedures around creating, modifying, and deleting user accounts. Sounds boring, is immensely important!
SSO & Federation: Making life easier for users (one login across many systems) while maintaining security.
The exam aims for a holistic understanding of IAM. Consider:

Risks vs. Costs: Perfect IAM could be suffocatingly restrictive. Finding the balance that aligns with your organization's risk tolerance is key.
Compliance: Regulations (GDPR, HIPAA, PCI DSS, etc.) often have specific IAM requirements. Security has to play nice with legal obligations.
The Human Factor: Even the best system fails if users are tricked into giving up their passwords or ignore security policies.
Get ready for a deep dive into the strategies and tools that keep your digital assets safe, and your business operations flowing smoothly!
Employee role changes are like a security tap dance – if your timing's off, sensitive data can trip and fall. Let´s break down the risks and how automation can smooth those transitions.

Common Pitfalls

- The Ghost of Access Past: Old permissions linger. Now the accountant-turned-marketer has unnecessary access to financial systems...uh oh!
- Inadequate Handoff: New role's access is set up, but revoking the old takes days (or gets forgotten entirely). This creates an overprivileged user.

- Who Did What, When?: If permissions aren't updated promptly, was it ex-accountant Steve or new-marketer Sally who messed with that sensitive file? Audit trail gets muddy.

Why This Matters

- Insider Threats: Whether malicious or accidental, lingering access makes misuse easier. Maybe old habits persist, or an angry ex-employee still has a way in.
- Compliance Headaches: Many regulations require prompt control adjustments based on job duties. Failing this = potential fines and reputation damage.
- Operational Overload: Manual permission tweaking is error-prone and sucks up IT's time, especially with frequent role changes.

Automation to the Rescue

1. Role-Based Access Control (RBAC): Define roles like "marketer", with pre-set permission bundles. The user gets put in the role, automatically gaining AND losing the right access.
2. Workflows: Upon role change, trigger workflows. Disable the old account immediately, start the creation of the new account (with the right RBAC role already attached).
3. Identity Governance Integration: Tie these workflows into your overall identity management system, ensuring centralized logging and oversight of the whole process.

Additional Benefits of Automation

- Speed: Reduces the window where someone is over-privileged, minimizing risk.
- Consistency: Less room for human error, your policies are ENFORCED by the system, not just hoped for.
- Audit-Friendliness: Clear logs of 'this role change triggered these access adjustments at this time', making investigations easier if something does go wrong.

Exams test your ability to see the big picture:

- Zero-Trust in Motion: Even internal role changes should be treated with some suspicion. Rapidly shrinking the old access minimizes opportunity.
- Scale Matters: Automation is a lifesaver in large organizations with frequent moves. Manual processes break down quickly.
- Change Management: The best tech is useless if you don't also have solid procedures around initiating and logging role transitions.

Let's Get Complex: An employee is on extended leave. We can't fully delete their account, but need to temporarily restrict access. How do we design this, both the policy and potential automation side? Handling employee leave while maintaining security requires a mix of careful policy and, ideally, some automation assists. Let's break it down:

Policy Foundations

Temporary Deactivation vs. Restriction: Do we fully disable the account (best security), or severely limit its permissions (may be needed for operational continuity)? This decision depends on the leave type, the employee's role, and your risk tolerance.

Define Triggers: What prompts this? Parental leave? Medical? Each may have different rules based on your company policies and any local regulations impacting protected leave categories.

Handoff Plan: Often, work-in-progress must be transferred. Your policy mandates HOW this occurs: temp access for a colleague, manager review, etc. Security can't happen in a vacuum, here's where it meets HR procedures.

Audit & Reactivation: How do we track the account is in 'leave' state? When the employee returns, who initiates the reactivation process, and are there any additional security checks before restoring full access?

Automation to Help
Account Flagging: Ideally, your identity management system can flag an account with a "Temporary Leave" status, distinct from "Terminated" or "Active".
Role-Modification Workflows: Trigger a workflow that REMOVES most permissions, perhaps leaving only basic email. RBAC is your friend here!
Time-Based Reactivation: Some systems can automatically reactivate accounts after a set date. Use with caution – you might want a manual check-in first.
Logging, Logging, Logging: Every adjustment to the account due to leave status should be heavily logged for future audit purposes.
System Limitations: Not every organization has fancy automated tools. Sometimes it's a well-documented manual checklist, better than nothing!
Unexpected Returns: Employee comes back early? Your process needs to reverse things quickly! Design for this scenario, it WILL happen.
The 'Forgotten' Account: Biggest danger is accounts languishing in restricted mode long after they should be active. Regular audits are a must, even with automation.
Privacy Considerations: In some jurisdictions, monitoring an employee's email on leave is a legal minefield. Know your territory!
Deprovisioning isn't just about hitting the 'disable' button on an internal account. Think of it as untangling a web of digital access threads the employee has spun over time.

Beyond the Core: Where Access Lurks

- SaaS Galore: Cloud apps galore! Did that sales rep have access to Salesforce, a document collaboration tool, a project management system...? Each needs its own offboarding.
- Third-Party Vendors: Did they have logins to partner portals, systems for placing orders, etc.? Cutting these ties is easy to overlook, yet often represents a significant risk.
- The Non-Obvious: Wi-Fi access, physical building keycards, specialized equipment logins...deprovisioning needs to be holistic.

Data Classification's Role

Remember when we labeled our data by sensitivity? This comes roaring back during termination! Consider:

- High-Risk Access: An employee with access to top-secret data needs a more rigorous deprovisioning process than someone handling public marketing materials.
- Retention Requirements: Regulations might dictate how LONG we need to retain logs of certain terminated accounts related to sensitive data access.
- Backups and Archives: Did their account have data that's also subject to long-term retention? Deprovisioning can't just be deletion in those cases.

Why This is Harder Than it Seems

- Lack of Centralization: Knowing what someone had access to in the first place is often harder than it should be, especially in decentralized IT environments.
- SaaS Sprawl: New cloud tools get adopted ad-hoc, with no clear 'owner' who's brought into the offboarding process.

- Manual Processes: Spreadsheets + emails = gaps waiting to happen. Even diligent admins miss things under pressure.

The exam emphasizes the practical consequences:

- Lingering Access is a Breach Waiting to Happen: Angry ex-employees, compromised accounts sold on the dark web...the longer it lives, the riskier it gets.
- Audit Agony: Failing to properly deprovision users across all their touchpoints is a major compliance red flag.
- Zero-Trust in Reverse: Assume the user is now hostile. Treat their potential access with the same respect for risk as an external hacker!

Think of access reviews as security checkups for your accounts. Just like an annual physical vs. rushing to the doctor with a sudden fever, access reviews have both scheduled AND urgent varieties. Let's break it down:

Why Reviews Matter

Creep Happens: Over time, people gain permissions as projects shift. Those rarely get removed proactively, leading to over-privileged users.

Zombie Accounts: Employees leave, contractors finish...their accounts should be dead but sometimes linger on. Huge risk!

Auditing & Compliance: Many regulations require periodic attestation that user access is still appropriate for job roles.

Access Tiers and their Frequency

High-Risk Accounts:

Privileged admins (those who control the systems themselves): Quarterly reviews, at least, maybe monthly if your risk tolerance is low.

Access to Top Secret Data: Depends on regulations, but tighter review is better.

Finance Systems: Those who can move money warrant frequent scrutiny.

General Users:

Knowledge Workers: Annual to semi-annual reviews may suffice, barring anything suspicious.

Basic External Portals: Maybe a yearly review is enough IF the data exposed is low-risk.

Triggers for Unscheduled Reviews

Role Change: IMMEDIATELY review, with an eye towards removing old, now irrelevant permissions.

Project Termination: If access was granted for a specific project, check if it's still needed when the project ends.

Suspicious Activity: Account used from odd locations, at weird hours, etc. warrants a deep dive ASAP.

Security Incidents: Was an account part of a breach? Even if accidental, its permissions need immediate review to limit further damage.

Risk-Driven, Not Check-Box: A nuclear plant needs tighter reviews than a blog. Your threat model dictates the right level of paranoia.

Automation Helps: Tools scanning for dormant accounts, or those with odd permission combinations, make the reviewer's job more focused.

Don't Forget the 'Who': Who conducts these reviews? Ideally, someone outside the immediate manager chain (to avoid conflicts of interest).

Let's get specific: A small healthcare clinic has a cloud-based records system. Suggest an access review strategy, being mindful of their likely resource constraints.

Since small healthcare clinics have the dual challenge of sensitive data and often limited IT staff, let's focus on a streamlined and efficient access review strategy.

Tiering It Out

- Tier 1: The Critical Few
    - Doctors, Nurses, Key Admins: These accounts touch the most sensitive data. Quarterly reviews, even if somewhat less formal due to the small team.
    - Billing/Insurance Access: Handling financial data warrants similar scrutiny.
- Tier 2: Regular Staff
    - Receptionists, General Office Workers: Annual review may be sufficient, with an emphasis on any accounts touching the records system EVEN with limited permissions.
- Tier 3: The Odds & Ends
    - Maintenance Contractors (Physical Access): Yearly may be fine, but have a procedure tied to their sign-in upon visiting the clinic.
    - Temporary Staff: At the END of their contract, immediate review and disablement unless a strong extension justification exists.

Streamlined Process

1. No Fancy Software Needed: A well-maintained spreadsheet can be the starting point. List accounts, associated roles/systems, and a review schedule.
2. Manager + One: Have the direct manager attest to the access necessity, BUT a senior person (office manager, lead doctor) adds a secondary approval layer.
3. "Did it Change?" Focus: The review isn't a re-granting of access each time, but specifically looking for anything that has changed since the last review.

Triggers Remain Key

- Termination: Delete immediately, no lingering access, period.
- Security Red Flags: Any odd activity from even the receptionist's account warrants immediate review, resources don't excuse ignoring potential breaches.
- The Power of Questions: Train staff to ask "Should [Name] still have access to [System]?" when projects end or roles visibly shift.
- Regulations Matter: HIPAA has some say in this. The simple strategy may still be compliant, but it's worth verifying.
- Growth Changes Things: As the clinic gets bigger, so will the need for more formalized processes and potentially automation tools to aid them.
- Outsourced Help?: Even a few hours of a security consultant yearly might be affordable, adding an expert eye to the review process.

Let's add a wrinkle: The clinic adopts a small telehealth platform. How does this impact their access review plan?

Picture privileged access creep like this: you start with a tidy backpack for a day hike, but end up hauling a giant, overstuffed suitcase up a mountain. It might've made sense at SOME point, but now it's just dangerous.

How Creep Happens

The "Temporary" That Sticks: Project needs urgent access? Fine. Project ends, access lingers... now we have an over-entitled user.

Job Shifts are Messy: New responsibilities piled on, but the old permissions never get removed. Frankenstein accounts are born!

Lack of Clarity: "Admin" gets used loosely. Is that server admin, app admin, minor "manage your own team's data" admin...?

Consequences

Breach Delight: Attackers LOVE an overprivileged account. Compromise one, and they've got the keys to the kingdom.

Insider Threat Made Easier: Disgruntled employee with too much access can cause far more damage than if their permissions were tightly scoped.

Audit Nightmare: Explaining why the intern has database admin rights will be...unpleasant.

Mitigation: Balancing Act

Role-Based is Your Buddy: Define roles clearly (not just by person). "Sales Rep - Tier 2" gets a specific bundle of permissions.

Request & Approval Workflow: Getting additional access shouldn't be the wild west. Formal process with justification.

Temporary Access Timebombs: Can the permission grant automatically expire after a set period? If more time is needed, it forces a re-review.

Access Reviews Tied to Changes: New project? Role change? Those are review triggers.

Least Privilege is the Goal, Reality is Compromise: Sometimes, we WILL grant a tad more access than ideal for the sake of getting work done. Vigilance in reviewing is what compensates.

Tooling Evolves: Privileged Access Management (PAM) solutions exist, making this technically easier in large orgs. The core principles matter even without fancy tech.

The Human Factor is HUGE: All the tech in the world won't stop a manager rubber-stamping access requests for their team without thinking.

Let's get messy: In a fast-paced startup, rigid processes are hated. How do we nudge towards better access control without causing a revolt?

Startups thrive on a certain degree of chaos, and imposing rigid security often backfires spectacularly. So, let's get sneaky in the name of eventual good habits!

Small Wins, Not Big Battles

- Focus on the 'Why': Don't lead with "CISSP says..." Instead, frame it as "Remember when that temp worker nearly leaked our client list? Let's avoid THAT again."
- Make It Easy to do Right: A convoluted form to request access will be ignored. A Slack channel where the right approver lives is more startup-speed.
- Champions Matter: Find that one dev who IS security-minded. Their endorsement of a process change is worth 10x your nagging.
- The Post-Mortem Advantage: Something DID go wrong? Use that as the catalyst: "Okay, we need a slightly more organized way..."

Specific Tactics

- "What If I Lose My Laptop?" Exercise: Get them thinking in terms of what an overprivileged account enables an attacker to do.
- Temporary Access by Default: Can some permissions be okay for 24 hours, forcing a re-request if truly needed long-term? Less friction.
- "Dormant" Account Spotchecks: Publicly gamify it... "Congrats, Sarah has the least overprivileged account this month! Pizza for her team!" Subtle peer pressure.
- Tie to a Real Threat: Not abstract "risks", but "Competitor X is known to do these dirty tricks, let's not make it easy for them."

- Culture Beats Thou-Shalt-Nots: If the startup rebels, you've lost before you even start. Build the security mindset organically.
- Acceptable Risk: Perfect won't fly. Focus on reducing the most egregious cases of privilege creep first. Small victories pave the way for later, larger ones.
- This WILL Change: As the startup succeeds (hopefully!), they face more scrutiny, more regulations. Nudging them early makes that transition smoother later.

Introducing real-world scenarios is where the CISSP mindset truly shines! Let's craft multi-factor authentication (MFA) strategies, balancing security needs with the context of each situation.

A Note: 'MFA' doesn't mean every scenario needs the exact same solution

Scenario (a): Remote Access VPN

- Threat Level: High. Our VPN is a gateway into the internal network. A breach here is potentially catastrophic.
- Users: Varied. Full-time employees, contractors, maybe third-party vendors needing occasional access.
- Justification:
  - Hardware Tokens: Provide strong, one-time passcodes. Good for core employees, reduces reliance on personal devices.
  - Push-Based App: Easier for contractors/vendors. "Approve" on their phone adds a layer beyond just the password.
  - SMS as Last Resort: It's the weakest OPTION, but better than nothing for truly rare/emergency access grants.

Scenario (b): Cloud-Based HR System

- Threat Level: Medium-High. Sensitive employee data is here! Compromise leads to identity theft risks, not just system disruption.
- Users: Primarily internal employees, from a mix of company-provided and personal devices.
- Justification:
  - Authenticator App: Widely adopted, strikes a good balance of strong MFA without overly complex management.
  - FIDO Keys: Consider these for HR admins themselves, or as an optional upgrade for security-conscious employees. Highly phishing resistant.
  - Careful about SMS: It's more vulnerable, but could be an 'only if absolutely necessary' fallback for lost devices.

Scenario (c): Critical Server Admins

- Threat Level: VERY High. If admins are compromised, it's game over. Assume sophisticated attackers will target them specifically.
- Users: Small, defined group. Security awareness should (hopefully) be higher.
- Justification:
  - FIDO Keys are KING Here: Phishing resistance is vital, and the minor inconvenience is acceptable for privileged users.
  - Hardware Tokens as Backup: Redundancy matters! If the FIDO key is lost, a token ensures they're not locked out.
- Usability vs. Strength: We always trade-off! Admins get the strongest MFA, since risk justifies it. Regular users need something less cumbersome.
- Cost: Hardware tokens scale poorly for large orgs. Apps are cheaper long-term.

- Don't Forget the Basics: MFA is amazing, but useless if the passwords themselves are weak. Policy around complexity still matters!

Let's make it messy: The company has many non-tech-savvy employees. How might this impact your MFA choices, particularly for the HR system?

When your users struggle with tech, overly complex MFA becomes a recipe for frustration and security workarounds (think passwords on sticky notes!). Let's adjust our HR system MFA strategy accordingly:

Prioritizing Usability
Push-Based Reigns Supreme: "Approve this login?" on their phone is the most intuitive option. Limit complex authenticator apps to a small subset of truly tech-comfortable users.
Voice Call as Backup: For those who REALLY struggle, a simple "Press 1 to approve" automated call can work, though it's a bit less secure than other methods.
Extended Training & Support: MFA rollout HAS to include hands-on help for the non-tech-savvy, and ongoing helpdesk readiness for questions.
Security Awareness is Key: Emphasize why we're doing this. "Keep YOUR paycheck data safe" resonates more than abstract threats.
Trade-offs We Might Make

SMS Acceptance: We'd prefer not to, but it may be the only realistic option for some users. Strict policies on when it's allowed (device lost, etc.) are key.
Longer "Remember Me" Periods: If MFA on every login is too painful, extend the trusted device window slightly. Risk trade-off, but better than workarounds.
Exemptions with a Caveat: Truly extreme cases might be stuck with password-only IF they access zero truly sensitive data AND have strong monitoring on their account.
Additional CISSP Thoughts

Risk Isn't One-Size-Fits-All: Maybe HR admins DO get stricter MFA, even if the average HR rep has a simpler setup.
Threat Landscape Changes: If phishing attacks spike, we might tighten things up even if it causes grumbling.
"Good Enough" is Sometimes Acceptable: Perfect MFA is useless if it drives users to unsafe practices. Getting SOME additional protection is a win.
User Education is an Ongoing Battle: It's not just initial rollout, but constant reinforcement of why this matters, making it part of the company culture.
Let's get tricky: A new regulation requires that for certain actions within the HR system, the user's physical identity must be verified. How could we potentially achieve this in conjunction with MFA?

Let's dive into the wonderful, yet perilous, world of Single Sign-On (SSO)! Think of it like a master key to various websites and applications.

SSO Benefits: User Experience

- One Login to Rule Them All: Users love not juggling a zillion passwords. Fewer forgotten passwords = less IT helpdesk load.
- Seamless Flow: SSO aims to whisk users between systems they're authorized for without constantly re-proving who they are.
- Can Reduce Password Reuse: If SSO is done well, users are less tempted to reuse the same weak password everywhere out of laziness.

SSO Mechanisms (SAML & OAuth)

- SAML: The older standard, XML-heavy. Often used in enterprise environments for SSO into internal systems.
- OAuth: Modern, designed for web APIs. You've seen it with "Login with Google/Facebook" on countless sites. More about authorization delegation than just SSO.

The Dark Side: IdP Compromise

If the Identity Provider (the system vouching for the user's identity) gets breached, things gets ugly FAST. Here's why:

- The Domino Effect: All those systems trusting the IdP are now potentially wide open to anyone holding those stolen IdP credentials.
- Attacker's Delight: They breach ONE system, gain access to MANY. This is why IdPs are prime targets.
- It's Not Just Hacking: Angry ex-admin at the IdP? Disgruntled insider? Those threats become far scarier with SSO.
- Detection is Tricky: Unauthorized logins may look legit if they're coming through the trusted IdP, making breaches harder to spot.

CISSP Approach

- Defense in Depth is Key: SSO won't save you if your password at the IdP itself is "123456". Strong authentication there is non-negotiable.
- Zero-Trust Even Within: Relying systems shouldn't blindly accept anything from the IdP. Some level of additional checks (device posture, unusual behavior) are wise.
- Incident Response Plans: Assume your IdP will be breached someday. How quickly can you revoke trust, cut off access, and start damage assessment?
- User Awareness Matters: Phishing the IdP is often easier than attacking it directly. Users need to be wary of those login screens!

Let's imagine a scenario: A healthcare system uses SSO for doctors and nurses to access patient records, billing, etc. How does the risk of IdP compromise change our security posture?

When we introduce SSO into a healthcare system, the stakes skyrocket. Let's explore how IdP compromise becomes an especially scary scenario and how we'd adjust our security posture accordingly.

Amplified Risks

- Data Sensitivity: Patient records are among the most tightly regulated and sensitive data out there. A breach is reputationally and legally devastating.
- Impact on Critical Operations: If SSO goes down, can doctors even access the records they need to provide care? Downtime isn't just lost productivity, it could endanger lives.
- Targeting the Doctors: Spear-phishing attacks aimed specifically at healthcare providers to get their IdP credentials become far more appealing to attackers due to the potential payoff.

Adjusting Our Security Posture

- Hardening the IdP to the Extreme: This becomes our crown jewel to protect. Strict access controls, auditing, multi-level monitoring are a must.
- Privileged Account Vigilance: IdP admins should have separate, highly-monitored accounts for their normal work. Only for IdP changes do they use the scary powerful ones.
- Break Glass Procedures: If the IdP goes down, a pre-planned way to SEVER that trust and let systems fall back to local auth (even if clunky) is vital.
- User Training Goes Beyond Phishing: Teaching healthcare staff to spot suspicious login prompts, even on their internal network, becomes crucial.

- Zero-Trust Intensifies: Relying systems shouldn't JUST trust the IdP. Unusual doctor login from a weird location at 3am? That warrants extra scrutiny.
- Regulations Matter A LOT: HIPAA (in the US) and similar laws elsewhere likely dictate specific safeguards around any system handling patient data.
- Compliance vs. True Security: Checking regulatory boxes might not be enough. Attackers don't care about your compliance if they can breach the IdP.
- Outsourcing the IdP: SaaS IdPs exist. Does that relieve our risk? Not really, it just shifts where the responsibility and scrutiny needs to lie.

Unfortunately, MFA isn't a magical shield. Attackers have gotten clever, and real-time phishing coupled with session hijacking can sometimes sidestep it. Let's dissect how this works and what defenses we have beyond the purely technical.

The Attack Chain

Phishing Lure: Not just generic spam, this is targeted. Maybe the attacker knows an employee uses a certain bank, crafting a fake "urgent fraud alert" email from that bank.
Fake Login, But Real MFA: The victim, panicked, lands on a convincing replica of their bank's login. They enter their username, password, AND diligently use their MFA authenticator app/token.
The Hijack: The attacker's tool is proxying this in REAL TIME. The second the MFA code is used, the attacker races to the real bank site, armed with the stolen credentials AND that valid one-time code.
Victim is None the Wiser: Maybe the fake site shows an error (whoops!), the victim assumes they mistyped, and goes about their day. Meanwhile...
Attacker Empties the Account: Now in, they transfer funds, change passwords, etc. Depending on their goal, discovery might take far too long.
Countermeasures: Beyond Tech

User Awareness Strikes Back: Train users on highly targeted phishing (spear-phishing). Stress that urgency + fear is the attacker's weapon of choice.
Pause & Verify: Can we instill a 'if it feels weird, call the real bank/IT/etc. on a known good number to double-check' culture?
MFA Type Matters: FIDO keys are inherently more phish-resistant. Push adoption where it makes sense (high-risk users!).
Device Posture Checks: Is the login from a familiar device/location? Some MFA systems allow this as an extra factor, making hijacking riskier.
Anomaly Detection: Sudden flurry of activity on an account right after login? Even without fancy AI, some basic behavior analysis can raise red flags.

Defense in Depth: MFA is amazing, but attackers adapt. We layer defenses to compensate if one falls.
The Human is ALWAYS the Weakest Link: Tech can only do so much. Security awareness targeted to how attacks actually work is crucial.
Reporting is Key: If users feel safe reporting they almost got phished (no punishment!), we catch attacks sooner, potentially while the hijacker is still fumbling around.
Let's ditch the humble password and dive into the world of biometrics! Picture this: you are your authentication, whether it's your fingerprint, your face, or even something more exotic like the pattern of veins in your hand.
Biometric Basics

- Enrollment: First, the system captures your unique characteristic (fingerprint scan, etc.). This is turned into a template (a mathematical representation, not a literal picture).
- Authentication: When you come back, you scan again. A new template is generated and compared to the stored one. Close enough match? You're in!

The Error Rates: FAR vs. FRR

- False Acceptance Rate (FAR): The likelihood the system lets in an imposter. Think of this as security failing.
- False Rejection Rate (FRR): The chance you, the legit user, get denied. Think of this as usability failing.

The Security Balancing Act

- High-Security Zone: NuNuclear plant access? We crank the accuracy settings to minimize FAR, even if that means employees occasionally need to re-scan their finger. Security is paramount.
- General Office: FRR becomes more annoying. We might slightly increase the FAR, making it a bit easier to get in, for the sake of smooth workflow.
- Context Matters: Is the user also swiping a badge and entering a PIN? Biometrics then become one layer, not the sole gatekeeper.

CISSP Considerations

- Accuracy Isn't Static: Injury, dirt on the sensor, etc. all impact it. Systems should adapt over time to an individual's baseline for better user experience.
- Liveness Detection: Fake gummy finger won't fool a good system! High-security implementations must ensure it's a real, live body part being scanned.
- Isn't Foolproof: High-res photos, detailed masks...biometrics can be defeated with effort. It's about raising the attacker's work factor.
- Privacy Concerns: How is biometric data stored? Can it be reverse-engineered into a usable image of my face/fingerprint? This needs careful consideration, especially as regulations vary globally.

Let's get specific: A bank considers using facial recognition for ATM access. Is this a good idea? Consider FAR/FRR trade-offs and the unique context.

Let's analyze the bank's ATM facial recognition idea – convenience vs. risk needs a careful balance!

Potential Advantages

Could Reduce Card Skimming: If no card is needed, those physical skimming devices become less useful. A potential security win.
Faster Transactions: In theory, you walk up, it recognizes you, and you're ready to withdraw cash. Nice for customer experience.
"Cool" Factor: Being tech-forward has marketing value, though that shouldn't drive security decisions alone.
The FAR/FRR Challenge

Outdoor Environment: Lighting changes, sunglasses, hats, winter scarves...these all mess with facial recognition accuracy. High FRR likely frustrates users.
Imposter Risk: Is casual fraud (someone grabbing a lost card) the big threat, or organized crime? FAR needs to be tuned to the threat model.

System Complexity: Adding robust cameras to countless ATMs, designed for harsh conditions, gets expensive.
Opt-In, Not Mandatory: Forcing this on customers is a PR nightmare. It needs to be an alternative to the card, not the only way.
Anti-Spoofing is KEY: Can someone trick the camera with a high-quality photo of the account holder? Robust liveness detection is a must.
Regulatory Hurdles: Biometric usage is often regulated. The bank can't just do this willy-nilly, legal implications abound.
Is the Problem Worth Solving?: How much loss does the bank actually incur from ATM card skimming today? Does the fancy tech cost justify the risk reduction?
Alternatives?

Could two-factor with the card be enough of an improvement? Chip + PIN was already a big leap.
Behavioral Biometrics: How a user interacts with the ATM (typing speed, etc.) can be an additional check, less reliant on external factors than pure facial recognition.
This scenario shows how the exam isn't about 'right' answers, but justifying your decision based on the specific context, risks, and trade-offs!
Think of access control models like choosing a bouncer for a party. DAC is the friendly one, MAC is the rule enforcer, and ABAC is the bouncer with an incredibly detailed guest list. Let's see where each fits best!

DAC (Discretionary Access Control): When Flexibility Is Okay

- Collaborative Project: Teammates need to easily share files, edit each other's work...DAC's focus on the owner setting permissions is perfect.
- "Low-Stakes" Data: Company picnic photo album? If leaked, it's annoying, not catastrophic. DAC is probably fine.
- Legacy Systems: Sometimes, we're stuck with what we have. Old systems might only support simple DAC, upgrading is too costly.

MAC (Mandatory Access Control): Strict Enforcement

- Classified Environments: Top Secret data gets labeled "Top Secret", and only those with the right clearance can EVER touch it. MAC is made for this.
- Need-To-Know: Even within the same clearance, maybe only the project team should see those files. MAC lets us slice and dice access by labels.
- Highly Regulated Industries: Some regulations mandate strict controls, regardless of whether the company thinks it's necessary. MAC helps prove compliance.

ABAC (Attribute-Based Access Control): Finesse Is Key

- "It Depends": Access okay during business hours, from the office, on a managed device? ABAC combines subject, object, and environment factors for nuanced rules.
- Dynamic Permissions: Doctor can view ANY record at the hospital...but during an emergency, location restrictions are lifted. ABAC can adjust on the fly.
- External Sharing Made Safer: Allow partners to view ONLY project files related to their contract with dynamic expiration dates. ABAC's strength lies in this granularity.
- No One Model Is Perfect: Large systems often use a mix! Highly sensitive core is MAC-land, while the user-facing stuff has more flexible elements.
- Complexity Is the Enemy: ABAC is amazing, but with too many rules it becomes a management nightmare. Balance power with usability.
- It's About Data, Not Tech: Start by classifying the data by sensitivity. This classification then drives your sensible access control choices.

Scenario: A university stores academic records (grades, etc.) along with research data of varying sensitivity. Which access control models likely play a role in a well-designed system?

Picture RBAC (Role-Based Access Control) as the bridge-builder in a diverse tech landscape. Let's see how it takes messy real-world permissions and brings some sanity, especially when Windows and Linux need to play nice.

The Challenge: Systems Speak Different Languages
Windows Active Directory: Built around the idea of groups, themselves having rights. User "Alice" is in the "Accounting" group, so she inherits their file share access.
Linux File Systems: Owner, group, and everyone else permissions. That critical file is owned by the "Finance" group, and world-readable (uh oh, maybe that's not ideal).
RBAC to the Rescue

Forget the Nitty-Gritty: We don't directly map AD groups to Linux file ownership. Instead, focus on what they mean – an "Accountant" role exists abstractly in BOTH systems.

Mapping Roles: Somewhere (could be a dedicated RBAC tool, could be custom scripts), we maintain a table:
"Accountant" Role = Member of the "Accounting" AD group
"Accountant" Role = Member of the "Finance" Linux group
Enforcing the Abstraction: When Alice logs in, the systems don't ask "What AD groups is she in?". They ask "What ROLES does she have?".

Benefits
Centralized Logic: If we change what an "Accountant" should access, we update the RBAC mapping, not permissions on countless files.
Reduced Drift: No more hand-tweaking Linux groups to kinda-sorta match an AD change. RBAC keeps them in sync.
Audits Get Easier: Asking "Does Alice have the right role?" is more human-answerable than combing through file permissions.

This Takes Effort: That mapping isn't magic! Initial setup and ongoing maintenance as roles change has a real cost.
Imperfect Translation: Maybe Linux has finer-grained options than AD groups allow. Some compromise in the name of manageability is likely.
Least Privilege Still Rules: Just because we CAN map roles across systems doesn't mean we SHOULD give everyone access to everything.
Tools Help, but Don't Solve Everything: Large orgs might use an Identity Governance solution with fancy RBAC features. Mid-sized might have clever scripts. The core principle is the same.

Let's illustrate when RBAC's simplicity shines, and when ABAC's nuanced rules become necessary.
Scenario 1: RBAC In Its Element – A Hospital

- Well-Defined Roles: Doctor, Nurse, Lab Tech, Billing Admin... their job duties are fairly consistent, access needs predictable based on that role.
- Centralized Control: IT loves this. New nurse joins? Drop them in the "Nurse" role, and their access to medical records, drug dispensing systems, etc. is instantly provisioned.

- Simplicity for the User: Doctors don't want to think about granular permissions, they want to see the patient's chart! Role-based makes this seamless.

Scenario 2: ABAC's Time to Shine – Global Manufacturing

- The "Engineer" Problem: Do they get unfettered access to all factory machine controls? Only if they're:
    - On-Site: Remote access is too risky for production equipment.
    - On-Shift: No messing with the machines after hours without approval.
    - Project Specific: Engineer for line A shouldn't touch line B's controls.
- External Consultants: They get temporary, read-only access to designs IF connecting from company-approved devices and during the contract period.
- "Break Glass" Overrides: Major equipment failure? ABAC can temporarily suspend the usual restrictions in a documentable way for approved engineers.
- Real-World Isn't Neat: Job titles don't always equal what someone needs to do their job. This is where pure RBAC starts to creak, and some ABAC elements likely creep in.
- Change is the Enemy: RBAC works if roles are stable. Constant re-orgs mean frequent permission rework, which gets messy.
- Auditability: With ABAC, be sure you can answer "WHY did this person have access at that moment?" for compliance and forensics.

Let's complicate things: Even in the hospital, a traveling specialist doctor comes on board temporarily. How might we handle this without abandoning RBAC entirely?
The traveling specialist throws a wrench into our neatly organized hospital RBAC model! Let's explore ways to accommodate them without throwing the whole system out the window.

Options with Varying Complexity
The Temporary Role: Create a "Specialist" role. Mimics core doctor permissions, but with a built-in expiry date to match their contract. Clunky, but workable for short stints.

RBAC + A Sprinkle of ABAC: Augment the "Doctor" role with a 'Visiting' attribute. Systems with awareness of this attribute then layer on additional restrictions (maybe read-only on certain systems initially, requiring sign-off for any changes).

"Supervised" Mode: Maybe the specialist never gets direct system access. Instead, a permanent staff doctor acts as their 'proxy', performing actions upon the specialist's instruction after due diligence. Extreme, but high-security if warranted.

Factors Influencing Our Choice
How Sensitive is the Data?: Regular doctor permissions on the general patient records? Perhaps option 1 is fine. Specialty data with tighter controls might necessitate option 2.
How Long is "Temporary"?: A few weeks? The pain of role creation might be worth it. Six months? We better invest in a more sustainable solution.
Risk Tolerance: How worried are we about the specialist abusing access? Past track record with similar arrangements plays a role.
Regulatory Constraints: Some rules might limit what a "temporary" person can do, forcing a more restrictive approach.
Least Privilege Still Reigns: We grant the bare minimum needed for them to function, even if it's a bit inconvenient at first.

Audit Trail Is Key: Whatever we do, it should be clear WHY the specialist had the access they did, and when.
This Isn't Just a Tech Problem: Policy needs to exist around when using a "temporary" model is justified, and what oversight is required. RBAC is the tool, not the entire solution.
Think of IAM policy enforcement as a series of security checkpoints a user must pass through to get to their destination (the data). Let's figure out where these checkpoints exist.

Enforcement Points
The Database Itself: For highly structured data, the database may have its own permissions. Ideal for fine-grained control within that dataset, but can become a management headache in a large environment.

API Gateway: Modern applications often have APIs as their gatekeeper. The enforcement point here checks if the user is authenticated AND if their role/attributes allow them to make that specific API call.

Application Code: In less structured setups, access logic may be embedded in the application itself. This is more flexible, but beware of it leading to inconsistent enforcement across different parts of the app.

Why It Matters Where We Enforce
Centralization vs. Sprawl: Database-level is amazing for a single database, terrible if we have dozens of them. API gateways help centralize some logic.
Performance Bottlenecks: Checking permissions on every single database query can slow things down. Sometimes, bulk permissions upfront with more casual checks later are a trade-off.
"Before" vs. "After": Can the API gateway outright block the obviously bad request, or does the app need to do the work and THEN respond with "unauthorized"? This impacts user experience.
Flexibility: What if we need to change a permission frequently? Updating database-level stuff is often more disruptive than tweaking an API policy.

Defense in Depth: Even if the API gateway checks, a critical app might have an extra layer of enforcement in its own code, just in case.
Zero-Trust in Motion: Don't assume initial login is the only check. Maybe a long-lived session accessing highly sensitive data re-prompts for certain actions.
Auditability Varies: Some systems make it easier to answer "who accessed what, when" based on where the enforcement point lies. This matters for compliance!

Vetting SaaS vendors for their IAM practices is no mere box-ticking exercise! Think of yourself as a detective digging into how well they handle the keys to their kingdom (which holds your data, too!).

Key Areas to Probe

Account Management:

How do THEY provision/deprovision employee accounts? Lax practices here are a massive risk.
Password Policies: Strong enough? MFA that's actually enforced?
Privileged Accounts: Are these tightly inventoried and monitored, or is it the wild west for their admins?
Access Controls:

Least Privilege: Do they REALLY adhere to it? Can we see role definitions to assess if they're granular?
Segregation of Duties: Do they prevent conflicts of interest, with checks and balances within THEIR systems?
Incident Response:

Do they even have a plan for an account breach? Do WE get notified...and how quickly?
Breach Tabletop: Can they articulate how they'd investigate and contain a potential IAM failure within their system?
Documentation Requests (Get Specific)

SOC 2 Type II: Recent, and with relevant sections (security, availability). What did the auditor REALLY look at?
Their IAM Policies: Not just a fluffy doc. We want password complexity rules, specifics on privileged access workflows.
Incident Playbook Excerpt: The section on compromised accounts tells us a lot about their maturity.
Red Flags to Watch for

Vagueness: If they dance around questions or provide generic docs, trust is eroded.
"It Depends": While some customization is fine, a core set of IAM practices should be non-negotiable.
Deflecting Blame: "Our users are the problem, they choose bad passwords..." shows an immature security culture.
Outdated Standards: Still referencing old NIST guidance instead of the latest shows they aren't keeping up.

Risk to US is Key: A minor password policy weakness may be acceptable if they handle low-sensitivity data for us. Highly sensitive data warrants tougher scrutiny.
Trust but Verify: Even with great docs, consider a right to audit clause. Knowing we COULD peek behind the curtain keeps them honest.
This is Ongoing: Vendors change. Re-assess IAM posture at renewal time, especially if they've had growth spurts or acquisitions (those are messy!).

Keeping tabs on third-party accounts requires continuous vigilance. Regular access reviews are a good start, but they only capture a snapshot in time. Let's discuss how to keep an eye on them throughout their lifecycle.

Methods for Continuous Monitoring

1. Logging & Alerting:

- Targeted Logging: Ensure your systems generate logs specifically for vendor accounts (may require some customization). Focus on privileged actions, accessing sensitive data sets, unusual login times/locations.
- Alerting Thresholds: "Vendor account modified critical system config" should trigger an immediate alert to your team, not get buried in a weekly report.
- Baseline Behavior: Over time, can you establish normal usage patterns for vendor accounts? Anomalies then warrant extra scrutiny.

2. Vendor's Internal Monitoring:

- Can They Help?: Ideally, THEY have solid logging of what their users do within their systems, potentially shareable with you under the right agreements.

- Joint Reviews: Even if you don't get raw logs, quarterly reviews of vendor-side activity reports can be eye-opening compared to what you're seeing in your own environment.

3. Technical Safeguards:

- Network Segmentation: Limit where vendor accounts can even reach within your network. Reduce the potential blast radius.
- API Restrictions: If their access is via API, lock down the specific calls they're allowed, making misuse more obvious.
- Jump Boxes: For 'hands-on' access, force them through a system you heavily monitor, making it harder to hide tracks.

Connecting Monitoring to Access Reviews

- Feeding the Review: Logs of vendor activity become a vital input when deciding if their access level is still justified. Did they use the privilege even once?
- Escalation Path: What if monitoring throws up red flags? The review process needs a clear path to either rapid containment (account disable) or an immediate ad-hoc review if the activity is suspicious, but not outright malicious.
- Privacy vs. Security: Monitoring what a PERSON does can get thorny, contractually and ethically. Focus on technical logs tied to accounts where possible.
- Cost of Vigilance: Heavier monitoring = more SIEM licenses, more analyst time, etc. Needs to align with the risk a given vendor poses.
- "Shadow Access": Beware of informal relationships where your employees grant access directly, bypassing the formal vendor process...this breaks all monitoring!

When a vendor's IAM system gets breached, it's like a fire alarm going off in someone else's building that might also hold your valuables. Let's craft a response plan emphasizing swift action and communication, since we're partially at their mercy.

Incident Response Playbook Outline

1. Immediate Actions:

- Trigger Internal Incident Response (IR): Don't wait for vendor confirmation, the second there's credible suspicion, kickstart your own IR process.
- Identify Impacted Systems: Where did this vendor have access? This dictates the initial scope of your investigation and containment.
- Isolate (If Possible): Can you sever their access points quickly? This buys time, but may disrupt operations (consider this trade-off carefully).

2. Vendor Engagement:

- Designated Point of Contact: Pre-agreement on WHO at the vendor you talk to in a crisis is vital, avoids comms chaos.
- Information Needs:
    - Breach Scope: What systems, what kind of access was gained?
    - Timeline: When did it happen, how long were they in?
    - Containment: What have THEY done already?
- Push for Transparency: This is where a strong contract helps. You need timely, honest updates, not downplaying the situation.

3. Internal Investigation:

- Log Analysis: Forensic focus on any vendor-related account activity. Look for anomalies, data exfiltration attempts, etc.
- Credential Resets: Any passwords, tokens, etc. used in the vendor system get rotated IMMEDIATELY on your side.

- System Scrutiny: If they had 'deep' access, a broader compromise assessment might be needed, depending on your trust level in the vendor's containment.

4. Communication

- Internal: Board-level briefing may be needed if impact is severe. Impacted business units MUST be kept in the loop, even if information is incomplete at first.
- External: Regulatory reporting obligations? Do customers need to be notified? Legal and PR get involved here.

CISSP Considerations

- Playbook Testing: Tabletops where the vendor breach is the scenario are essential to uncover friction points.
- Breach the Contract: Vendor's actions (or inaction) may violate your agreements. Be prepared to escalate this, but carefully, as it could hinder the technical response.
- "Assume the Worst": Until proven otherwise, operate as if your data IS exposed. Better to overreact initially than be caught flat-footed.

VI. Security Assessment and Testing

Imagine your network, your applications, your entire digital infrastructure as a fortress. Security assessments and testing are like hiring a team of expert attackers – sometimes friendly, sometimes adversarial – to find those cracks in the walls, those unlocked windows, and those hidden escape tunnels before the REAL bad guys do.

Why Assessments & Testing Matter

Proactive Defense: Don't wait for a breach to tell you something's wrong. Assessments proactively hunt for vulnerabilities so you can fix them.
Compliance Check: Many regulations (PCI DSS, HIPAA, etc.) mandate regular testing to demonstrate you're doing your security due diligence.
Beyond the Scanner: Automated vulnerability scans are great, but they don't catch logic flaws, complex attack chains, or the human factor, which is where skilled testers excel.
Evolving Threats: Attack techniques change constantly. Assessments make sure your defenses aren't stuck in the past, protecting against yesterday's attacks.
Types of Assessments & Testing (A Taste)

Vulnerability Scanning: The automated workhorse. Tools look for known weaknesses in your systems, giving you a patching to-do list.
Penetration Testing: Ethical hackers mimicking real attackers. They try to break in, showing you where your fancy defenses crumble in practice.
Social Engineering: Preying on the human element! Phishing campaigns, pretexting phone calls...testing if employees are your weakest link.
Configuration Reviews: Secure systems can be undone by sloppy settings. These reviews look for misconfigurations that create openings.
Code Reviews: For custom software, looking for security flaws baked into the code, things automated scanners won't find.
CISSP Approach

Testing isn't a "check-box" activity. Here's where exam questions will try to trick you:

Risk-Driven: What are we MOST worried about? Protecting crown jewel data warrants a deeper dive than a public brochure website.
Scope is Everything: A poorly scoped pen test is a waste of money. Get specific about what's in bounds, what techniques are allowed, etc.
Findings to Action: A report listing vulnerabilities is useless. The assessment needs to drive a remediation plan with clear priorities.

Think of a vulnerability scan report like a doctor's checkup result. You've got a bunch of test results, but what needs URGENT attention versus what's just "good to be aware of"? Let's dissect a scan output (Nessus is a popular choice) and make sense of how to prioritize this potential firehose of findings.

Sample Nessus Output Snippet

- Vulnerability 1: Apache Struts - Remote Code Execution (CVE-2022-1234)
  - CVSS Score: 10.0 (Critical)
  - Affected Host: Production Web Server A
  - Exploit: Metasploit module available
- Vulnerability 2: SMB Legacy Protocol Enabled (Not a CVE)
  - CVSS Score: 3.5 (Medium)
  - Affected Hosts: 150 Workstations
  - Exploit: Potential for lateral movement
- Vulnerability 3: WordPress Plugin Out-of-Date (CVE-2021-9876)
  - CVSS Score: 7.8 (High)
  - Affected Host: Marketing's Public Blog
  - Exploit: Proof-of-concept online, no fancy tools needed

Prioritization Factors

1. CVSS Score: The baseline. Critical (9-10) usually screams "patch me NOW!". But, this score is in a vacuum...other factors adjust that initial panic level.
2. Asset Criticality: A remote code execution on a domain controller is game over. Same vuln on a developer's rarely-used workstation? Less urgent.
3. Exploit Availability: A theoretical vuln is bad. A Metasploit module makes it script-kiddie accessible, increasing the urgency.
4. Network Placement: Internet-facing box with the spicy vulnerability? Yikes. Internal-only, behind layers of firewalls? Gives us some breathing room.
5. Compensating Controls: Maybe we CAN'T patch that production server instantly. Strong network segmentation, WAF rules, and HIPS mitigate (but don't remove!) the risk, buying us time.

CISSP Approach

- Don't Blindly Trust the Scanner: False positives happen. A skilled analyst is still needed to weed out noise and understand the true landscape.
- Business Impact Matters Most: A "High" vuln that affects nothing sensitive may actually be lower priority than fixing a "Medium" impacting your financials.
- Trends Over Snapshots: Is a system consistently riddled with critical vulns? That paints a bigger picture of neglect, needing systemic change, not just a patch Tuesday rush.

Let's play security strategist: What if ALL three vulnerabilities affect the same production web server? How does our remediation plan change?

With all three vulnerabilities targeting the same production web server, the alarm bells are deafening! Let's shift from prioritization to a rapid-response remediation plan.

Factors Heightening the Urgency

- Criticality Stacking: Individually, some of these vulnerabilities might have been manageable. Together, they create a 'perfect storm' scenario for attackers.
- Public-Facing: This isn't some obscure internal system, it's directly exposed to the internet. Attacker motivation is sky-high.
- Exploit Variety: From script kiddie-ready Metasploit exploits to a potential zero-day in the old WordPress plugin, we're vulnerable to multiple attacker skill levels.

Our Remediation Plan Outline

1. Contain the Bleeding (If Possible)

- Can the Server Be Taken Offline?: Ideal, but likely disrupts the business. Might be necessary if it's a truly dire situation.
- WAF to the Rescue?: If you have a web application firewall, emergency rules blocking exploit attempts for the known CVEs might buy some time.
- Aggressive Segmentation: Can you quickly firewall it off from anything truly critical, limiting the damage if it does get popped?

2. Parallel Patching

- Apache Struts: Top priority. Vendor patch likely exists, but needs testing before going live. Any workaround instructions get evaluated ASAP.
- SMB Legacy: Disabling is usually painless. Remediation plan if we discover it's truly needed by something.
- WordPress Plugin: Update if available. If not, can the plugin be entirely disabled if its functionality isn't mission-critical?

3. Assume Compromise, Then Breathe

- Forensic Focus: Even if we patch, was the server already breached using an unknown exploit? Logs need deep analysis, and potentially a full system rebuild.
- Hyper-Monitoring: Scrutinize this server's traffic even after patching is done. Attackers may return if they think they successfully got in the first time.
- Incident Response Might Kick In: Severity may trigger your full IR process – containment, forensics, comms, etc. Patching becomes just one aspect.
- Root Cause is Key: Why was this server in such a bad state? Outdated patching process? Poor dev practices? Fixing THAT is what prevents the next crisis.
- "Perfect" is the Enemy of "Good Enough": Sometimes, quick and dirty mitigation is needed to avert disaster. Elegance can come later once the fire is out.

Let's imagine a scenario where a vulnerability scan flags the same critical vulnerability on three different web servers. Why our response might be wildly different for each highlights the importance of context in security!

Our Network
Server A: Public-facing e-commerce site. Processes credit card data, the lifeblood of the company.
Server B: Internal HR system. Stores employee info, nothing directly internet-accessible.
Server C: DMZ web server. Hosts marketing materials, allows some limited file uploads.
The Feared Vulnerability

The scanner screams about a remote code execution (RCE) flaw in a common web framework used on all three servers. CVSS score is a perfect 10.0, exploit code is readily available.

Why Responses Diverge
Server A (E-commerce): Drop everything! Patching may disrupt the business, but a breach is infinitely worse. Emergency change procedures are invoked, potentially even taking the site offline if a fix isn't immediately deployable.
Server B (Internal HR): Urgent, but controlled. Patching likely needs to go through standard change processes, but those get fast-tracked. Focus on network segmentation – can we further isolate it while remediation is ongoing?

Server C (DMZ): Serious, with tricky choices. Can we disable the vulnerable feature? Aggressive monitoring for exploit attempts, and a deeper dive on what those file uploads are actually used for. Might need some quick risk acceptance if business impact of patching is too severe.
CISSP Considerations

Data Is King: The e-commerce server's function drives the extreme response. Storing sensitive data = being a high-value target.
Blast Radius: Breach of Server A likely has regulatory reporting obligations, while Server B might be an internal cleanup job...massive difference in fallout.
"Perfect" Patching Isn't Always Possible: DMZ servers often have messy mixes of software. We demonstrate due diligence through compensating controls and making the risk trade-offs explicit.
The Human Factor: Public-facing servers are more likely to be targeted by opportunistic attackers, while an internal HR system breach suggests a more sophisticated adversary already inside the network.

Picture asset discovery like a detective trying to map out a crime scene. Active methods are like knocking on doors and interviewing witnesses, while passive is like analyzing footprints and security camera footage. Let's dive into the techniques, when each shines, and why a perfectly accurate asset inventory remains that elusive unicorn.

Active Discovery Techniques

- Port Scanning (Nmap is king): Sends probes to systems, sees what ports are open. Tells you what services are *potentially* running.
- OS Fingerprinting: More nuanced, tries to guess the specific OS version based on response quirks. Helps tailor attacks if vulns are OS-specific.
- Credentialed Scans: If you have valid credentials, you can ask the system nicely "what are you?" and get a far more accurate picture.

When Active Is Ideal

- Initial Mapping: "What the heck is even out there?" is answered well by active scans. You need a baseline to work from.
- Targeted Assessments: Vulnerability scans against specific systems rely on active techniques to find those flaws.

Passive Discovery Techniques

- Traffic Sniffing (Wireshark is your friend): Capture those network packets! Source/destination IP, protocol patterns, etc., reveal what's communicating, sometimes even what the application is.
- DHCP Logs: When a device gets an IP address, it often leaves a record with hostname, tells you what kind of device it likely is.
- Wireless Probes: Even without connecting, devices shout out what networks they're looking for, revealing their existence.

When Passive Excels

- Avoiding Detection: Stealthy attackers (or Red Teams) may prefer passive. Less likely to set off alarms on a well-monitored network.
- Sensitive Devices: Blasting industrial control systems with port scans is a recipe for a VERY bad day. Passive is safer.
- Finding the Unexpected: Rogue device plugged in by a well-meaning employee? Passive often spots what active scans miss.

Challenge: The Ever-Changing Inventory

- Cloud Ephemerality: Spinning up/down instances constantly? Our asset map ages out FAST.

- IoT Explosion: Smart sensors, rogue BYOD devices...the network edge gets fuzzy.
- Shadow IT: Frustratingly common. Users spin up an AWS instance, forget to tell IT, creating a security blind spot.

CISSP exam

- Both are Needed: It's not an either/or! Active for the baseline, passive for continuous monitoring to catch drift.
- Inventory is an Ongoing Process, Not a State: The perfect map only exists for a fleeting second. Tools help, but success lies in the procedures around them.

Many organizations see vulnerability scans as a mere checkbox to satisfy some auditor or to hit a compliance target. Let's shift that mindset, focusing on how these scans are a weapon in our proactive security arsenal, even when we aren't under a strict mandate.

Vulnerability Scans as Part of Risk Management

Know Your Enemy: Scans reveal what an attacker might see – unpatched systems, misconfigurations, etc. This shapes our defensive priorities.
Quantifying the Problem: "We're kind of vulnerable" vs. "30 critical unpatched systems" gets leadership's attention in the RIGHT way, driving resource allocation.
Trendspotting: Are we getting better or worse? Scans over time show if our patching, configuration hardening, etc., are actually working.
Focusing Remediation: A huge scan report is overwhelming. Risk-based prioritization driven by these scans ensures we tackle the most critical issues FIRST.
Proactive Mitigation: Think Like an Attacker

Attack Surface Mapping: New project spins up servers? A scan ensures they don't accidentally expose something they shouldn't.
Zero-Days Loom: No patch yet, but your scan flags systems vulnerable to a recently publicized exploit. Time for compensating controls and hypervigilance!
"The Forgotten Systems": Legacy systems no one actively manages tend to fester with vulnerabilities. Scans drag them back into the light.
Merger Mayhem: Acquiring another company? Scan their network FAST to understand the risks you're inheriting before the ink dries on the deal.

Compliance is a Floor, Not a Ceiling: Meeting a regulation is good, but shouldn't be our ultimate security goal. Vulnerability assessments are part of true due diligence.
Scope and Frequency Matter: Scanning the whole org quarterly is pointless if high-risk systems should be scanned far more often, and with greater depth.
Integrate, Don't Just Report: Scan data feeds into ticketing systems, informs change management...it becomes part of how we RUN security, not a separate task.
Accept Limitations: Scans catch a LOT, but not logic flaws in custom code, or the risk of an employee clicking a well-crafted phishing link. Defense in depth remains key!
The scanner lit up like a Christmas tree! Alerts for "high" severity vulnerabilities plastered your dashboard. You braced for fallout, yet...the penetration testers waltz in and shrug. They can't break in using any of the supposedly glaring holes your scan highlighted. What's the deal?
Why the Disconnect?

- Vulnerability Scanners See Potential: A scan often flags known vulnerabilities based on software versions, like an outdated Apache server. It's saying "this *could* be exploited, theoretically."
- Penetration Testers Seek Reality: They try to actually chain exploits together, turning that theoretical into a real breach. Many factors can derail this in practice:
    - Mitigating Controls: Maybe there's a firewall rule you didn't know about, or an IPS, blocking the attack path the scanner couldn't see.
    - Configuration Nuances: The vulnerability might be real, but its severity depends on an obscure config setting on the system that lessens the impact.
    - False Positives: Scanners aren't perfect. Especially in complex environments, they sometimes misidentify things.

The Unique Value of Each

- Vulnerability Scanning: Breadth and Baseline. Great for that initial lay of the land, and tracking our patching hygiene over time. Casts a wide net, essential for our early warning system.
- Pentesting: Depth and True Adversary Simulation. Answers the "so what?" Can those vulnerabilities really lead to our data walking out the door? Forces us to think about how attacks chain together in the real world.
- Both Are Tools in the Box: Dissing the scanner because the pen test went well misses the point. We need both for a comprehensive security posture.
- Results Drive Action: Did the scan find vulnerabilities that, while unexploitable THIS time, highlight a systemic issue needing a fix? Absolutely valuable!
- Risk vs. Theoretical Exploitability: A low likelihood exploit on a system holding nuclear secrets is still probably a worse problem to have than an easily exploitable flaw on the cafeteria lunch menu website. Context always matters!

Picture the Rules of Engagement (ROE) document for a penetration test as a contract mixed with a safety manual. It's what keeps the exercise productive, avoids accidental outages, and prevents the pen testers from ending up in handcuffs! Let's dissect the key elements:

Core Components of an ROE

Scope: The playground, and the electrified fences. IP ranges, systems in-bounds, types of attacks allowed (DoS, data exfiltration, etc.). Crucial to get right!
Off-Limits: The crown jewels needing extra protection (or those ancient systems that'll fall apart if you breathe on them wrong). Might have time-of-day restrictions too.
Timing: Don't run a noisy scan during peak business hours. Maintenance windows might be ideal. Attack duration is defined to prevent endless, aimless poking.
Communication: Who gets alerted if the testers think they hit the jackpot? What about if something breaks mid-test? Clear escalation paths are vital.
When Tight Rules = Bad Results

"Don't Touch Anything Important": If we ONLY let them hit test systems, how do we assess the risk to the real deal? Overly restrictive scoping limits value.
Attack Path Censorship: Can't use phishing, can't exploit workstation 0-days...might miss how most attackers would actually get in, giving a false sense of security.
The Timing Trap: Attacks ONLY during off-hours doesn't show if our monitoring would even notice the intrusion during a normal business day.

It's a Negotiation, Not a Dictate: Pentesters help here -- they understand attacks. Security AND business needs must be balanced for the results to be actionable.
Lawyers Might Get Involved: Especially with cloud systems, contracts may limit what testers can do. Understand these constraints BEFORE the exercise.
"Success" Isn't Always Breaching: Maybe the goal is explicitly testing incident response, and a breach is guaranteed. Defining what a "win" looks like is key.
Social engineering is the art of hacking the human being, often bypassing all those fancy technical defenses. Let's explore common attack methods and how to use that knowledge to make our security training truly stick.

Attack Vectors: Bypassing the Tech

- Phishing: The classic! Disguised emails, targeting urgency (unpaid invoice!) or greed (free gift!), aiming to get the user to click a bad link or open a malicious attachment.
- Vishing: The same trickery, but over the phone. Pretending to be IT support, a frantic 'CEO' needing help, anything to build trust and get sensitive info or get the victim to take an action.
- Pretexting: The con artist approach. Attacker builds a false backstory, becoming a trusted supplier, a concerned coworker... the goal is to gain enough info for a later, more damaging, stage of the attack.

Tailoring Training to Hit Home

First, we need to understand our weaknesses:

- Are employees falling for phishing constantly? Training needs to dissect real examples, why they're convincing, and teach those subtle red flags.
- Recent incident exploited trust between departments? Training scenarios mimicking that exact dynamic make it real for people vs. generic anti-phishing videos.
- Industry-Specific Threats: Healthcare has different bait than a manufacturing plant. Current events (tax season scams, etc.) create timely training moments.

Beyond Just Awareness

- Make Reporting Easy & Blame-Free: Employees need a safe way to say "I almost screwed up", so we can learn and analyze close calls.
- Training Isn't One-and-Done: Regular, bite-sized refreshers are better than an annual marathon session people instantly forget.
- Test Them (Kindly): Simulated phishing campaigns (with no punishment for those who fall for it) show what sticks from the training... and where we need to focus next.

A pentest report can be a powerful weapon for getting security fixed... or it can be a dusty tome left unread on a manager's desk. How we communicate the results makes all the difference. Think of it like translating a complex medical diagnosis to a patient – we need to tailor the message to the audience.

The Executive Summary: Risk in Business Terms

- Avoid Jargon Overload: Don't start with CVSS scores and exploit names. Focus on what a successful attack could do: data stolen, operations halted, etc.
- Tie to Business Objectives: "Compliance gap" is less compelling than "this flaw opens us to fines harming our expansion plans".
- Prioritization is Key: Not all vulnerabilities are equal. What MUST we fix right now to avoid imminent disaster?
- A Visual Helps: A simple risk heatmap (likelihood vs. impact) can be more impactful than lengthy text for some executives.

The Tech Team Deep Dive: Remediation Roadmap

- Specificity Rules: Don't just say "patch Apache Struts", include version numbers, configuration changes if relevant, and potential workaround instructions if a patch is unavailable.
- Consider Exploit Availability: Something with a Metasploit module is more urgent to address than a theoretical vulnerability.
- Think Beyond the Obvious: Did the test reveal a process problem (poor firewall rules changing because of haphazard IT tickets)? Recommendations need to address that root cause.

Additional Considerations

- Who Presents Matters: Security staff explaining to their peers is different than the pentester delivering bad news to the executives who hired them.
- Don't Sugarcoat, But Offer Solutions: Scare tactics alone backfire. Present the risk, but ALSO a reasonable path to fix it.
- Tell a Story: If possible, narrate the attack path in plain English. This makes it real for non-technical folks who might struggle with the raw report.

When it comes to audits, let's ditch the idea that it's simply about internal teams being cozy with what they audit, while external equals true independence. Both approaches have their time and place. Here's how to know when to leverage which:

Internal Audit: Strength in Familiarity

Process Deep Dives: They know the 'why' behind how things are done (sometimes messy, but that's the reality). This can uncover risks an outsider, focused on textbook 'should', would miss.
Trend Spotting: Since they're embedded in the org, they see subtle shifts over time. Gradual control failures become visible, where a snapshot audit by an outsider might look OK.
Relationship Currency: Internal auditors often have existing rapport with those being audited. This can make sensitive conversations easier, though they must watch out for becoming too friendly.
The Cost Factor: They're on the payroll already! Leveraging their skills for audits outside their core area can be a win, if managed well to avoid the 'too close to the problem' pitfall.
External Audit: The Power of 'Fresh Eyes'

Unburdened by History: "We've always done it this way" blindness is hard to fight internally. Outsiders ask those 'dumb' questions that reveal shaky assumptions.
Specialized Expertise: Need to audit complex tech or a niche regulatory area? External firms likely have those skills, internal teams rarely do.
The Independence Stamp: Sometimes, it's less about what they find, and more that an outsider said it. This carries weight with boards, regulators, etc.
Benchmarking Opportunity: They've seen how others tackle similar problems, bringing in that industry context that may be lacking internally.
Scenario Time
Routine Process Audit: A core business process is due for its annual audit. Internal is probably the efficient choice, assuming enough segregation of duties exists.
Merger Mayhem: Just acquired another company. External auditors are a must to get a FAST and unbiased view of the risks you've just bought.
Suspected Fraud: Sensitive situation almost always warrants an external firm to ensure impartiality (and sometimes forensic expertise).

Important Note: The quality of the auditor, whether internal OR external, matters more than the mere label. Brilliant internal teams and clueless external consultants both exist! Picking the right person for the job is key.

Let's shift our thinking from audits being a painful chore to them being a tool PROACTIVELY supporting the business. By understanding the "why" behind specific audits, we ensure our effort is focused where it truly matters.

Mapping Security to What the Business Cares About

- PCI DSS: The Classic Example. Want to process credit cards? This isn't optional. Security goals here are non-negotiable, dictated by the standard to protect that payment data and avoid costly fines.
- SOC 2: Building Customer Trust. Especially for SaaS, proving you have good security around their data isn't just nice-to-have, it closes deals. The SOC 2 focuses on operational security, aligning with those client expectations.
- ISO 27001: The International Flavor. Doing business in Europe? This standard often carries weight. It's more holistic than PCI DSS, demonstrating an overall mature security program.
- Industry-Specific Regulations: Healthcare has HIPAA, finance has its own alphabet soup of regulations, etc. Compliance here directly ties to license to operate.
- "No One Asked, But We Should": Maybe you're building the next life-saving medical device. Proactively auditing against security best practices, even if not yet regulated, builds public trust AND protects your innovation.

Security Goals Driven by Business Objectives

Let's look at a hypothetical scenario to see how this plays out:

- Company X launches a new online marketplace. It wants to:
    - Attract both individual sellers AND big retailers
    - Scale rapidly, aiming for international growth
    - Minimize legal/compliance overhead early on

Security Audit Priorities Might Look Like:

1. URGENT: Hit the absolute minimum for processing payments securely (likely a scaled-back PCI assessment). Without this, no business happens.
2. SOON: SOC 2 Type 1 to show basic controls are in place. This will be vital for those bigger retail partners
3. LATER: Full blown PCI DSS when volume warrants, ISO 27001 down the road as they eye those international markets

Important Considerations

- Risk Appetite Matters: Highly risk-averse org may go above and beyond baseline compliance. Audits are a shield against the unknown.
- Don't Be Myopic: A narrow audit to check a box misses opportunity. Maybe findings SHOULD influence the biz (ex: being way overbuilt for the current customer base, so costs could be reduced)

Picture a toolbox packed with different auditing frameworks. Knowing which tool is right for the job is crucial! Let's figure out when to reach for each, and why one-size-fits-all is a recipe for a subpar audit.

Frameworks: A Quick Taste

ISO 27001/27002: The internationally recognized security management standard. Broad in scope, good when an overall "stamp of approval" is needed, with flexibility in how you meet the controls.

NIST CSF: Made for critical infrastructure but broadly useful. Prioritizes outcomes (Identify, Protect, etc.), giving organizations a roadmap, not just a checklist.
COBIT: The IT governance darling. If the goal is aligning security with broader business processes and risk management, COBIT's focus shines.
Scenario-Based Selection

Multinational SaaS Provider: Customers are demanding proof of solid security. ISO 27001 is likely the play here, carrying that global recognition.
US-Based Power Company: Protecting critical infrastructure is the name of the game. NIST CSF provides a structure aligned with that mission.
Financial Firm Struggling with Scattered IT: Security feels reactive and chaotic. COBIT can help bring order, mapping security efforts to the business goals that are clearly getting lost.
Internal Audit Needs a Boost: The team KNOWS things are sloppy, but lack a structured way to assess and report. Implementing ISO 27001/2 gives them a backbone, even if external certification isn't the near-term goal.
Important Notes

Frameworks Aren't Magic: A stellar audit against the wrong framework is useless. First define the need, THEN pick the tool.
Combo Platters Exist: Using NIST CSF for high-level structure, and ISO standards for specific controls, is perfectly valid.
Maturity Matters: A startup shouldn't dive headfirst into full-blown ISO 27001. Right-size the effort to avoid it becoming a soul-crushing burden.
Here's a discussion focused on creating effective remediation plans and turning audit insights into real improvements:

Beyond the Report: Proactive Remediation

A successful audit isn't just about the final report, but it's a starting point for driving meaningful change. Here's how to create a powerful remediation plan:

1. Prioritization: Not Everything Is Equal

- **Severity and Risk:** Focus on findings that could have the most significant impact on operations, compliance, or financial health. Consider potential consequences like data breaches, regulatory fines, or even reputational harm.
- **Feasibility:** Balance the complexity of the fix against the potential gains. Can it be addressed quickly with minor tweaks, or does it require a system overhaul? Prioritize actions with a good return on investment.
- **Dependencies:** Understand if certain changes need to be made in a sequence for success. For example, updating documentation might be dependent on a new process being put in place first.

2. Accountability: Ownership is Key

- **Clear Responsibilities:** Assign each remediation task to specific individuals or teams, not just departments. This promotes direct accountability.
- **Cross-Functional Collaboration:** Sometimes audit findings reveal issues that span teams or processes. Ensure that those involved in the change work together effectively, not in silos.
- **Empowerment:** Provide those responsible for remediation not only with tasks, but with sufficient authority, resources, and support they need to implement the changes successfully.

3. Timelines: A Sense of Urgency

- **Realistic Deadlines:** Don't leave things open-ended. Set achievable milestones and deadlines, while considering the severity of the findings and the effort required.
- **Progress Tracking:** Implement a way to track progress against milestones. This could be as simple as a shared spreadsheet or using project management tools for more complex remediation plans.
- **Communication:** Regular updates to stakeholders maintain focus and allow for addressing any roadblocks early on.

4. Continuous Improvement: Learning From the Lessons

- **Root Cause Analysis:** Don't just fix symptoms. Address the underlying reasons for the audit findings. Were there failures in process design, training, or oversight?
- **Preventive Measures:** Analyze how similar findings can be avoided in the future. Consider adjusting policies, procedures, training programs, or even technology solutions to mitigate future risks.
- **Knowledge Sharing** Share lessons learned across the organization, especially with teams engaged in similar processes. This fosters a culture of learning and proactive risk management.

Action Plan Example

Audit Finding	Priority	Owner(s)	Timeline	Remediation Steps
Inadequate inventory controls	High	Warehouse Manager	3 months	1. Revise inventory procedures 2. Implement cycle counts 3. Train staff on new process
Inconsistent data backup	Medium	IT Department	6 months	1. Review backup policy 2. Automate backups 3. Test restore procedures

Additional points to emphasize:

- **Executive Buy-in:** Remediation plans need leadership support to be a priority across the organization.
- **Communication:** Open channels of communication throughout the process ensure everyone understands their roles and deadlines, avoiding surprises.
- **Iteration:** View audit remediation as an evolving process. As situations and risks shift, the plan might need adjustments for successful long-term improvement.

Shift your mindset from audits being a compliance exercise to being a catalyst for organizational growth.

VII. Security Operations

In today's rapidly evolving threat landscape, robust security operations are critical to maintaining the confidentiality, integrity, and availability of an organization's assets. This section outlines the essential processes, methodologies, and technologies that underpin our approach to detecting, analyzing, responding to, and mitigating security incidents effectively. Our focus is on proactive risk reduction while ensuring operational efficiency and resilience.

Overview

This section will delve into the following key elements of our security operations:

- **Threat Detection and Monitoring:** We'll detail our mechanisms for continuous monitoring of networks, systems, and endpoints, highlighting the technologies and strategies used to identify anomalous activity and potential threats.
- **Incident Analysis and Response:** This includes our processes for investigating alerts, determining their severity, containing threats, and implementing remediation measures. We'll outline incident escalation procedures and collaboration with relevant stakeholders.
- **Vulnerability Management:** We'll illustrate how proactive vulnerability identification, assessment, prioritization, and patching practices are integrated into our security operations.
- **Metrics and Reporting:** We'll discuss the key security metrics and performance indicators monitored for insights into our security posture, enabling us to track progress and make informed decisions.
- **Continuous Improvement:** Building on the concept of security as an ongoing journey, this segment will explore our approach to continually learn from incidents, adapt our security controls, and improve organizational resilience.

By providing a comprehensive look at our security operations, this section serves to enhance transparency, demonstrate our commitment to proactive security, and instill confidence in our approach to protecting critical assets.

Let's break down a scenario and explore classification impacts:

Scenario: Suspicious Network Traffic

Monitoring tools have flagged unusual network traffic patterns:

- **Traffic Origin:** Multiple connections from unknown external IP addresses are attempting to access sensitive company servers.
- **Timing:** These connection attempts are happening outside standard business hours.
- **Behavior:** The traffic involves unusual port scans and attempted downloads of large data packets.

Incident Classification: The Power of a Label

1. **Potential Incident:**
    - **Focus:** Initial screening and investigation are prioritized. The objective is to rapidly determine if this activity presents a genuine threat.
    - **Actions:**
        - Triage: Analyzing traffic source, intended targets, and nature of activity.
        - Baseline Comparisons: Review typical traffic patterns to identify deviations.
        - Log Analysis: Scrutiny of system and firewall logs for related indicators.

- **Reporting:** Preliminary notification to stakeholders about a *potential* issue under investigation.

2. **Confirmed Breach:**
    - **Focus:** Containment and remediation take center stage after classification is upgraded. The scope expands to assessing damage and preventing further harm.
    - **Actions:**
        - Isolation: Severely limit or cut off compromised system communication.
        - Preservation of Evidence: Forensic protocols for logs, network data, and affected systems.
        - Root Cause Analysis: A deep dive into how the breach happened and exposed vulnerabilities.
    - **Reporting:**
        - Regulatory or contractual reporting mandates might be triggered.
        - Detailed reporting to executives, potentially including legal involvement.
        - Information sharing with external entities (e.g., law enforcement, vendors) if appropriate.

Why Classification Matters

- **Urgency and Scale:** A confirmed breach commands a greater sense of urgency and scaling of resources for containment and investigation.
- **Legal and Regulatory:** Breaches often trigger specific reporting requirements to authorities and affected parties. Starting these processes hastily before confirmation could lead to misreporting and unnecessary escalation.
- **Response Actions:** Isolation of compromised systems or network segments is often part of breach response, which is less disruptive if done after careful assessment.
- **Reputation and Customer Trust:** Prematurely announcing a breach before complete information is gathered can be damaging. Controlled, transparent communication after confirmation is critical.

Important Notes

- **Classification Is Dynamic:** As investigations progress, the classification of the incident may change, leading to adjustments in actions and reporting.
- **Thresholds:** Organizations should have well-defined criteria and processes for determining what constitutes a potential incident versus a confirmed breach.

Here's the study guide content focused on the importance of playbooks for complex incidents, designed to sound natural and offer the kind of depth that would help a professional excel on the CISSP exam:

Playbooks: Beyond the Basics

It's easy to dismiss playbooks as tools for newcomers, but seasoned security analysts benefit tremendously from them, especially in the face of complex threats like ransomware. Let's explore why detailed playbooks matter even for highly experienced teams:

Consistently Defeating Complexity: Ransomware attacks often involve a multi-pronged approach: initial compromise, lateral movement, data exfiltration, and finally, the crippling encryption stage. A comprehensive playbook provides step-by-step guidance for each of these stages, ensuring a structured and thorough response, even when multiple systems and technologies are affected.

Mental Bandwidth During Crisis: During a high-stress event, cognitive load can skyrocket, making even routine tasks feel overwhelming. A playbook offloads decision-making that can be pre-planned (like isolation procedures, communication channels, forensic actions), freeing mental resources for analyzing the unique aspects of the current attack.

Experience as a Playbook Ingredient: A mature playbook isn't static. It incorporates lessons learned from prior incidents – what succeeded, what went wrong, and where delays occurred. These insights refine isolation protocols, enhance collaboration, and pinpoint areas where automation can streamline responses in the future.

Crafting an Effective Ransomware Playbook

To further explore this concept, consider these key elements:

Stakeholder Involvement: Playbooks are most effective when developed with input from network engineers, system admins, legal, PR, and executives – reflecting their needs and constraints during an incident.

Specificity Isn't Rigidity: Include the essential actions, BUT allow for analyst judgment if the situation warrants deviation. It's about guidance, not blind adherence.

Regular Testing and Updates: Simulate parts of the playbook (like a tabletop exercise) to uncover gaps. Update it as technology, regulations, or internal processes evolve.

Playbooks are an analyst's superpower, especially in the fight against time-sensitive, complex attacks like ransomware.

Experience doesn't negate the need for a playbook; it fuels the creation of a better one.

A well-structured playbook translates into quicker containment, less damage, and a stronger security posture overall.

Let's break down the key elements of an incident communication plan designed to bring order to a potentially chaotic situation:

Incident Communication Plan: The Essentials

Pre-Defined Roles & Contact Lists: Knowing WHO is responsible for WHAT communication during an incident saves precious time. This includes:

Technical Leads: Communicating the nature of the issue, updates on containment, and remediation progress to relevant teams.

Executive Spokesperson: Owning high-level internal and external communication, often working alongside legal counsel to ensure proper messaging.

Customer-Facing Team: If services or data are affected, this pre-identified team manages customer notifications in a controlled manner.

External Considerations

Law Enforcement: Determine under what circumstances they should be contacted (ransomware, suspected data theft, etc.). Build pre-existing relationships for smoother interaction if needed.

Affected Customers: Establish a clear, honest, and proactive approach that balances providing information with minimizing panic or reputational damage. This could include status pages or pre-drafted holding statements.

Regulatory Bodies: Understand mandatory reporting requirements (such as breach notifications) and build those timeframes directly into your plan.

Templates, Not Rigidity: Have templates for common communications (e.g., initial incident confirmation, status updates, issue resolution). These provide a base while still allowing customization to suit the specific situation.

The Power of Practice: Run communication simulations alongside technical tabletop exercises. This highlights potential sticking points, like how to balance technical updates with clear messaging for executives.
An incident communication plan isn't just for PR; it's a core element of effective response.
Proactive and well-timed communication builds trust internally and externally throughout an incident.
Regulatory reporting mandates drive timelines. Don't get caught unprepared and potentially face additional penalties.
The Eradication and Recovery Balancing Act

Containment is just the first step. Thorough incident response demands complete malware eradication while paving the way for a swift, secure return to normal operations. Let's discuss the sometimes-competing priorities that shape this stage:

Malware Eradication: The Forensic Considerations:

Volatile Data: Identify and capture this (e.g., memory, running processes) before it's lost on power cycle. It can contain vital clues about the attack's methods.
System Image vs. 'Clean' Reinstall: If time permits, a full disk image preserves evidence at its most pristine. However, in severe cases, a scorched-earth reinstall can be necessary for faster recovery.
Chain of Custody: If legal action is likely, document every step of handling infected systems. This ensures evidence integrity.
Recovery: When Speed Isn't Everything

Data Restoration: Backups are crucial, but validate them before restoration. Avoid accidentally reintroducing the malware.
Patches and Mitigations: Identify the vulnerability that allowed the incident as part of the recovery. Don't rebuild only to be compromised again.
Phased Approach: If possible, restore services in a staged manner, monitoring closely for any signs of reinfection.
Regulatory Requirements Shape the Timeline

Incident severity isn't the sole factor driving the recovery approach. Consider these regulatory influences:

Breach Notifications: Deadlines for reporting confirmed data breaches to individuals and authorities introduce a hard time limit to your investigation and recovery efforts.
Financial or Healthcare Data: Regulations like PCI DSS or HIPAA may impose stricter restoration and monitoring requirements to demonstrate compliance has been regained.
Industry-Specific Standards: Critical infrastructure providers, for example, might have mandated processes for system restoration after an incident that go beyond simple technical fixes.
Eradication and recovery create a tension between the need for detail and the need for speed. Security analysts must balance them contextually.
Forensics can't be an afterthought. Plan for evidence preservation in ways that align with your organization's risk tolerance and legal obligations.

It's not JUST about restoring servers; it's about restoring trust by ensuring the attack won't simply recur.
et's unpack a chain of custody failure and why it spells trouble in court:

Scenario: Evidence Mishandling

Imagine a forensic analyst tasked with investigating potential data theft from a company laptop. Rather than using a write blocker when creating a forensic image of the hard drive, the analyst works directly on the seized device. Potential consequences of this misstep cascade from there:

Data Contamination: Imaging without a write blocker opens the door to modifying the drive's contents, even unintentionally. This undermines the integrity of the evidence. Were files deleted by the attacker to cover their tracks, or accidentally by the analyst?
Defense Attorney's Dream: Any skilled defense lawyer will zero in on this breach of protocol. They'll argue that since the prosecution can't definitively prove the evidence hasn't been altered, it can't be trusted as proof of their client's actions.
Doubt Cast on Findings: Even if the data DOES point to wrongdoing, the mishandling casts doubt. The prosecution now has to prove both the guilt of the defendant AND that the evidence remained pristine throughout the investigation.
Potential Inadmissibility: In severe cases, the judge could rule the evidence tainted beyond repair and declare it inadmissible. This might severely weaken or even collapse the prosecution's entire case.
Chain of Custody as a Shield (and a Sword)

Maintaining Trust: A rigorously maintained chain of custody is the security professional's proof that evidence is what it claims to be. Mishandling shatters this trust.
The Burden Shifts: Now, the analyst is under scrutiny on the witness stand, explaining their actions rather than the evidence they collected. This puts the defense in the driver's seat.
Beyond Technical Skills: The courtroom magnifies the importance of procedural adherence alongside technical expertise. A single mistake can discredit otherwise sound findings.

Chain of custody isn't just paperwork - it's a core element of any prosecutable investigation.
Proper evidence handling protocols are as vital to a successful case as the forensic findings themselves.
Court cases can hinge on the minutiae of procedure, elevating seemingly small actions to critical importance.

Let's explore how tools like EnCase and FTK transform raw data into actionable evidence:

Unmasking Evidence with Forensic Suites

Forensic tools are a digital detective's utility belt. Here's how tools like EnCase and FTK can be used to uncover critical evidence:

Acquiring a Disk Image (The Digital "Snapshot"):

Write Blockers: The Golden Rule: These hardware devices are essential, enforcing read-only access to prevent any accidental modification of the original drive.

Image Formats: Tools generate images in standard formats (like E01 or dd), ensuring both portability and compatibility with other forensic tools for later analysis.
Hashing for Integrity: Calculate a cryptographic hash at acquisition. Any changes to the image will produce a different hash, signaling possible tampering.
Recovering Deleted Files (Finding the Hidden):

The Art of "Un-deleting": Files marked as "deleted" often still have data on the disk until that space is overwritten. Forensic suites carve through file systems and unallocated space to find recoverable files.
The Nuances of Fragmentation: If a file is fragmented, tools might only recover portions. However, even partial recovery can be valuable (e.g., email fragments, metadata).
Timeline Artifacts (When Things Happened):

File System Timestamps: Each file has creation, modification, and access timestamps (MAC times). These help piece together user actions or malware activity.
Beyond Filesystems: Tools might parse browser history, log files, and system artifacts to assemble a comprehensive timeline of a user's activity.
Tool Validation: Trust but Verify

Known Test Data: Before deploying a tool on real evidence, verify its behavior using drives with known content. Do file hashes match expectations? Does it accurately recover what should be recoverable?
Standards and Validation Suites: Organizations like NIST provide test data and guidelines to help validate forensic tools for accuracy and reliability.
Staying Current: Forensic tool development is ongoing. Updates often address new file systems, encryption types, and address vulnerabilities found in earlier versions.

Forensic tools bring structure and scientific rigor to complex investigations.
Understanding a tool's capabilities AND limitations is crucial to interpret findings accurately.
Validation isn't a one-time event; it's an ongoing process to ensure tools keep pace with evolving technology.
Let's explore how tools like EnCase and FTK transform raw data into actionable evidence:

Unmasking Evidence with Forensic Suites

Forensic tools are a digital detective's utility belt. Here's how tools like EnCase and FTK can be used to uncover critical evidence:

Acquiring a Disk Image (The Digital "Snapshot"):

Write Blockers: The Golden Rule: These hardware devices are essential, enforcing read-only access to prevent any accidental modification of the original drive.
Image Formats: Tools generate images in standard formats (like E01 or dd), ensuring both portability and compatibility with other forensic tools for later analysis.
Hashing for Integrity: Calculate a cryptographic hash at acquisition. Any changes to the image will produce a different hash, signaling possible tampering.
Recovering Deleted Files (Finding the Hidden):

The Art of "Un-deleting": Files marked as "deleted" often still have data on the disk until that space is overwritten. Forensic suites carve through file systems and unallocated space to find recoverable files.
The Nuances of Fragmentation: If a file is fragmented, tools might only recover portions. However, even partial recovery can be valuable (e.g., email fragments, metadata).
Timeline Artifacts (When Things Happened):

File System Timestamps: Each file has creation, modification, and access timestamps (MAC times). These help piece together user actions or malware activity.
Beyond Filesystems: Tools might parse browser history, log files, and system artifacts to assemble a comprehensive timeline of a user's activity.
Tool Validation: Trust but Verify

Known Test Data: Before deploying a tool on real evidence, verify its behavior using drives with known content. Do file hashes match expectations? Does it accurately recover what should be recoverable?
Standards and Validation Suites: Organizations like NIST provide test data and guidelines to help validate forensic tools for accuracy and reliability.
Staying Current: Forensic tool development is ongoing. Updates often address new file systems, encryption types, and address vulnerabilities found in earlier versions.
Key Takeaways
Forensic tools bring structure and scientific rigor to complex investigations.
Understanding a tool's capabilities AND limitations is crucial to interpret findings accurately.
Validation isn't a one-time event; it's an ongoing process to ensure tools keep pace with evolving technology.

Let's discuss how transforming isolated log entries into a full-fledged attack narrative is a critical skill for security analysts:

Piece by Piece: Assembling the Attack Puzzle

A single alert rarely tells the whole story. True root cause analysis often involves connecting these dots:

- **Firewall Logs: The Initial Signal:** These logs might flag unusual traffic patterns - a surge of connections from unknown IPs on unusual ports, connections to unexpected destinations, or attempts at exploiting known vulnerabilities. This is often your first red flag.
- **System Logs: Where Things Get Interesting:** Look for signs corresponding to the firewall activity. Attempted login failures (especially to privileged accounts), process launches from unusual locations, or file modifications within critical system directories point to potential intrusion.
- **User Activity Logs: The Human Factor:** Are the unusual login attempts tied to a legitimate user's account (possible compromise)? Do the suspicious processes match something that user would normally execute (or definitely *shouldn't* be running)?

Piecing Together Attack Stages

By correlating logs, the attack narrative often takes form:

1. **Initial Foothold:** Did the attacker exploit a vulnerability (firewall log shows that port, system logs confirm successful process launch)? Or did they trick a user with social engineering (unusual login, activity outside that user's norm)?
2. **Lateral Movement:** Logs might reveal system-to-system connections, remote administration tool usage, or attempts to access sensitive data shares, mapping out the attacker's spread within the network.

3. **Exfiltration:** Look for large data transfers to external IPs, file compression activities (preparing stolen data), or the introduction of new external storage devices that don't match typical usage patterns.

- Individual logs are breadcrumbs; correlation is the trail showing the attacker's full journey.
- Time synchronization between log sources is crucial for accurate sequencing of the attack.
- Understanding "normal" activity baselines for different user roles makes spotting anomalies during analysis easier.
- Centralized log aggregation tools (SIEM) massively streamline this kind of multi-source investigation.

Let's explore the legal maze that is eDiscovery, especially when it ventures beyond your familiar territory:

Data Preservation: When "Delete" Isn't an Option

Duty to Preserve: In civil litigation, once a lawsuit seems reasonably likely, organizations have a legal obligation to preserve potentially relevant evidence, including electronic data. This means no selective deletion or overwriting of data – even routine activities might need to be halted.
Scope Is Broad: It's not just the smoking gun email that needs preservation. It includes drafts, backup tapes, system logs, potentially even employees' personal devices if they were used in ways relevant to the case.
Penalties for Non-Compliance: If data destruction occurs (intentional or accidental) after the duty to preserve arises, courts can impose sanctions. These might be fines, damaging rulings on the merits of the case, or even contempt of court findings.
Cross-Border Nightmares: Conflicting Laws

When incident data resides in other countries, things get really complex:

Data Privacy Protections: The EU's GDPR, for example, limits what personal data can be collected, how long it's retained, and with whom it's shared. These limits can clash with the broad preservation requests common in US-based litigation.
Data Sovereignty: Some countries have laws restricting the transfer of certain data types outside their borders entirely, potentially putting international companies in a no-win situation.
Conflicting Orders: You could face a US court order to produce data that a foreign regulator insists cannot leave their jurisdiction.
Strategies for Navigating the Chaos

Proactive Legal Counsel: Early legal involvement is crucial for developing preservation strategies with cross-border risks in mind.
Data Mapping: Knowing where your data is physically stored (and the laws governing each location) is essential to make informed choices if litigation strikes.
International Agreements: The US has mechanisms like Mutual Legal Assistance Treaties (MLATs) for formal legal cooperation. These are slow, but awareness of them can help with planning.
Privacy-Focused Solutions: If dealing with EU data, explore solutions like anonymization or pseudonymization to meet discovery requests while minimizing the transfer of sensitive personal data.
Key Takeaways

eDiscovery is rarely purely technical; legal requirements drive your actions.

Cross-border incidents require specialist expertise to balance compliance with obligations in different jurisdictions.
Proactive planning – knowing where data resides, understanding applicable laws – is vital for minimizing risk when the litigation storm inevitably arrives.
Let's break down how different log types provide unique windows into suspicious activity or potential incidents:

The Investigative Power of Different Logs

Windows Event Logs (The System's Diary)

Strength: Breadth of information on operating system and application activity. Logon/logoff events, process creation, file access, service status changes, and much more.
Focus: Detecting intrusions on the host itself, anomalous user behavior, privileged account activity, software malfunction or attempted tampering.
Caveats: Can be verbose, requiring sifting for the important events. Local logs are easily tampered with if the attacker gains sufficient control of the system.
Web Server Access Logs (The Web's Footprints)

Strength: Record each HTTP request to the server. Reveals client IP addresses, pages accessed, timestamps, user-agent strings (browser/device info), and error codes.
Focus: Pinpointing web-based attacks (SQL injection attempts, probing for vulnerabilities), unusual access patterns, potential data exfiltration, or website defacement.
Caveats: Can be limited if the attacker manipulates request headers. Primarily focused on server-side activity, may not fully reflect the end-user's experience if client-side attacks are involved.
DNS Query Logs (Mapping the Network's Conversations)

Strength: Log DNS resolution activity, showing which domain names systems queried, and the responses received.
Focus: Unmasking malicious infrastructure (connections to known bad domains), detecting data exfiltration disguised as DNS traffic, identifying malware command-and-control activity.
Caveats: Many legitimate connections to CDNs and external services create noise. Encrypted DNS protocols (like DoH) may limit your visibility.

No single log type is a silver bullet. Correlating across them is where the true analytical power lies.
The most valuable log source depends on the nature of the suspected attack.
Understanding the "normal" patterns in each log type for your environment is crucial to make anomalies stand out.
Scenario: Logs Collected, Value Missing

Picture this: your organization diligently invested in a SIEM solution. Logs from servers, firewalls, and endpoints pour in. Yet, when a breach does occur, that mountain of data offers frustratingly little insight during the investigation. Here's what could be going wrong:

Lack of Normalization: Log formats are wildly inconsistent. Windows Event IDs, Apache's access log syntax, and vendor-specific firewall logs all speak their own language. Without normalization, attempts to correlate data across systems fail.

Noisy Correlation Rules: The SIEM tries its best with out-of-the-box rules. However, these are generic, resulting in a flood of low-value alerts about "suspicious" activity that's actually normal for your environment. Analysts drown in false positives.
Missed Thresholds: Some events may be genuinely important but lost in the noise. Without fine-tuned alerting, a single failed login is treated the same way as a brute-force attack overwhelming a public-facing server.
Transforming Data into Actionable Intelligence

Here's how to turn your SIEM into a powerful security weapon:

Normalization is Key: Map logs to a common schema (IP addresses, timestamps, user identifiers, etc.). This ensures that different systems can be understood in a unified way.
Custom Correlation Rules: Think like an attacker. What sequence of events signifies a true threat in your environment? Build rules that detect that specific pattern, reducing the irrelevant noise.
Alert Thresholds Matter: Multiple failed logins within a short period? That's a higher priority alert than a single failure. Fine-tune your system to prioritize events that truly signify risk.

A SIEM isn't a magic box; it's a tool that needs skilled configuration to match your environment.
Log collection is the starting line, not the finish. True value comes from turning data into actionable insights about potential threats.
Regular review of SIEM rules and thresholds is essential. Environments and threats evolve, so should your detection capabilities.
Let's unpack the complementary roles of automation and human-driven hunting, and then touch on the complexities of insider threat monitoring:

Automated Alerts: The Strengths and Limits

- **SIEM Dashboards & Pre-Built Logic:** These tools shine in detecting known threats. Pre-defined rules watch for patterns like malware signatures, vulnerability exploitation attempts, or typical post-compromise activities.
- **The Power of Speed:** Automated systems operate at machine scale, alerting analysts to suspicious events as they happen, crucial for rapid containment and minimizing the impact.
- **Weaknesses:** They're inherently 'backward-looking'. They miss attacks designed to fly under the radar of existing rules. Zero-day exploits or attackers closely mimicking legitimate user actions can go undetected.

Proactive Threat Hunting: When Humans Lead

- **Hypothesis-Driven:** Threat hunters don't wait for alerts. They posit an attack scenario (what would data exfiltration look like?) and comb logs for subtle traces.
- **Connecting the Unusual Dots:** Finding unusual process execution chains, minor anomalies across several logs... these may be insignificant individually, but a skilled hunter spots them as potential pieces of a larger attack.
- **Tool of Choice:** Threat hunting may leverage SIEM platforms, but also involve raw log analysis and custom scripts to dig deeper.

Monitoring for the Enemy Within

Detecting malicious insider activity poses unique challenges:

- **Baseline Deviations:** What's "normal" varies wildly between users. An accountant accessing HR files is suspicious, a system admin accessing them might not be. This makes anomaly detection tricky.

- **Granular Monitoring, Privacy Risk:** To spot subtle signs of data gathering or sabotage, you need deep visibility into user actions. This raises ethical concerns and potential legal risks if employee monitoring crosses the line.
- **Technical Limitations:** Insiders might know exactly how to avoid triggering alerts, carefully blending in with their regular work patterns.
- Automated alerts and threat hunting are NOT competitors, but complementary elements of a robust security strategy.
- Novel attacks often demand the human ability to find unusual patterns, ask "what if..." questions, and look beyond the obvious.
- Insider threat detection requires careful balancing between security needs, privacy rights, and technical limitations.

Let's dive deep into the world of patching, where things are rarely as simple as clicking "install updates":

Patch Prioritization: Risk and Realities

Severity is Only Part of the Puzzle: Yes, critical vulnerabilities demand quick action. BUT, also consider exploit availability in the wild, the asset's exposure (internet-facing vs. internal), and potential business impact of the patch itself.
Dependencies: The Tangled Web: Sometimes fixes for one system rely on others being patched first. Or, a patch breaks a legacy application crucial to business operations. Understanding these relationships is vital for a smooth rollout plan.
The Importance of Backout Plans: Even well-tested patches can have unforeseen consequences in production. Having a documented rollback procedure (system image back, older-version installer readily available) is the safety net that allows for aggressive patching.
Baselines: The Foundation for Sanity

Golden Images: A secure baseline is a known-good starting point for OS and application installations. Deviations from this signal potential tampering, misconfiguration, or unauthorized software.
File Integrity Monitoring (FIM): Tools watch critical files and directories. Changes, whether malicious or just an accidental config edit, trigger alerts, allowing for rapid investigation and correction.
Registry Hardening: Especially on Windows, registry bloat and insecure settings are common. Tools compare registry state to a secure baseline and assist in hardening, reducing the attack surface.
Tools of the Trade

Patch Management Suites: Often part of vulnerability scanners or endpoint protection. These automate patch deployment but also aid reporting and dependency mapping, which are crucial.
FIM Specialists: Solutions like Tripwire or OSSEC excel in detecting file-level changes. Integration with your ticketing system is key for streamlined response.
Group Policy & Template Tools: If managing many Windows systems, Group Policy provides centralized control. Template solutions help author and distribute secure registry baselines efficiently.

Patching without a plan is a recipe for chaos, especially in complex environments.
Risk-based prioritization makes patching efforts efficient and reduces business impact.
Knowing what "good" looks like is the first step toward detecting unauthorized changes.

Let's map out the journey from vulnerability discovery to a more secure environment:

Vulnerability Remediation Workflow

1. **Discovery:** Vulnerability scanners (Nessus, Qualys, etc.) provide a starting point, but their output is raw data, not a finished plan.
2. **Prioritization:** Weigh risk factors (CVSS score, exploit availability, asset criticality) to determine which vulnerabilities warrant the most urgent attention.
3. **Remediation Options:**
    - **Patching:** The ideal scenario, but vendor patches need to be tested in a non-production environment first.
    - **Compensating Controls:** If a patch isn't available, network segmentation, tighter firewall rules, or disabling vulnerable services temporarily reduce the attack surface.
    - **Risk Acceptance:** Rarely a good long-term option, but for low-risk vulnerabilities on legacy systems, it might be the most practical approach, alongside close monitoring.
4. **Deployment:** Automate patching where possible. Thoroughly document manual fixes done for audit and reproducibility purposes.
5. **Validation:** Re-scan after remediation! This confirmation step is crucial to ensure the fix truly addressed the issue and didn't introduce new problems.

The Challenge of the Unfixable

- **Vendor Delays and Zero-days:** When a patch is absent, having strong compensating controls and monitoring is essential while awaiting a solution.
- **Broken Functionality:** Sometimes patches cause worse problems than the vulnerability. This is where testing environments are invaluable. Partial patching (e.g., only on internet-facing systems) might be a temporary compromise.

Authenticated Scans: Going Deeper

- **Beyond Perimeter Defenses:** Firewalls only reveal so much. Authenticated scans log into systems like an admin, mapping out missing patches, weak configurations, and vulnerable software versions more comprehensively.
- **The Keys to the Kingdom:** These scans need privileged credentials. Secure credential storage is paramount; consider a Privileged Access Management (PAM) solution to protect them.
- **Noise vs. Value:** Authenticated scans can be noisy. Tailor them carefully to avoid useless alerts about benign configuration differences between machines.

Key Takeaways

- Vulnerability management isn't about perfect scores; it's about demonstrably reducing risk over time.
- "Fixed" doesn't always mean "solved." Contingency plans when things go wrong are a sign of a mature security process.
- Look beyond external scans: Attackers will gladly exploit weaknesses inside a network that perimeter defenses miss.

Let's dissect the different focus areas of DR and BCP, and then explore how to test those plans in a way that reveals hidden weaknesses:

DR: The Technical Heartbeat

Focus: Disaster Recovery plans obsess over restoring IT systems and data. This includes servers, network infrastructure, databases, and critical applications.

Metrics: DR success is measured in RTO (Recovery Time Objective – how quickly systems should be back online) and RPO (Recovery Point Objective – how much data loss is acceptable).
Tech is King: DR often leans heavily on solutions like replication, backups, failover sites, and cloud-based recovery.
BCP: Keeping the Business Alive
Focus: Business Continuity Planning takes a wider view. It asks, "What core business functions MUST keep operating, and how do we make that happen, even with a major disruption?"
Beyond IT: BCPs consider non-technical dependencies: facilities, supply chains, communication channels, personnel availability, and alternate work procedures.
Business-Oriented Metrics: It's less about server uptime, more about preserving customer service levels, order fulfillment capability, or reputation.
Contrasting Downtime Impacts

Call Center: Disruption here means lost customer interactions. Potential consequences are mounting wait times, abandoned calls, and damaged brand perception.
E-commerce Site: Every minute offline is lost revenue. Secondary effects include delayed shipments, frustrated customers, and a potential surge in support cases once the site is restored.
Tabletop Exercise: Stress-Testing the Plan

Let's make your exercise count:
Scenario: Tailor it to your organization's risks (fire, ransomware, etc.). Make it impactful enough to disrupt core operations significantly.

Participants: Include not just IT, but business unit stakeholders, executives, and even external parties (vendors, PR specialists, etc.).

Injecting Chaos:
Key Personnel Unavailable: That lead database admin is on vacation when the disaster happens.
Cascading Failures: Your backup data center...is also in the path of a hurricane.
Communication Breakdowns: Phone lines are down, now how does your incident response team coordinate?
Beyond the "Happy Path:" Force participants to grapple with hard choices. Rationing resources, explaining delays to customers, and facing ethical dilemmas create the pressure that exposes plan weaknesses.

DR is a vital component of BCP, but not the whole picture.
Impact is measured differently for various parts of the business. Understanding those pain points drives better planning.
Tabletop exercises can't be scripted. It's the unexpected complications that truly test your team's adaptability.
Now ill unpack the critical metrics that shape data protection strategies and explore the impact of the cloud:

RTO vs. RPO: The Time and Data Balance

- **Recovery Time Objective (RTO):** This asks, "How quickly must systems be back online?" Minutes matter for critical business processes. Hours of downtime might be tolerable for others.
- **Recovery Point Objective (RPO):** This asks, "How much data loss is acceptable?" Zero is ideal but expensive. Some businesses can tolerate losing a few hours of data, others can't.

Scenarios Dictate Urgency

- **Near-Real-Time Replication:** Ideal for systems where any data loss or significant downtime hurts badly:
    - Financial Trading: Lost transactions = lost money
    - E-commerce Websites: Each missed order impacts revenue
    - Healthcare: Up-to-date patient data is crucial
- **Nightly Backups Acceptable:** Where some data loss and a few hours' recovery time make sense:
    - Internal Report Storage: Re-generating reports may be inconvenient but not a disaster.
    - Development Environments: Losing a day's work is annoying, but not business-critical.

Cloud-Based DR: Game Changer?

- **Benefits**
    - Scalability on Demand: Spin up recovery resources only when needed, not maintaining a dormant data center.
    - Geographic Redundancy: Cloud providers replicate your data across regions easily for geographic protection.
    - Outsourcing Complexity: They handle infrastructure headaches, you focus on configuring the DR process.
- **Considerations**
    - Cost: Pay-as-you-go models can be unpredictable if you're suddenly restoring terabytes of data.
    - Network Bandwidth: Recovering large data sets over the Internet might take longer than expected.
    - **Testing is Everything:** Simply replicating to the cloud won't do. Did your complex applications *actually* start up correctly in the new environment?

Key Takeaways

- There's no one-size-fits-all backup solution; business needs drive RTO and RPO.
- The cloud offers DR flexibility, but it doesn't magically eliminate planning and testing requirements.

VIII. Software Development Security

In a world driven by software, secure code is no longer a "nice-to-have"... It's a shield for your organization's data, reputation, and the foundation of customer trust. This section dives into the practices that transform development from a source of vulnerabilities into a pillar of your security posture.
We'll explore:

- **Threat Modeling for Developers:** Shifting from reactive patching to proactively identifying risks during the design stage.
- **Secure Coding Standards:** Mastering the principles that prevent common vulnerabilities like injection attacks and buffer overflows.
- **Code Review as a Security Tool:** Integrating peer reviews and automated analysis to catch flaws before they reach production.
- **DevSecOps: Agility with Guardrails:** How to integrate security seamlessly into development processes without stifling innovation.

Remember, every line of code is an opportunity to build resilience or create an entry point for attackers. This section equips you to make the right choices.

Let's compare development approaches and the security challenges they introduce:
Traditional (Waterfall) vs. Agile: A Security Perspective
Waterfall: Security Checkpoints
Design, development, then dedicated testing/security phases.
Pros: Structured gates allow in-depth security review at each stage.
Cons: Vulnerabilities found late are costly to fix; security seen as a blocker, not baked-in.
Agile: Sprinting with Vulnerabilities?

Short iterations (sprints) deliver features incrementally.
Pros: Faster feedback loop means potential issues surfaced earlier.
Cons: Constant change makes comprehensive security hard; testing often focuses on functionality, not attack vectors.
Trade-Offs in a Nutshell
Waterfall offers the illusion of control, but late-stage discoveries often derail schedules. Agile is faster but could result in a pile of insecure features shipped at pace.

SecDevOps: The Search for the Middle Ground
Culture, Not Just Automation: Security can't be an afterthought bolted onto DevOps speed. It's about:
Developers owning security as part of quality, not an adversary to fight.
Security folks providing tools integrated into dev workflows, not just reports at the end.
Shared Metrics: Success measured by security AND feature delivery velocity.
Changing the Mindset:

"We build it, we secure it": Developers don't throw code "over the wall" for later fixing.
Threat modeling as part of design: Asking "what can go wrong?" alongside "how does this work?".
Process Changes

Automated security testing in build pipelines: Catching flaws early.
Treating security work as user stories: They're prioritized in sprints, not mythical future "hardening" phases.

No model is inherently 'secure'. Each demands tailored methods to balance agility and security. SecDevOps aims for security at speed, but the biggest hurdle is often changing how people think and work.
let's explore both the promise and the complexities of "shift left" security:

"Shift Left" – The Logic

- **Prevention vs. Cure:** Much like in medicine, finding issues early is cheaper and less disruptive. Flawed design choices are harder to fix in production than during a whiteboard session.
- **Security as an Enabler:** Early security input helps bake security into features instead of trying to layer it on top later. This reduces refactoring and delays.
- **Empowered Developers:** When developers have security tools and knowledge, they take more ownership of preventing problems in the first place.

"Shift Left" – The Challenges

- **Skillset Gap:** Not all devs are security pros. Training, good tooling, and a security team ready to *collaborate* are essential.
- **Balance with Agility:** Too many rigid gates slow you down. It's about smart, frequent checkpoints, not recreating Waterfall in disguise.
- **Context Matters:** Early threat models are less accurate than those done on a working system. Striking the right balance of 'early enough' vs. 'meaningful' is key.

Hybrid Models: When Regulation Calls

Some industries can't embrace full Agile due to compliance needs or the risk profile of the systems they build:

- **Healthcare / Finance:** Extensive documentation and audit trails may be mandated.
- **Critical Infrastructure:** Rigorous testing phases before deployment, as the cost of failure is too high.

Hybrid Approaches Might Feature:

- **Agile-Inspired Sprints:** Building features iteratively within a regulated larger framework.
- **Documentation as Code:** Keeping design docs up-to-date via tools integrated into dev workflow eases compliance burden.
- **Formal Security Gate Reviews:** At key milestones, but fueled by in-sprint security work to avoid last-minute surprises.
- "Shift Left" is a philosophy, not a rigid prescription. The goal is to make security proactive and less disruptive.
- Regulation doesn't mean giving up on speed entirely. Finding the right blend for your industry is crucial.

Now ill dissect three prevalent OWASP Top 10 vulnerabilities, demonstrating their dangers with code snippets and real-world impact:

1. Injection (A01:2021)

- **Code Vulnerability (SQL Injection):**

```
username = input("Enter your username: ")
query = "SELECT * FROM users WHERE username = '" + username + "'"  # Unsafe!

```

- **Attacker Exploitation:** A malicious user enters ' OR '1'='1 as input, resulting in the query becoming: SELECT * FROM users WHERE username = '' OR '1'='1'. Since '1'='1' is always true, the attacker gains access without needing a valid username.
- **Real-World Impact:** The infamous 2011 breach of Sony's PlayStation Network exposed millions of user records due to SQL injection flaws.

2. Broken Authentication (A02:2021)

- **Code Vulnerability (Weak Password Storage):**

```
public void registerUser(String username, String password) {
    // Storing passwords in plain text - very bad!
    db.store("users", username, password);
}
```

- **Attacker Exploitation:** If the database is breached, attackers have plain text passwords. Credential stuffing attacks (trying these leaked combos on other sites) become rampant.
- **Real-World Impact:** The LinkedIn breach in 2012 resulted from poor password hashing, leading to millions of credentials leaked and easily cracked.

3. Cross-Site Scripting (XSS) (A03:2021)

- **Code Vulnerability (Unescaped User Input):**

```
<p>Search Results: <?php echo $_GET['query']; ?></p>
```

- **Attacker Exploitation:** An attacker crafts a link like: https://vulnerableapp.com/search?query=<script>alert('XSS')</script>. A victim clicking this has the script execute in their browser.
- **Real-World Impact:** XSS is used for session hijacking, website defacement, and injecting malicious redirects to spread malware.
- These classic vulnerabilities persist due to seemingly minor coding oversights.
- The impact is far-reaching: data theft, credential abuse, and damage to brand reputation.
- **Secure Coding Principles:** Input validation, strong password hashing, and output sanitization are fundamental defenses.

Now....Input validation and sanitization seem similar but have crucial differences:

Input Validation: The Gatekeeper

- **Focus:** Checking if input FITS the expected format:
  - Is an email address structured correctly?
  - Does a "Date" field contain only numbers and separators?
- **Actions:**
  - Rejecting outright invalid input.
  - Providing clear error messages to the user for correction.
- **Strength:** Prevents a wide range of malicious input from entering the system.

Sanitization: Cleaning Up for Context

- **Focus:** Modifying input to make it SAFE for the intended use.
- **Context is King:**
  - **HTML Sanitization:** Removing dangerous tags (<script>, <iframe>, etc.) to prevent XSS.

- **SQL Sanitization:** Escaping special characters to neutralize injection attempts.
- **Strength:** Allows accepting somewhat messy input, fixing it to align with the system's needs.

When Validation Isn't Enough

- **Legacy Databases:** You can't fix bad schema design with validation alone. Sanitization is a must to protect flawed systems.
- **Rich User Input:** Think comment fields on a forum. You may want to allow some HTML tags (<b> for bold) while stripping dangerous ones.
- **Internal Data Flows:** Even between trusted systems, sanitization at boundaries reduces the risk of misinterpretations causing bugs.
- **Defense in Depth:** Use validation as the first line, and sanitization as the context-aware backup.
- **Data Integrity:** Overly aggressive sanitization can corrupt *valid* data. Balance security with functionality.
- **Libraries over Hand-Rolling:** Use well-tested sanitization libraries whenever possible to avoid reinventing the wheel (and introducing your own vulnerabilities!).

Error Handling: When Help Hurts

- **Verbose Errors: The Enemy Within** Stack traces, internal object structures, or debugging dumps in error messages provide a goldmine for attackers. They reveal system architecture, technologies used, and potentially paths for deeper exploitation.
- **Logs Are Watching (Especially When They Shouldn't Be)** While crucial for debugging for developers, excessively detailed logging can leak sensitive data:
    - User inputs echoed back into logs on failure.
    - Database queries logged in plain text.
- **User-Facing Slipups** Generic "An error occurred" is unhelpful for users, but overly specific messages might expose backend logic, guiding attackers in their next moves.

Fix the Leak, Not Just the Bug:

- **Error Handling Tiers:**
    - Developers see the nitty-gritty details (locally, never in production)
    - System logs get enough to troubleshoot, but not raw user data.
    - Users get helpful but vague prompts to retry, contact support, etc.
- **Centralized Error Handling:** Prevents inconsistent error messages across different code sections from leaking varying levels of information
- **Think Like an Attacker:** When a feature fails, ask yourself, "What does this inadvertently tell someone trying to break the system?"

Secure Coding Tools: Preventing Mistakes

- **IDE Plugins: Your First Line of Defense** Plugins for popular IDEs analyze code as you type, highlighting potential issues like:
    - Catching the use of unsafe functions prone to error-handling leaks.
    - Flagging places where user input flows into error messages unsanitized.
- **Static Code Analyzers (SAST): Deep Inspection** These dissect your codebase, even parts you're not currently editing. They find patterns that hint at error handling logic flaws across your project.
- **Not a Magic Bullet:** Tools are great at spotting common patterns, but won't catch every logical mistake in how you structure error handling.
- Error handling IS a security issue. Consider it during design, not just when bugs happen.
- Balance usability and security in error messages. Users need guidance, not a developer roadmap.

- Tools empower developers to catch issues *before* they become vulnerabilities in production code.

Let's break down the pros, cons, and ideal use cases for the security testing trio: SAST, DAST, and IAST.

SAST (Static Application Security Testing)

- **The Microscope:** Analyzes source code without running the app. Looks for known vulnerability patterns, bad coding practices, etc.
- **Strengths:**
    - Early in the SDLC: Integrates with IDEs to give developers rapid feedback.
    - Finds 'deep' flaws: Not just at the web layer, but in how data is processed throughout the application.
- **Blindspots:**
    - Misses runtime issues: Great for bad code, less so for zero-day attacks or configuration mistakes.
    - Can be noisy: Might generate some false positives, requiring tuning.

DAST (Dynamic Application Security Testing)

- **The Black Box Attacker:** Scans a running web app, probing for vulnerabilities like XSS, injections, misconfigurations. Doesn't need the source code.
- **Strengths:**
    - Simulates real attacks: Finds what an attacker could exploit, regardless of how it was built.
    - Configuration Check: Good for spotting weak server settings, TLS issues, etc.
- **Blindspots:**
    - Less 'inside' visibility: Harder to pinpoint the root cause in the code.
    - Coverage Challenges: Might miss whole sections of an app if it can't guess the right inputs.

IAST (Interactive Application Security Testing)

- **The Hybrid:** Has an agent inside the running app, providing more context than DAST, but without full source code analysis like SAST.
- **Strengths:**
    - Blends speed and insight: Faster than SAST, more accurate than DAST.
    - Great for complex apps: Tracks data flow to connect the dots between inputs and vulnerable behavior.
- **Blindspots:**
    - Setup overhead: Requires instrumenting your app with the agent.
    - Still less coverage than SAST at the pure code level.

When To Use Which

- **Early Development:** SAST is king. Catching flaws while coding saves headaches later.
- **Staging/QA:** SAST + DAST combo is ideal. Find code-level issues and runtime problems.
- **Production:** DAST for regular checks. IAST considered valuable here if feasible to continuously monitor the running app.
- No single tool is perfect; they complement each other.
- "Shift left" with SAST, but don't neglect DAST for the external attacker's perspective.
- IAST is gaining popularity as a balance, but its suitability depends heavily on the specific application.

Now ill illustrate the hidden dangers lurking within third-party code and how SCA tools can help:

Scenario: The Vulnerability Domino Effect

Imagine your web application uses a popular JavaScript library for frontend form handling. A new vulnerability is discovered in that library, allowing attackers to inject malicious code when users submit forms (XSS attacks). Consequences can include:

- **Data Theft:** User form submissions, potentially containing sensitive information, are siphoned off to attacker-controlled servers.
- **Session Hijacking:** Attackers gain control of authenticated user sessions, impersonating legitimate users.
- **Defacement:** The application's appearance is modified for malicious purposes.
- **Reputation Damage:** Even if the breach is contained, the loss of customer trust can be significant.

SCA Tools: Shining a Light on Dependencies

1. **Dependency Mapping:** SCA tools scan your project, building an inventory of directly used libraries AND the libraries those libraries in turn depend on (transitive dependencies).
2. **Vulnerability Matching:** This inventory is cross-referenced against known vulnerability databases (like the NVD). You get a clear report of which components are affected, and the severity.
3. **Integrated Updates:** Good SCA tools suggest safer alternative versions, or even generate automated pull requests to update your dependencies.

Integrating SCA into Your Workflow

- **Build Pipelines:** SCA checks become a step that can fail your build if high-risk vulnerabilities are found.
- **Developer Notifications:** Don't just generate reports, push alerts directly to the responsible teams to prioritize fixes.
- **Regular Scans:** Even if your code doesn't change, new vulnerabilities in dependencies are discovered continuously.
- You inherit the vulnerabilities of the libraries you use. Proactive dependency management is essential.
- SCA isn't just about finding problems, it's about enabling rapid remediation within your SDLC processes.
- A single vulnerable dependency can undermine the security of your entire application.

Let's get into the value of fuzz testing and the challenges of rapid patching in Agile environments:

Fuzz Testing: Breaking Things on Purpose

Fuzzing's Goal: To trigger unexpected application behavior by bombarding it with invalid, malformed, or random inputs.
Going Beyond the Expected: Fuzzers stress-test input validation, error handling, and parsing logic. They aim to cause crashes, hangs, or leaks that might hint at deeper exploitable flaws.
Zero-Day Hunting: Because fuzzing doesn't rely on known attack signatures, it's a valuable tool for uncovering brand-new vulnerabilities attackers haven't yet figured out.
The Role of Fuzzing in Mature Security

Proactive Defense: Fuzzing during development finds bugs before deployment, reducing the attack surface.

Supplement to Other Testing: Fuzzing complements SAST/DAST, finding what those tools might miss due to their focus on known patterns.
Continuous Approach: Integrating fuzzing into CI/CD pipelines means your testing keeps up with the pace of code changes.
Vulnerability Remediation: Agile vs. Urgent

When a critical vulnerability hits, it throws a wrench into neatly planned sprints. Here's how to balance speed and stability:

Risk-Based Prioritization: Not every vuln warrants dropping everything. Severity, ease of exploitation, and the asset's exposure drive your response.
Dedicated Hotfix Branches: If possible, address the fix in a branch, allowing testing and release without derailing the main sprint.
Regression Testing is KEY: A rushed fix breaking existing features is worse than the original problem. Have a quick-turnaround test suite focused on the affected area.
Communication is Everything: Transparency with stakeholders about the delay and risks is essential to buy time for doing the fix right.

Fuzzing is a powerful tool for proactive security, especially in uncovering the unknown vulnerabilities.
Mature security isn't just about finding flaws, but about how you respond to them.
Agile environments demand adaptable patching processes, emphasizing testing and communication.
Let's explore how code signing becomes more than just verifying origin, and then dive into the nuts and bolts of system hardening.

Code Signing: Trust and Accountability

- **The Basics:** Using digital signatures, code signing allows users and systems to verify two essential things:
  - **Authenticity:** The code came from the expected developer (protects against malicious substitution).
  - **Integrity:** The code hasn't been modified in transit (supply chain attack defense).
- **Non-Repudiation:** A developer can't easily deny authorship of signed code. This is valuable in legal contexts or when enforcing strict change control.
- **Malware Check:** If a malicious actor compromises a build environment and injects code, the broken signature will immediately flag the software as tampered with.
- **Key Management is Crucial:**
  - Secure storage of signing keys: If these are leaked, an attacker can sign their own malicious code.
  - Revocation Processes: A mechanism to revoke compromised keys and to update software with a new signature.

Hardening: Trimming the Fat

The principle of 'least functionality' is your guiding star in hardening:

1. OS Hardening: * **Patching Religiously:** Focus on security updates, not just the latest features. * **Disable Unused Services:** Every network port open or service running is a potential attack vector. * **User Accounts:** Enforce least privilege, strong passwords, and remove default admin accounts. * **File system Permissions:** Lock down access to sensitive system files.

2. Web Server Hardening: * **Software Updates:** The web server itself (Apache, Nginx) and related libraries need security patches. * **Remove Default Files:** Cleanup sample pages and scripts attackers know about. *

HTTP Headers: Configure to minimize information leakage (X-Powered-By, etc.). * **Input Filtering:** Strong server-side input validation helps even if the application code has flaws.

3. Database Hardening: * **Network Exposure:** Restrict access to only the systems that need it. * **Account Lockdown:** Remove defaults, enforce least privilege for application-specific DB accounts. * **Encryption:** Sensitive data should be encrypted at rest within the database.

- Code signing provides a layer of trust throughout the software lifecycle.
- Hardening is an iterative process: Start with a secure baseline, then continually refine based on the specifics of your application stack.
- 'Attack Surface' is the keyword: Every bit of functionality is an opportunity for attackers.

Let's explore why secrets management is essential for automation security, and then break down the complexities of container security:

Secrets Management in CI/CD: The Keys to the Kingdom

- **Hardcoded Credentials: A Recipe for Disaster** Build scripts, deployment configs, etc., often end up containing passwords, API keys, or database connection strings. This leads to:
    - Leaks in Source Control: Devs often accidentally commit these secrets.
    - Broad Exposure: Scripts need the credentials in plaintext, creating a wider attack surface if the script is compromised.
- **Secrets Vaults: A Saner Approach**
    - Centralized Storage: Secrets are stored encrypted within the vault. Access is role-based.
    - Dynamic Injection: CI/CD pipelines retrieve secrets at runtime from the vault, inserting them into processes or configurations only when needed.
    - Audit & Revocation: You gain visibility into how secrets are used and can easily rotate them if needed.

Popular Secrets Management Tools

- HashiCorp Vault
- AWS Secrets Manager
- Azure Key Vault
- Kubernetes Secrets (basic solution, often paired with third-party vaults for stronger controls)

Container Security: It's Not Just About the Image

- **Image Scanning: The First Line of Defense** Similar to SCA for dependencies, these tools analyze containers for known vulnerabilities and outdated packages. They integrate with your image registry.
- **Runtime Security: Enforcing Behavior** Even a "clean" image can be exploited if the container is misconfigured. Tools help:
    - Isolate Containers: Restrict network access and available system calls.
    - Detect Anomalies: Monitor for suspicious activity (file changes, unexpected processes) inside running containers.
- **Kubernetes Hardening: Orchestration Woes** K8s adds its own attack surface:
    - RBAC Gone Wrong: Excessive permissions on service accounts are a huge risk.
    - Network Policies: Isolate pods and namespaces to limit lateral movement blast radius.
    - "Secure by Default" mindset: Avoid running containers as root whenever possible.
- Secrets sprawl is a side effect of automation. Centralized secrets management is essential for CI/CD hygiene.
- Container security requires a layered approach: image content, runtime behavior, and the orchestration platform all matter.

IX. Study Strategies & Additional Resources

Conquering the CISSP exam is a significant milestone, but it's also where your true learning journey intensifies. This section provides guidance on fine-tuning your study methods and arms you with resources to stay ahead in the ever-evolving world of cybersecurity.
We'll cover:

- **Reinforcing Your Knowledge:** Techniques to move beyond memorization and achieve true understanding.
- **Staying Sharp:** How to maintain your expertise and keep up with new threats and technologies.
- **Community Power:** Leveraging the wisdom of the cybersecurity community for support and insights.

Remember, cybersecurity is a marathon, not a sprint. The tools and habits you develop here will pave the way for lasting success in the field. Let's get started!

Let's talk about how turning knowledge into visuals can enhance your understanding of the CISSP's sprawling landscape.

Why Knowledge Mapping Works

- **Breaks Down Silos:** The CISSP domains can feel like separate islands. Mapping forces you to explicitly show how they connect. Example: how does Identity and Access Management (Domain 2) impact Incident Response (Domain 6)?
- **Reveals Your Weak Spots:** Struggling to visually link two topics? That's a sign you need to revisit the material and bridge that gap.
- **Engages Different Learning Styles:** Not everyone thrives on textbooks. Visualizing concepts taps into spatial and diagrammatic thinking.

How to Put it into Practice

- **Mind Maps:** Start with a central concept (e.g., "Risk Management"). Branch out into subtopics (risk assessment, treatment, etc.). Further subdivide with related concepts and even cross-domain links.
- **Concept Diagrams:** More structured than mind maps. Use boxes and arrows to show relationships clearly:
    - Entity Relationship Diagrams: To map data structures and how data flows through a system (relevant for software security and data protection).
    - Flowcharts: To visualize security processes, making decision points and conditional logic explicit.

Example: Mapping a Vulnerability Management Workflow

A mind map could explode outwards from "Vulnerability Management" to include branches for scanning tools, risk prioritization, patching processes, communication channels, and links to change management concepts from other domains.

Tools

- **Pen and Paper:** The low-tech option is still powerful!
- **Mind Mapping Software:** XMind, MindNode, etc., offer features like color-coding and easy rearrangement.
- **Collaborative Whiteboards:** Miro, Conceptboard – great for group study sessions.

- Knowledge mapping is an ACTIVE learning process. The act of creating the diagrams forces deeper thinking.
- Your maps are unique to YOU – they reflect how your brain makes connections.
- Revise your maps as you study – they are a living reflection of your progress.

Moving along, Now ill dissect how seemingly minor word changes in CISSP questions can lead to drastically different interpretations.

Sample Question: Original

In developing a disaster recovery plan, which of the following should be the FIRST priority?

- (A) Resumption of business operations
- (B) Protection of human life
- (C) Safeguarding critical data
- (D) Communicating with stakeholders

Variation 1: "most important" instead of "FIRST"

In developing a disaster recovery plan, which of the following is the MOST IMPORTANT consideration?

- **Impact of Change:** This version is broader. Yes, protecting life is paramount, but if it's not the top priority within the DR plan itself, another answer might be better (e.g., resuming business operations to prevent financial ruin is often more critical in DR planning than data, which may have backups).

Variation 2: Adding "after a major incident"

In developing a disaster recovery plan, which of the following should be the FIRST priority after a major incident occurs?

- **Impact of Change:** Now, the question is squarely about immediate response, not the pre-written plan. Protecting human life (option B) becomes the likely best answer.

How to Spot the Nuances

- **Scope Words:** "First," "most important," "primary" all subtly change how you weigh the options.
- **Context Specifiers:** Did the question suddenly include "during development" vs "after an incident"? This alters your frame of reference.
- **Absolutes vs. Relatives:** Watch out for "ALWAYS" or "NEVER". CISSP questions are more likely to ask for the "BEST" or "MOST LIKELY" approach.
- Slow down and analyze each word: CISSP questions deliberately test your ability to parse the details.
- Think from the perspective of the "ideal" security professional, not just your specific job role.
- If you're torn between two answers, it's likely because the question is subtly ambiguous – look for the word that makes one choice a slightly better fit.

In the world of the CISSP, the "best" answer often hides among several partially correct ones. Let's practice weighing those subtle differences.

Sample Question Scenario

Your organization's web application handles sensitive customer data. You've discovered a vulnerability that could allow unauthorized access. Which of the following is the BEST response?

- (A) Immediately deploy a newly released vendor patch, even though it hasn't been thoroughly tested in your environment.
- (B) Disable the web application entirely until the vulnerability can be properly fixed.
- (C) Implement a temporary workaround using your web application firewall (WAF) to block exploit attempts.
- (D) Rate the risk as low based on the difficulty of exploitation and continue to monitor network traffic.

Evaluating The Trade-offs

None of these choices are inherently *wrong*, but they represent different priorities:

- **(A) Speed vs. Stability:** A patch fixes the root cause, but untested software brings the risk of unforeseen bugs, potentially disrupting the service.
- **(B) Security vs. Availability:** Shutting down the app is secure, but might have an unacceptable business impact.
- **(C) Mitigation vs. Fix:** A WAF lowers risk, buying time for a proper solution, but may not block all attack variations.
- **(D) Acceptance vs. Action:** Doing nothing is rarely ideal, but if the exploit is truly difficult and resources are limited, focused monitoring might be justifiable short-term.

"Best" Depends on Context

- **Criticality of the App:** Is this revenue-generating or does it support essential operations? This elevates the importance of option (A) or (C) to keep things running.
- **Risk Tolerance of the Org:** Some are more conservative – option (B) might be preferred, even with the impact.
- **Maturity of Processes:** Do you have rapid patch testing and rollback procedures? This makes (A) less risky.
- Look for the choice that balances security with operational realities.
- Consider the specific organization's context – there's no universally perfect answer.
- "Best" often means the most well-managed risk, not necessarily complete risk elimination.

Ok.. now well decode the hidden language within CISSP questions and how they guide your thinking towards the right answer.

Keywords as Signposts

- **Defense in Depth:** This phrase signals layered security. Look for answers involving multiple controls (firewalls AND IDS, etc.), not just a single solution.
- **Least Privilege:** Focus narrows to access controls. The best answer will be the one that grants the minimum necessary permissions to perform a task.
- **Confidentiality:** Integrity and availability matter, but less so in this question. Prioritize answers about encryption, access controls, and data classification.
- **Integrity:** Data must be accurate and unaltered. Look for solutions focused on hashing, change/version controls, and tamper detection.
- **Availability:** Uptime is key. Answers involving redundancy, load balancing, and failover mechanisms are likely strong contenders.
- **Risk-Based Approach:** Rarely about absolutes. Weigh the cost of security controls against the likelihood and impact of the threat they are designed to mitigate.

Beyond Buzzwords: Question Intent

- **"MOST effective" / "BEST course of action":** There's likely trade-offs. Analyze the potential impact of each answer choice on security AND business operations.
- **"PRIMARY objective" / "FIRST step":** Forces prioritization. Consider foundational concepts (e.g., secure configuration before deploying fancy tools).
- **"EXCEPT" / "NOT":** Easy to miss these! They reverse the question – you're looking for the one INCORRECT statement among plausible ones.
- Keywords offer clues to the question's core focus, steering you away from less relevant options.
- Understanding the intent behind the question helps you evaluate seemingly correct answers in a more nuanced way.

Let's turn your practice exams into strategic battlegrounds where you maximize your score.

Round 1: The Blitz

- **Goal:** Build confidence and momentum.
- **Method:** Sweep through, answering only the questions that jump out as immediately obvious. Mark the rest for later rounds.
- **Mindset:** Don't overthink! Trust your gut on these.

Round 2: Focused Analysis

- **Goal:** Chip away at the tougher questions.
- **Method:** Tackle those you marked earlier. Break down the wording, eliminate obviously wrong answers. If it'll take serious time to puzzle out... mark it AGAIN, but with a different symbol to distinguish it.
- **Mindset:** Be methodical, but don't let a single question stall you out.

Round 3: Educated Guesses (If Time Allows)

- **Goal:** Minimize unanswered questions.
- **Method:** Return to those extra-tricky questions. Look for patterns in the answers, the possibility to eliminate at least one or two bad options.
- **Mindset:** It's a calculated risk, not blind guessing. .

Adapting as You Go

- **Energy Management:** Tired and losing focus? Rearrange the rounds – tackle some "easy" ones to refresh.
- **The Dreaded Wall:** If you hit a string of questions you just don't know, don't panic. Mark them, move on, and regain composure. They may spark a memory later.
- **Time is Your Enemy:** Keep a close eye on the clock. Adjust the rounds if you're running behind – better to have educated guesses on everything than perfect answers on half the exam.
- This strategy maximizes your points per minute. The CISSP isn't just about knowing the answers, it's about knowing how to tackle the exam itself.
- Don't leave questions blank! A reasoned guess is better than no answer at all.
- Practice with full-length mock exams to refine this technique and build your time-sense.

The CISSP exam is not just a test of your knowledge, but of your endurance. Let's optimize your mental stamina alongside your technical preparation:

Deep Breathing: Your Secret Weapon

The Science: Slow, deep breaths activate the parasympathetic nervous system, counteracting the stress response. This improves focus and clears brain fog.
In Practice: Take a few seconds even mid-question: 4-count inhale, hold briefly, 6-count exhale. Repeat a few times.
Bonus: This doubles as a way to discreetly reset if you hit a frustrating question.
Posture Matters

The Slump Trap: Hunching compresses your chest, impacting breath, and signals fatigue to your brain.
Conscious Correction: Roll your shoulders back, straighten your spine, and lift your chin slightly. Do this periodically throughout the exam.

Benefits: Improved oxygen flow helps you think more clearly, and the physical act sends a "power pose" signal to your mind.
Strategic Breaks

When, Not If: The exam allows breaks. Use them wisely, even if you don't feel you need one yet. Get up and move around.
Mini-Resets: Brief walks, stretching in the restroom stall – they re-oxygenate your blood.
Avoid Burnout: It's harder to regain focus than to maintain it. Breaks are preventative.
Additional Tips

Hydration: Even mild dehydration impacts cognition. Sip water throughout, not just giant gulps during breaks.
Light Snacks: Choose brain-friendly options (nuts, fruit) if permitted, to avoid blood sugar crashes.
Eye Strain: Consciously look away from the screen every few minutes to rest your eyes.

Your mind is part of your toolkit for the exam. Train it alongside the technical knowledge.
Small acts matter: These aren't about huge gains, but preventing the gradual decline that makes the last questions feel impossible.
Practice during mock exams: Discover which techniques work best for you.

Let's turn those last few agonizing question choices into strategic victories. When you're down to two options, it's time to go beyond memorization and tap into your analytical side.

Dissecting the Wording

Absolutes vs. Relatives: Watch for words like "ALWAYS", "NEVER", "BEST", or "MOST". These choices are often either wrong or need careful context.
Specific vs. Vague: Is one option overly detailed, while the other is a broad generalization? The specific one is more likely correct, especially if it aligns with a concept from your studies.
Hidden Assumptions: Does one of the answers only seem true if you make additional assumptions that weren't in the question itself? This should raise a red flag.
Leveraging Your Domain Knowledge

Think Like an Ideal Practitioner: If you were responsible for implementing the solution, which option aligns with secure practices, even if it's slightly more difficult or expensive?
Prioritize Core Principles: Recall fundamentals like least privilege, confidentiality, and defense-in-depth. The option that best embodies these is often (though not always) the right one.
When In Doubt, Keep it Simple: Overly complex solutions are less common in the CISSP world than well-executed, straightforward ones.
Example: Let's say your choices are:

A: Deploy a next-gen firewall with integrated AI-powered threat detection.

B: Restrict the service to specific source IP addresses using existing firewall rules.

Analysis: While 'A' sounds fancy, is it the most appropriate given the limited question info? 'B' is simpler, aligns with the principle of least privilege, and leverages tools likely already present. B is often a better CISSP-world answer.

Don't despair when you don't know for sure. Making educated guesses is an essential CISSP skill.
The question writers are giving you clues, often unintentionally! Learn to spot them.
Your broader security knowledge is your compass when the specifics get hazy.

Accuracy through deliberate practice is sometimes more important than raw speed. Let's design some drills to sharpen your focus and catch those sneaky errors before submitting your exam.

Drill 1: The Slow Burn

- **Materials:** A set of practice questions (15-20 to start). Set aside MORE time than you'd normally take for this amount.
- **Process:**
    - Answer each question methodically, talking yourself through your reasoning out loud (if you won't disturb others).
    - After you have an answer for each, DON'T check the solutions yet.
    - Re-read each question AND answer choice as if seeing them fresh. Could you argue AGAINST your chosen answer?
- **Goal:** Forces you to slow down and fully consider the wording. Breaks the habit of "snap judgments" that lead to careless mistakes.

Drill 2: Pattern Recognition

- **Materials:** Your incorrectly answered questions from past practice sets.
- **Process:**
    - Don't look at the correct answers just yet. Instead, only look at the questions and your own wrong choices.
    - Categorize your errors: Did you misread the question? Miss a keyword? Get fixated on a minor detail while missing the bigger picture?
- **Goal:** You'll start to see the types of traps you personally fall for. This awareness prevents repeating them on exam day.

Drill 3: The Backwards Blitz

- **Materials:** A practice set where you're already confident in most answers.
- **Process:**
    - Start with the LAST question and work your way upwards.
    - Immediately AFTER each question, check the answer key.
- **Goal:** Breaks your predictable thinking patterns. You're more likely to catch careless errors when questions are out of order.
- These drills are for when you're NOT under immediate time pressure. They train the habits that carry over to the exam.
- Be honest about your error patterns. Are you rushing? Overcomplicating? Fix those weaknesses deliberately.
- Review is not just about right vs. wrong: It's about HOW you arrive at those answers.

Below is a collated list of crucial terminologies and acronyms:

General Security Concepts:

- **AAA**: Authentication, Authorization, and Accounting
- **ACL**: Access Control List
- **AES**: Advanced Encryption Standard
- **APT**: Advanced Persistent Threat
- **BIA**: Business Impact Analysis
- **BYOD**: Bring Your Own Device
- **CIA Triad**: Confidentiality, Integrity, Availability
- **DLP**: Data Loss Prevention
- **DRP**: Disaster Recovery Plan
- **EOL**: End Of Life
- **IAM**: Identity and Access Management
- **IDS/IPS**: Intrusion Detection System / Intrusion Prevention System
- **IoT**: Internet of Things
- **ISMS**: Information Security Management System
- **PKI**: Public Key Infrastructure
- **RSA**: Rivest-Shamir-Adleman (encryption algorithm)
- **SAST**: Static Application Security Testing
- **SIEM**: Security Information and Event Management
- **SOC**: Security Operations Center
- **SSL/TLS**: Secure Sockets Layer / Transport Layer Security
- **WAF**: Web Application Firewall

Security Domains:

- **IAM**: Identity and Access Management
- **SDLC**: Software Development Life Cycle
- **SoD**: Separation of Duties
- **TCSEC**: Trusted Computer System Evaluation Criteria

Regulatory Frameworks:

- **GDPR**: General Data Protection Regulation
- **HIPAA**: Health Insurance Portability and Accountability Act
- **ISO 27001/2**: International Organization for Standardization Information Security Standards
- **NIST**: National Institute of Standards and Technology
- **PCI DSS**: Payment Card Industry Data Security Standard
- **SOX**: Sarbanes-Oxley Act

Cryptography and Network Security:

- **DH**: Diffie-Hellman (key exchange protocol)
- **ECC**: Elliptic Curve Cryptography
- **MAC**: Message Authentication Code
- **PAP**: Password Authentication Protocol
- **SHA**: Secure Hash Algorithm
- **VPN**: Virtual Private Network
- **WEP/WPA/WPA2/WPA3**: Wireless Encryption Protocols

Incident Response and Management:

- **CSIRT**: Computer Security Incident Response Team
- **MTBF/MTTR**: Mean Time Between Failures / Mean Time To Repair

Risk and Security Management:

- **RA**: Risk Assessment
- **RMF**: Risk Management Framework
- **ROI**: Return on Investment
- **RTO/RPO**: Recovery Time Objective / Recovery Point Objective
- **TVM**: Threat and Vulnerability Management

Security Architecture and Design:

- **DAC/MAC/RBAC**: Discretionary Access Control / Mandatory Access Control / Role-Based Access Control
- **DMZ**: Demilitarized Zone
- **NGFW**: Next-Generation Firewall
- **SDN**: Software-Defined Networking

Software and Application Security:

- **API**: Application Programming Interface
- **CI/CD**: Continuous Integration / Continuous Deployment
- **DAST**: Dynamic Application Security Testing
- **OWASP**: Open Web Application Security Project
- **SCA**: Software Composition Analysis
- **XSS**: Cross-Site Scripting
- **CSRF**: Cross-Site Request Forgery

Business Continuity and Disaster Recovery:

- **BCP**: Business Continuity Planning
- **COOP**: Continuity of Operations Planning
- **DRP**: Disaster Recovery Planning
- **MTD**: Maximum Tolerable Downtime

Legal, Regulations, and Compliance:

- **CFAA**: Computer Fraud and Abuse Act
- **ECPA**: Electronic Communications Privacy Act
- **FOIA**: Freedom of Information Act
- **GLBA**: Gramm-Leach-Bliley Act

Physical Security:

- **CPTED**: Crime Prevention Through Environmental Design
- **EAC**: Electronic Access Control
- **HVAC**: Heating, Ventilation, and Air Conditioning
- **MANTRAP**: A physical security device or enclosed space used to control access to a secure area

Cryptographic Methods and Standards:

- **HMAC**: Hash-Based Message Authentication Code
- **IV**: Initialization Vector
- **KDF**: Key Derivation Function
- **PFS**: Perfect Forward Secrecy
- **PGP**: Pretty Good Privacy

Cloud Computing and Virtualization:

- **IaaS/PaaS/SaaS**: Infrastructure as a Service / Platform as a Service / Software as a Service
- **VLAN**: Virtual Local Area Network
- **VM**: Virtual Machine
- **VPC**: Virtual Private Cloud

Here's a comprehensive glossary that includes key terms across various domains of cybersecurity:

A

- **Access Control**: The selective restriction of access to a place or other resource.
- **Advanced Persistent Threat (APT)**: A prolonged and targeted cyberattack in which an intruder gains access to a network and remains undetected for an extended period.
- **AES (Advanced Encryption Standard)**: A symmetric block cipher chosen by the U.S. government to protect classified information.

B

- **Backup**: A copy of data that can be used to restore the original after a data loss event.
- **Biometrics**: Authentication techniques that rely on measurable physical characteristics that can be automatically checked.

C

- **CIA Triad**: An information security model that comprises three main components: confidentiality, integrity, and availability.
- **Chain of Custody**: A process that tracks the movement of evidence through its collection, safeguarding, and analysis lifecycle.
- **COOP (Continuity of Operations Planning)**: A plan that ensures the continued operation of critical business functions during a wide range of emergencies, including localized acts of nature, accidents, and technological or attack-related emergencies.

D

- **DAC (Discretionary Access Control)**: A type of access control system in which owners or administrators of the protected system, data, or resource set the policies defining who or what is authorized to access the resource.
- **DLP (Data Loss Prevention)**: A strategy for making sure that end users do not send sensitive or critical information outside the corporate network.
- **DRP (Disaster Recovery Planning)**: A documented, structured approach with instructions for responding to unplanned incidents.

E

- **Encryption**: The process of converting information or data into a code, especially to prevent unauthorized access.
- **ECC (Elliptic Curve Cryptography)**: An approach to public-key cryptography based on the algebraic structure of elliptic curves over finite fields.

F

- **Firewall**: A network security device that monitors and filters incoming and outgoing network traffic based on an organization's previously established security policies.

G

- **GDPR (General Data Protection Regulation)**: A regulation in EU law on data protection and privacy in the European Union and the European Economic Area.

H

- **Hashing**: A function that converts an input (or 'message') into a fixed-size string of bytes, typically for security or data retrieval purposes.
- **HIPAA (Health Insurance Portability and Accountability Act)**: United States legislation that provides data privacy and security provisions for safeguarding medical information.

I

- **IDS (Intrusion Detection System)/IPS (Intrusion Prevention System)**: Systems that monitor network traffic for suspicious activity and issue alerts when such activity is discovered.
- **IAM (Identity and Access Management)**: A framework of business processes, policies, and technologies that facilitate the management of electronic or digital identities.

K

- **Key Derivation Function (KDF)**: A cryptographic algorithm that derives one or more secret keys from a secret value such as a master key, a password, or a passphrase using a pseudo-random function.

L

- **Least Privilege**: The principle of providing the bare minimum levels of access – or permissions – needed to perform a job function.

M

- **MAC (Mandatory Access Control)**: A type of access control by which the operating system constrains the ability of a subject or initiator to access or generally perform some sort of operation on an object or target.

N

- **NIST (National Institute of Standards and Technology)**: A physical sciences laboratory and a non-regulatory agency of the United States Department of Commerce.

P

- **PKI (Public Key Infrastructure)**: A set of roles, policies, hardware, software, and procedures needed to create, manage, distribute, use, store, and revoke digital certificates and manage public-key encryption.

R

- **Risk Assessment**: The overall process or method where you identify hazards and risk factors that have the potential to cause harm.
- **RSA**: An algorithm used by modern computers to encrypt and decrypt messages.

S

- **SIEM (Security Information and Event Management)**: Software that provides real-time analysis of security alerts generated by applications and network hardware.
- **SAST (Static Application Security Testing)**: A method of security testing that examines the code to find software flaws.

T

- **TCSEC (Trusted Computer System Evaluation Criteria)**: A United States Government Department of Defense (DoD) standard that sets basic requirements for assessing the effectiveness of computer security controls.

V

- **VPN (Virtual Private Network)**: A service that allows you to connect to the Internet via an encrypted tunnel to ensure your online privacy and protect your sensitive data.

W

- **WEP/WPA/WPA2/WPA3**: Protocols for securing wireless networks.
- **Waterfall**: A linear and sequential design approach for software development.

X

- **XSS (Cross-Site Scripting)**: A security vulnerability typically found in web applications that enables attackers to inject client-side scripts into web pages viewed by other users.

PRACTICE SECTION

You've studied the theory, dissected the terminology, and maybe even had a few "aha!" moments. Now, it's time to see how it all translates to tackling CISSP-style exam questions. Remember, the CISSP isn't just about knowing facts; it's about applying them in complex and sometimes ambiguous scenarios.

Each question is followed immediately by the answer and a detailed explanation. This immediate feedback loop is crucial for several reasons:

1. **Instant Reinforcement**: Learning research shows that immediate reinforcement after answering a question helps to solidify retention. It allows you to immediately understand your mistakes and learn the correct information.
2. **Contextual Learning**: Seeing the answer in the context of the question helps you understand why that answer is correct. This deepens your comprehension and aids in the application of knowledge, rather than mere memorization.
3. **Efficient Review**: Having the answer right after the question saves time and maintains your focus. You won't lose momentum by flipping pages to seek out answers.
4. **Targeted Learning**: If you get a question wrong, you can immediately learn the correct information. This targeted learning approach helps to quickly fill knowledge gaps.

How to Use This Section

- **Test Your Instincts:** Read the question, try to answer it before peeking at the choices. This hones your gut feeling for what a "secure" solution should look like.
- **Analyze, Don't Memorize:** The explanations provided aren't meant for you to simply memorize. They train you to think like the exam writers, understanding why one answer is superior to others.
- **Time Yourself... Or Don't:** Depending on your study stage, use some sets timed, others untimed to focus on deep analysis.
- **Embrace the Wrong Answers:** Sometimes learning what *isn't *right is as valuable as knowing the correct solution.

A Note on Difficulty

The questions here will range in difficulty, mirroring what you might expect on the real exam. Don't be discouraged if some feel impossible at first – your goal is to consistently improve your analysis skills.

Important: These practice questions are a tool, not a guarantee. The real CISSP exam may still surprise you – the key is adaptability.

Ready to Dive In?

Let's see how your preparation translates into action. Remember, focus on the "why" behind each answer as much as the answer itself!

You're designing a multi-factor authentication system for a highly sensitive application. Which combination provides the strongest level of security?

a. Password + Fingerprint scan
b. Password + Security Question
c. Smartcard + One-Time Password (OTP)
d. Password + Picture chosen from a set

Answer: c. Smartcard + One-Time Password (OTP). Explanation: This combines "something you have" (the smartcard) with "something you know" (OTP), which is harder to compromise than two knowledge factors, or a possession factor combined with biometrics which can sometimes be spoofed.

Your organization suspects an insider is exfiltrating data. You've been asked to recommend a tool to help detect this activity. Which type of solution is best suited?
a. Network-based Intrusion Detection System (NIDS)
b. Data Loss Prevention (DLP)
c. Vulnerability scanner
d. Firewall with deep packet inspection

Answer: b. Data Loss Prevention (DLP). Explanation: DLP solutions are specifically designed to monitor data in motion, at rest, and in use, looking for patterns of unauthorized access or transmission that might indicate exfiltration.

A critical vulnerability in a widely used software library was just announced, but no vendor patch is available yet. Which risk response strategy would be the most appropriate?
a. Risk Mitigation
b. Risk Acceptance
c. Risk Avoidance
d. Risk Transference

Answer: a. Risk Mitigation. Explanation: You can't eliminate the risk entirely (avoidance), and accepting it unaddressed is too dangerous. Mitigation means taking temporary measures to reduce the likelihood or impact of the exploit – this could include network segmentation, increased monitoring, etc., until a patch exists.

A healthcare startup is developing a mobile app for patients to manage chronic conditions. They've consulted you as a security expert during the design phase. Which of the following is the MOST critical risk factor to address?
a. Unauthorized access to the app's backend database
b. The use of cloud storage for patient data
c. Vulnerabilities in the app's code
d. The possibility of users installing malware on their devices

Answer: b. The use of cloud storage for patient data. Explanation: Healthcare data is subject to stringent regulations (e.g., HIPAA). The cloud introduces compliance complexities regarding data location, encryption, and ensuring the provider meets the required security standards. While other options are risks, they're less fundamental to the core business model.

Your security team has a backlog of vulnerabilities to address. You need to prioritize them based on risk. Which of the following combinations carries the HIGHEST potential risk?
a. Vulnerability with a CVSS score of 7.5, located on an internet-facing system, with a known exploit
b. Vulnerability with a CVSS score of 5.0, affecting internal systems, with no readily available exploit
c. Vulnerability with a CVSS score of 9.0, on an isolated test system with no production data
d. Vulnerability with a CVSS score of 3.0, affecting a legacy system with complex dependencies

Answer: a. Vulnerability with a CVSS score of 7.5, located on an internet-facing system, with a known exploit. Explanation: This combines a high severity score (potential impact), ease of exploitation, and direct exposure making it the most urgent to address.

During a ransomware incident, it's discovered that backups were also encrypted. The IT Director insists on paying the ransom to recover quickly. What major risk factor should you highlight to dissuade them?
a. There's no guarantee that paying will result in data recovery or prevent further attacks.
b. Paying the ransom increases the risk of regulatory fines due to non-compliance.
c. Paying the ransom may inadvertently fund criminal organizations or terrorist activities.
d. The incident response team could face lowered morale due to this decision.

Answer: a. There's no guarantee that paying will result in data recovery or prevent further attacks. Explanation: While other options are valid concerns, this directly undermines the value proposition of paying. Attackers aren't known for their reliability, and payment offers no protection from being targeted again.

Your company is outsourcing software development to a firm overseas. What's a key step to mitigate risks in this relationship?
a. Require the vendor to use your company's VPN, granting them access to internal systems.
b. Include robust service level agreements (SLAs) for security patching and incident response
c. Mandate the use of specific encryption algorithms to protect all data in transit.
d. Conduct a background check focused on the individual developers who will be assigned to the project.

Answer: b. Include robust service level agreements (SLAs) for security patching and incident response. Explanation: SLAs contractually obligate the vendor to maintain security standards and define how they'll respond to breaches. This is more proactive than solely focusing on technology or individual-level checks.

During an enterprise risk assessment, a security manager identifies a potential threat that could exploit a vulnerability in the company's public-facing web application, leading to data loss. What is the FIRST step the manager should take in addressing this threat?
a. Patch the vulnerability immediately without further analysis.
b. Conduct a business impact analysis to determine the potential consequences.
c. Implement a web application firewall (WAF) as a preventive control.
d. Review and update the incident response plan to include this specific scenario.

Answer: b. Conduct a business impact analysis to determine the potential consequences. Explanation: Before taking action on a potential threat, it's critical to understand the impact it could have on the business. This enables the security manager to prioritize responses based on the severity and likelihood of the threat materializing, ensuring that resources are allocated effectively.

A multinational corporation is considering expanding its operations into a new country. What factor is LEAST likely to influence the company's decision from a security and risk management perspective?
a. The geopolitical stability of the country.
b. The maturity of the country's IT infrastructure.
c. The personal preferences of the company's CEO regarding the country.
d. The regulatory environment related to data protection in the country.

Answer: c. The personal preferences of the company's CEO regarding the country. Explanation: While the CEO's preferences might influence business decisions, they are not a relevant factor in a risk management context. Decisions should be based on objective criteria such as geopolitical stability, IT infrastructure maturity, and regulatory requirements, which directly impact the company's risk exposure and security posture.

In the context of security governance, which of the following best describes the principle of 'separation of duties'?
a. Dividing tasks among multiple employees to prevent fraud and errors.
b. Ensuring that every employee has a backup to cover their duties during absences.
c. Assigning security-related tasks to a dedicated department within the organization.
d. Outsourcing certain tasks to third-party vendors to reduce operational risk.

Answer: a. Dividing tasks among multiple employees to prevent fraud and errors. Explanation: Separation of duties is a key governance principle aimed at mitigating the risk of fraud or errors by ensuring that no single individual has control over all aspects of a critical process. This division of responsibilities helps in the detection and prevention of malicious activities within an organization.

When evaluating the effectiveness of a security awareness training program, which metric is MOST indicative of success?
a. The number of employees who have completed the training.
b. A decrease in the number of security incidents reported.

c. Positive feedback received in post-training surveys.
d. The amount of time employees spend on training modules.

Answer: b. A decrease in the number of security incidents reported. Explanation: While completion rates, time spent, and positive feedback are important, the ultimate goal of security awareness training is to reduce security incidents. A measurable decrease in incidents following training indicates that employees are applying what they learned, making it the most direct indicator of the program's effectiveness.

A company implements an aggressive password policy requiring changes every 30 days, complex compositions, and prohibiting the reuse of previous passwords. Employees start writing down passwords or using simplified patterns. What principle has been overlooked in this scenario?
a. Confidentiality
b. Integrity
c. Availability
d. Usability

Answer: d. Usability. Explanation: While the password policy aims to enhance security, neglecting the usability aspect leads to counterproductive behaviors that can compromise security. Balancing security measures with usability is crucial to ensure that security controls do not encourage insecure workarounds by users.

In the aftermath of a data breach, a forensic analysis reveals that the breach was facilitated by an outdated and unpatched server. Which of the following would have been MOST effective in preventing this breach?
a. A comprehensive patch management program.
b. Increased security awareness training for employees.
c. Implementation of stronger network perimeter defenses.
d. More stringent password policies.

Answer: a. A comprehensive patch management program. Explanation: Regularly updating and patching systems is a fundamental aspect of maintaining security. A comprehensive patch management program ensures that vulnerabilities are addressed promptly, reducing the window of opportunity for attackers to exploit outdated systems.

During a contract negotiation with a cloud service provider, which clause is MOST critical to ensure the protection of an organization's data?
a. Service Level Agreement (SLA) specifying uptime and availability.
b. Right to audit the provider's security practices and compliance.
c. Indemnification clause for losses due to service outages.
d. Provision for unlimited data storage capacity.

Answer: b. Right to audit the provider's security practices and compliance. Explanation: While SLAs and indemnification clauses are important, the right to audit the provider's security and compliance practices is crucial for ensuring that the provider maintains a security posture that meets the organization's standards and regulatory requirements.

A security consultant recommends the implementation of multi-factor authentication (MFA) across the organization. Which of the following is NOT a typical factor used in MFA?
a. Something the user knows (password).
b. Something the user has (security token).
c. Something the user is (biometric verification).
d. Something the user dislikes (aversive conditioning).

Answer: d. Something the user dislikes (aversive conditioning). Explanation: MFA involves combining at least two of the following factors: something the user knows, has, or is. Aversive conditioning is not a recognized factor in authentication processes and does not contribute to the verification of a user's identity.

In the development of a new application, a project manager decides to use open-source components to speed up the development process. What is the MOST important security consideration in this context?
a. Ensuring that the open-source components are widely used and popular.
b. Verifying that the components are compatible with the company's existing IT infrastructure.
c. Conducting a thorough security review of the open-source components before integration.
d. Obtaining approval from senior management for the use of open-source software.

Answer: c. Conducting a thorough security review of the open-source components before integration. Explanation: While popularity and compatibility are relevant factors, the security of open-source components is paramount. Conducting a security review helps to identify and mitigate potential vulnerabilities that could be introduced into the application, ensuring the security of the final product.

A company's security policy mandates the encryption of all sensitive data at rest. However, an audit reveals that several older databases are not encrypted. What should be the security team's IMMEDIATE action?
a. Encrypt the identified databases at the earliest opportunity.
b. Assess the risk and potential impact of not encrypting these databases.
c. Develop a new security policy that accounts for exceptions.
d. Train database administrators on the importance of encryption.

Answer: b. Assess the risk and potential impact of not encrypting these databases. Explanation: Before taking action, it's important to assess the risk and impact associated with the unencrypted databases. This assessment will guide the prioritization and implementation of remedial actions, such as encryption, based on the sensitivity of the data and the potential exposure risk.

A manufacturing company relies on a critical piece of machinery controlled by an embedded system running Windows XP. This system can't be upgraded or patched due to vendor limitations. Which approach would BEST address the security risk?
a. Implement strict network segmentation to isolate the system.
b. Deploy a host-based intrusion prevention system (HIPS) on the legacy device.
c. Accept the risk, since the system is critical to production.
d. Purchase cyber insurance to cover the potential impact of a compromise.

Answer: a. Implement strict network segmentation to isolate the system. Explanation: Segmentation limits the blast radius if the device is compromised, preventing attackers from easily pivoting elsewhere in the network. While HIPS might help, it's unlikely to be supported on XP, and neither insurance nor risk acceptance truly mitigate the threat.

An organization is struggling to implement their data classification scheme. Employees find it confusing, and often misclassify sensitive data. What's a key step to improve this process?
a. Increase the number of classification levels to provide more granularity.
b. Provide clear definitions, examples, and mandatory training on the classification process.
c. Automate the classification process as much as possible, using data tagging tools.
d. Penalize employees for misclassification to increase compliance.

Answer: b. Provide clear definitions, examples, and mandatory training on the classification process. Explanation: At its core, classification is a human process. Before automation or complex schemes, users need a strong understanding of the "why" behind the categories and practical guidance to apply them consistently.

Your company develops cutting-edge robotics technology. Which data security control is MOST crucial to protect your core intellectual property (IP)?
a. Strong encryption for data at rest and in transit.
b. Digital Rights Management (DRM) solutions integrated into design documents.
c. Regular security awareness training for all research and development staff.
d. A comprehensive non-disclosure agreement (NDA) for employees and contractors.

Answer: b. Digital Rights Management (DRM) solutions integrated into design documents. Explanation: While all options are valid, DRM directly embeds controls into the IP itself, limiting copying, distribution, or modification even if the data leaks.

Your organization is migrating sensitive workloads to a public cloud provider. What's a key consideration for ensuring data security in this environment?
a. Replicating your on-premises firewall rules into the cloud provider's security groups.
b. Understanding the shared responsibility model and the provider's security offerings.
c. Insisting on using only the cloud provider's proprietary encryption mechanisms.
d. Maintaining full control over the physical servers where your data will be stored.

Answer: b. Understanding the shared responsibility model and the provider's security offerings. Explanation: Cloud security is a partnership. You can't simply lift your existing controls, and must understand what the provider handles vs. your own obligations.

Your company has a mix of on-premises data centers and cloud-based applications. You need a way to enforce consistent access controls regardless of where resources are located. Which solution is BEST suited for this?
a. Implement a VPN to tunnel all traffic back to the on-premises network for authorization.
b. Deploy a Cloud Access Security Broker (CASB) to act as an intermediary and control policy.
c. Federate identity across providers using a standard like SAML (Security Assertion Markup Language).
d. Migrate everything to the cloud and utilize the provider's built-in identity management.

Answer: c. Federate identity across providers using a standard like SAML (Security Assertion Markup Language). Explanation: Federation establishes trust between identity providers, enabling single sign-on (SSO) and seamless access across environments. While CASBs have their role, they focus more on visibility and data-level controls, not core authentication.

You're enhancing physical security at a facility. Barriers and access controls are in place. Which addition would BEST exemplify defense in depth?
a. Installing brighter lighting in the parking lot to improve visibility for security cameras.
b. Adding a security alarm system that triggers if both perimeter doors are opened simultaneously.
c. Hiring an additional overnight security guard to increase patrols.
d. Posting more warning signs about the presence of security measures.

Answer: b. Adding a security alarm system that triggers if both perimeter doors are opened simultaneously. Explanation: Defense in depth involves diverse, overlapping controls. The alarm adds a different type of protection – detecting anomalies even if a single barrier fails, unlike the others which enhance existing layers.

Your organization is adopting Zero Trust principles. Which change would MOST directly reflect this approach?
a. Investing in multi-factor authentication (MFA) for all remote access.
b. Disabling legacy protocols and focusing on micro-segmentation of the network.
c. Implementing continuous monitoring and authorization checks based on context.
d. Encrypting all data at rest, regardless of its sensitivity classification.

Answer: c. Implementing continuous monitoring and authorization checks based on context. Explanation: Zero Trust is about dynamic authorization based on user, device, data, and the current situation. While the other options are good security practices, they're not as core to the Zero Trust philosophy.

Your company needs to securely exchange large, highly sensitive files with an external partner. Which cryptographic approach is generally BEST suited for this?
a. Symmetric encryption using a pre-shared key, for maximum speed.
b. Hashing the files to ensure integrity, and transmitting them unencrypted for efficiency.
c. Asymmetric encryption using the partner's public key, combined with symmetric encryption for the data itself.
d. Steganography to conceal the sensitive files within image or audio files.

Answer: c. Asymmetric encryption using the partner's public key, combined with symmetric encryption for the data itself. Explanation: It offers the best blend of security and efficiency. Asymmetric handles secure key exchange, symmetric encrypts the data quickly, and hashing can be added for integrity.

When designing a secure network architecture, which of the following is the MOST critical factor to consider for protecting data in transit?
a. Implementing network segmentation and zoning.
b. Employing strong end-to-end encryption protocols.
c. Enforcing strict access controls on all network devices.
d. Regularly updating antivirus software on all endpoints.

Answer: b. Employing strong end-to-end encryption protocols. Explanation: While all the listed measures are important for network security, employing strong end-to-end encryption protocols is crucial for protecting data in transit, ensuring that data remains confidential and integral as it moves across the network.

In selecting a cryptographic algorithm for securing sensitive company communications, which factor should be the LEAST important?
a. The algorithm's resistance to cryptanalysis.
b. The computational efficiency of the algorithm.
c. The popularity of the algorithm in the industry.
d. The length of the cryptographic key.

Answer: c. The popularity of the algorithm in the industry. Explanation: The security of a cryptographic algorithm doesn't solely rely on its popularity but on its resistance to cryptanalysis, key length, and computational efficiency. An algorithm should be chosen based on its proven security and suitability for the organization's specific needs.

For a critical infrastructure system, which security principle is MOST important to ensure operational continuity in the event of a system failure?
a. Confidentiality
b. Integrity
c. Availability
d. Non-repudiation

Answer: c. Availability. Explanation: For critical infrastructure systems, ensuring operational continuity is paramount, making availability the most crucial security principle. These systems must remain operational even in adverse conditions to maintain the essential services they provide.

When implementing security controls in a software development environment, what is the primary purpose of adopting a Secure Development Lifecycle (SDLC) methodology?
a. To ensure compliance with regulatory standards.
b. To facilitate faster development and deployment cycles.
c. To integrate security considerations throughout the development process.
d. To reduce the cost of software development.

Answer: c. To integrate security considerations throughout the development process. Explanation: The primary purpose of adopting a Secure SDLC methodology is to integrate security considerations at every phase of the development process, ensuring that security is not an afterthought but a fundamental component of the software.

In a multi-tenant cloud environment, what mechanism is MOST effective in isolating different tenants' data and workloads?
a. Data encryption with unique keys for each tenant.
b. Physical separation of each tenant's data on separate hardware.
c. Virtualization-based logical isolation with robust access controls.
d. Regularly scheduled data backups and redundancy.

Answer: c. Virtualization-based logical isolation with robust access controls. Explanation: In a multi-tenant cloud environment, virtualization-based logical isolation is most effective for separating tenants' data and workloads, allowing multiple tenants to share the same physical hardware securely.

For an organization subject to GDPR, what is the PRIMARY purpose of implementing pseudonymization techniques on personal data?
a. To enhance the data's integrity and reliability.
b. To reduce data storage and management costs.
c. To minimize the impact of data breaches on data subjects.
d. To increase the speed of data processing and analysis.

Answer: c. To minimize the impact of data breaches on data subjects. Explanation: Under GDPR, pseudonymization is recommended to protect personal data by decoupling it from identifying information, thereby minimizing the impact on data subjects in the event of a data breach.

When integrating an Intrusion Detection System (IDS) into a network, what is the PRIMARY goal of placing it behind a firewall?
a. To reduce the volume of traffic the IDS needs to analyze.
b. To ensure the IDS can prevent attacks in real-time.
c. To encrypt the traffic before it reaches the IDS.
d. To facilitate easier maintenance and updates of the IDS.

Answer: a. To reduce the volume of traffic the IDS needs to analyze. Explanation: Placing an IDS behind a firewall reduces the volume of traffic to be analyzed by filtering out a significant amount of malicious traffic at the firewall level, allowing the IDS to focus on more sophisticated threats.

In a distributed system architecture, which of the following is the MOST effective way to ensure data consistency across multiple locations?
a. Implementing a centralized database with periodic synchronization.
b. Utilizing a blockchain-based ledger system for transactions.
c. Employing a distributed database with a strong consistency model.
d. Relying on manual reconciliation processes at regular intervals.

Answer: c. Employing a distributed database with a strong consistency model. Explanation: A distributed database with a strong consistency model ensures that all nodes in the system reflect the same data state simultaneously, providing the most effective way to maintain data consistency across multiple locations.

When designing a secure system, why is it important to implement the principle of least privilege?
a. To ensure that users have access to all the resources they need.
b. To minimize the potential impact of a system or user compromise.
c. To make the system easier to use and navigate for end-users.
d. To comply with all industry-standard security regulations.

Answer: b. To minimize the potential impact of a system or user compromise. Explanation: The principle of least privilege restricts users' access rights to only what is necessary for their role, minimizing potential damage in the event of a compromise by limiting the attacker's access to essential systems and data.

In securing a web application, which of the following is the MOST effective control to protect against Cross-Site Scripting (XSS) attacks?
a. Implementing strict session management controls.
b. Enforcing HTTPS for all application traffic.
c. Validating and sanitizing all user input.
d. Requiring multi-factor authentication for users.

Answer: c. Validating and sanitizing all user input. Explanation: The most effective way to protect against XSS attacks is by validating and sanitizing all user input, ensuring that malicious scripts are not injected into the application, which could be executed in the context of a user's session.

Your organization detects a network breach. After containment, which step is MOST critical for preserving potential legal evidence?
a. Creating a full disk image of compromised systems for forensic analysis.
b. Changing passwords on all potentially affected user accounts.
c. Documenting the incident in detail, following the chain of custody for all logs and artifacts.
d. Immediately rebooting the affected servers to clear out any malware from memory.

Answer: c. Documenting the incident in detail, following the chain of custody for all logs and artifacts. Explanation: Legal action may follow, making meticulous documentation essential. Disk images are valuable, but if their collection process isn't defensible in court, they're less useful. Password changes and reboots might disrupt evidence.

A security analyst is reviewing firewall logs and notices unusual traffic patterns using non-standard ports. Which OSI layer should they initially focus their investigation on?
a. Layer 2 (Data Link)
b. Layer 3 (Network)
c. Layer 4 (Transport)
d. Layer 7 (Application)

Answer: c. Layer 4 (Transport). Explanation: Transport layer protocols (TCP, UDP) deal with port numbers. Non-standard ports might indicate an attempt to evade typical firewall rules, while Layers 2 and 3 are less concerned with port specifics.

Your company is mandating the use of a VPN for all employees working remotely. What is the PRIMARY security benefit this provides?
a. It encrypts traffic, preventing eavesdropping on public Wi-Fi networks.
b. It blocks malware from automatically infecting employee devices.
c. It authenticates users before granting them access to the internal network.
d. It prevents distributed denial-of-service (DDoS) attacks against employee devices.

Answer: a. It encrypts traffic, preventing eavesdropping on public Wi-Fi networks. Explanation: VPNs are primarily about secure tunneling. While they can play a role in authentication and access control depending on the configuration, their core function is protecting data in transit.

A security bulletin is released detailing a zero-day vulnerability in a web server your company uses. However, no working exploit code is publicly available yet. How should you prioritize your response?
a. Immediately patch the web server, even if the patch is still in beta testing.
b. Implement compensating controls around the web server, such as stricter WAF rules.
c. Monitor exploit development closely and prepare for rapid patching once an exploit is released.
d. Accept the risk, as the lack of an exploit makes an attack unlikely.

Answer: c. Monitor exploit development closely and prepare for rapid patching once an exploit is released. Explanation: Zero days are high-risk, but without an exploit, it's less urgent. Compensating controls and monitoring for exploit activity is the most balanced approach while awaiting a patch.

In designing a network infrastructure for a new data center, what is the MOST critical consideration to ensure secure communication between servers within the center?
a. Implementation of VLANs to segment network traffic.
b. Use of redundant network paths to ensure availability.
c. Deployment of hardware-based firewalls between each network segment.
d. Encryption of data in transit between servers using TLS.

Answer: d. Encryption of data in transit between servers using TLS. Explanation: While VLANs, redundancy, and firewalls are important for network security and availability, encrypting data in transit using TLS is critical to ensuring that sensitive information remains confidential and integral as it moves across the network, protecting against eavesdropping and tampering.

When evaluating the security of a VoIP system, which of the following vulnerabilities should be considered FIRST?
a. Insufficient bandwidth causing call quality degradation.
b. Lack of support for older VoIP protocols.
c. Eavesdropping on unencrypted VoIP traffic.
d. Incompatibility with traditional analog phone systems.

Answer: c. Eavesdropping on unencrypted VoIP traffic. Explanation: While bandwidth, protocol support, and system compatibility are important considerations, the primary security concern with VoIP systems is the potential for eavesdropping on unencrypted traffic, which could lead to unauthorized access to sensitive communications.

For an organization implementing IPv6, what is a key security feature that should be utilized to protect the integrity and confidentiality of network traffic?
a. Stateless Address Autoconfiguration (SLAAC) for dynamic address assignment.
b. Mandatory use of IPsec for network communication.
c. Network Address Translation (NAT) to hide internal IP addresses.
d. The use of Anycast addresses to distribute load among servers.

Answer: b. Mandatory use of IPsec for network communication. Explanation: IPv6 inherently supports IPsec for encrypting and authenticating network traffic, providing a robust method to ensure the integrity and confidentiality of communications. Unlike IPv4, IPsec is an integral part of IPv6, making it a key security feature to be utilized.

In the context of wireless network security, why is WPA3 considered more secure than WPA2?
a. WPA3 uses a longer encryption key length.
b. WPA3 introduces individualized data encryption.
c. WPA3 mandates the use of stronger password requirements.
d. WPA3 employs a more efficient packet routing protocol.

Answer: b. WPA3 introduces individualized data encryption. Explanation: WPA3 enhances wireless security by introducing individualized data encryption, which provides personalized encryption for each device on the network, protecting against certain types of attacks that were possible on WPA2, such as KRACK (Key Reinstallation Attack).

In deploying a site-to-site VPN, what is the PRIMARY security concern that must be addressed to protect data across the Internet?
a. Ensuring that all devices support the same VPN protocols.
b. Authentication of each site before establishing the VPN tunnel.
c. The physical security of the VPN endpoint devices.
d. Selection of a suitable dynamic routing protocol.

Answer: b. Authentication of each site before establishing the VPN tunnel. Explanation: While protocol support, physical security, and routing are important, the primary concern with site-to-site VPNs is ensuring that each site is authenticated before establishing the VPN tunnel, thereby ensuring that data is only shared between trusted entities.

When configuring a firewall to protect an e-commerce application, what is the MOST effective approach to minimize exposure to web-based attacks?
a. Implementing default deny rules and allowing only necessary traffic.
b. Placing the e-commerce application in a DMZ.
c. Enabling deep packet inspection on all inbound traffic.
d. Regularly updating the firewall's firmware to the latest version.

Answer: a. Implementing default deny rules and allowing only necessary traffic. Explanation: While DMZs, deep packet inspection, and firmware updates are important, the most effective firewall configuration to protect an e-commerce application is to implement a default deny policy, explicitly allowing only the necessary traffic to and from the application, minimizing the attack surface.

For securing remote access to a corporate network, which of the following provides the BEST combination of security and usability?
a. An SSL VPN with multi-factor authentication.
b. A hardware-based dial-up token system.
c. A site-to-site VPN with static IP addressing.
d. SSH with password-based authentication.

Answer: a. An SSL VPN with multi-factor authentication. Explanation: An SSL VPN with multi-factor authentication offers a secure and user-friendly method for remote access, providing encryption, ease of use through a web browser, and strong authentication by requiring multiple forms of verification.

In a distributed denial-of-service (DDoS) attack, what is the PRIMARY target of the attackers?
a. The integrity of the data on the target system.
b. The confidentiality of communications to and from the target system.
c. The availability of the target system or network.
d. The authentication mechanisms of the target system.

Answer: c. The availability of the target system or network. Explanation: The primary objective of a DDoS attack is to compromise the availability of the target system or network, overwhelming it with a flood of traffic to disrupt normal services and deny access to legitimate users.

When considering the security of SNMP (Simple Network Management Protocol) within an enterprise network, what is a key practice to enhance its security posture?
a. Using SNMPv3 which includes mechanisms for authentication and encryption.
b. Limiting SNMP use to read-only operations.
c. Deploying SNMP traps for all network devices.
d. Ensuring that SNMP is accessible over the Internet for remote management.

Answer: a. Using SNMPv3 which includes mechanisms for authentication and encryption. Explanation: SNMPv3 is considered more secure than its predecessors because it introduces features for authentication and encryption, addressing the security weaknesses in SNMPv1 and SNMPv2, such as the transmission of community strings in clear text.

In the implementation of a new wireless LAN for a corporate office, which of the following measures would MOST effectively reduce the risk of unauthorized access?
a. Disabling SSID broadcasting.
b. Implementing MAC address filtering.
c. Enabling WPA3 encryption on the wireless network.
d. Decreasing the power levels of the wireless access points.

Answer: c. Enabling WPA3 encryption on the wireless network. Explanation: Enabling WPA3 encryption provides robust security by encrypting the data transmitted over the wireless network, significantly reducing the risk of unauthorized access compared to measures like SSID hiding, MAC filtering, or power level adjustments, which provide minimal security improvements and can be easily circumvented.

Your organization has a mix of legacy on-premises systems and a growing suite of SaaS applications. You've been asked to streamline the user provisioning and de-provisioning process. Which technology would BEST support this goal?
a. Role-based access control (RBAC) to manage permissions consistently.
b. A single sign-on (SSO) solution to reduce password fatigue for users.

c. A privileged access management (PAM) tool to control administrative accounts.
d. A cloud-based identity and access management (IAM) platform.

Answer: d. A cloud-based identity and access management (IAM) platform. Explanation: Modern IAM platforms are designed for hybrid environments, offering centralized provisioning, SSO, and integrations with various systems. While RBAC, SSO, and PAM are all important, they're components of a broader IAM solution.

A web application development team is using an attribute-based access control (ABAC) model. Which of the following is MOST likely to be used as a factor for making authorization decisions?
a. The user's job title within the organization.
b. The specific resource the user is trying to access.
c. The time of day that the access request is made.
d. The device's current security posture (patch levels, antivirus status).

Answer: c. The time of day that the access request is made. Explanation: ABAC is highly dynamic. Time, location, device status, etc., can all be factors alongside traditional attributes like roles. This flexibility is its core strength.

A new helpdesk employee is struggling because they can't resolve most of the tickets assigned to them. Upon investigation, you discover their permissions are too restrictive. Which security principle has been applied too strictly in this case?
a. Separation of duties
b. Need-to-know
c. Least privilege
d. Defense in depth

Answer: c. Least privilege. Explanation: Least privilege is about the bare minimum, but not so little as to prevent work from happening. In this case, it's impeding operational efficiency.

Your company recently implemented multi-factor authentication (MFA) using SMS-based one-time codes. Which attack could still compromise this system?
a. Credential stuffing attacks against user accounts.
b. SIM swapping attacks to intercept the SMS codes.
c. Shoulder surfing to observe a user entering the code.
d. Password spraying attacks to slowly guess weak passwords.

Answer: b. SIM swapping attacks to intercept the SMS codes. Explanation: This attack targets a weakness in the second factor itself, not the primary password. It highlights the importance of choosing strong MFA methods (authenticator apps, hardware tokens).

In the context of an IAM framework, which of the following best ensures the principle of least privilege is adhered to?
a. Implementing role-based access control (RBAC) with defined permissions for each role.
b. Enforcing periodic password changes for all users.
c. Utilizing single sign-on (SSO) for all enterprise applications.
d. Requiring biometric authentication for system administrators.

Answer: a. Implementing role-based access control (RBAC) with defined permissions for each role. Explanation: RBAC is a method of restricting system access to authorized users based on their roles within an organization. By defining specific permissions for each role, RBAC ensures that users have only the access necessary to perform their job functions, adhering to the principle of least privilege.

When integrating a new application into an existing SSO solution, what is the MOST critical security consideration?
a. The compatibility of the application with the SSO protocol in use.
b. The impact of the new application on the overall performance of the SSO system.
c. The method by which the application stores session tokens.
d. Ensuring the application supports the same authentication mechanisms as the SSO system.

Answer: d. Ensuring the application supports the same authentication mechanisms as the SSO system. Explanation: For seamless integration and to maintain security, it's crucial that the new application supports the same authentication mechanisms as the existing SSO solution. This ensures that the SSO system can effectively manage user authentication and access control for the application without compromising security.

In a federated identity management system, what role does the Security Assertion Markup Language (SAML) play?
a. It encrypts data in transit between identity providers and service providers.
b. It serves as a protocol for exchanging authentication and authorization data between parties.
c. It manages user sessions across different domains.
d. It provides a directory service for storing user credentials.

Answer: b. It serves as a protocol for exchanging authentication and authorization data between parties. Explanation: SAML is an XML-based standard used in federated identity management to exchange authentication and authorization data between an identity provider and a service provider, enabling single sign-on and secure data sharing across different systems.

What is the PRIMARY advantage of implementing Multi-Factor Authentication (MFA) in an organizational security infrastructure?
a. It reduces the need for complex password policies.
b. It provides a backup authentication method if one factor is compromised.
c. It significantly enhances security by requiring more than one form of verification.
d. It simplifies the user authentication process.

Answer: c. It significantly enhances security by requiring more than one form of verification. Explanation: MFA enhances security by requiring users to provide two or more verification factors to gain access to a resource such as an application, online account, or a VPN, making it more difficult for unauthorized persons to gain access.

For an organization using a Public Key Infrastructure (PKI), what is the role of the Certificate Authority (CA)?
a. It generates public and private key pairs for users.
b. It acts as a trusted third party to issue, revoke, and manage digital certificates.
c. It provides secure storage for users' private keys.
d. It encrypts data using the public key of the recipient.

Answer: b. It acts as a trusted third party to issue, revoke, and manage digital certificates. Explanation: In a PKI, the CA is a trusted entity that issues digital certificates to validate the ownership of a public key by the named subject of the certificate, thereby facilitating secure communication and transactions online.

When considering access control models for a highly secure environment, why might an organization choose Mandatory Access Control (MAC) over Discretionary Access Control (DAC)?
a. MAC allows users full control over the permissions of the objects they own.
b. MAC provides a higher level of security by enforcing policies based on classifications and clearances.
c. DAC is more complex and difficult to manage.
d. MAC is easier to implement in a decentralized environment.

Answer: b. MAC provides a higher level of security by enforcing policies based on classifications and clearances. Explanation: MAC is considered more secure than DAC because it restricts access based on security policies determined by the system, not by the users, using classifications and clearances. This makes it suitable for environments requiring a high level of security.

In an IAM system, what is the primary purpose of provisioning?
a. To monitor and record user activities within systems.
b. To grant, modify, or revoke user access to systems, networks, and applications.
c. To authenticate users at the time of login.
d. To encrypt user data stored within the system.

Answer: b. To grant, modify, or revoke user access to systems, networks, and applications. Explanation: Provisioning in IAM refers to the process of managing user access rights, including the creation, modification, and deletion of user accounts and permissions, ensuring users have appropriate access to organizational resources.

What is the most significant risk associated with not implementing timely deprovisioning of user access rights in an IAM system?
a. Increased administrative overhead for managing access rights.
b. Potential unauthorized access by former employees or contractors.
c. Inefficiencies in allocating system resources.
d. Conflicts in role-based access controls.

Answer: b. Potential unauthorized access by former employees or contractors. Explanation: Failure to promptly deprovision user access rights can lead to former employees or contractors retaining access to sensitive systems and data, posing a significant security risk by potentially allowing unauthorized access and data breaches.

In a large enterprise, what is the PRIMARY benefit of using a directory service, such as LDAP, for IAM?
a. It decentralizes user management, spreading the load across multiple servers.
b. It allows for the encryption of data at rest and in transit.
c. It centralizes the management of user information, simplifying administration and access control.
d. It replaces the need for traditional authentication mechanisms like passwords.

Answer: c. It centralizes the management of user information, simplifying administration and access control. Explanation: LDAP and similar directory services provide a centralized location to store and manage user information, making it easier to administer access controls, manage user attributes, and streamline authentication and authorization processes across the enterprise.

When designing an IAM system for an organization with high employee turnover, what feature is MOST important to ensure security and operational efficiency?
a. Automated provisioning and deprovisioning of user accounts.
b. Regular mandatory password changes for all users.
c. Biometric authentication for access to all systems.
d. Comprehensive logging of all user actions for audit purposes.

Answer: a. Automated provisioning and deprovisioning of user accounts. Explanation: Automated provisioning and deprovisioning streamline the process of granting and revoking access to new and departing employees, respectively, ensuring that access rights are always up to date and reducing the risk of unauthorized access, which is crucial in environments with high employee turnover.

Your organization's highly sensitive data resides in a cloud-based SaaS application with limited configuration visibility. You're concerned about potential misconfigurations. Which type of security assessment is MOST appropriate?
a. Penetration testing to attempt exploitation of vulnerabilities.
b. Vulnerability scanning using an authenticated credentialed scan.
c. Code review to analyze the SaaS application's source code.
d. Configuration review against the vendor's security baselines and industry best practices.

Answer: d. Configuration review against the vendor's security baselines and industry best practices. Explanation: Since you have limited control of the app itself, focusing on the settings you CAN manage is critical. Pen tests and code reviews are less useful, and credentialed vulnerability scans might be limited by the SaaS model.

You're interpreting the results of a network vulnerability scan. Many findings are marked as "Informational." How should you prioritize these?
a. Ignore them entirely, as they don't represent immediate risks.
b. Address them after higher-severity vulnerabilities have been fixed.
c. Investigate them, as they might reveal configuration details useful to an attacker.

d. They are likely false positives and can be safely disregarded.

Answer: c. Investigate them, as they might reveal configuration details useful to an attacker. Explanation: "Informational" doesn't mean irrelevant! These findings can provide a roadmap for an attacker to build upon, leading to more serious compromise later.

Your organization is planning a tabletop exercise to test its incident response plan. Which of these objectives is this type of exercise BEST suited for?
a. Identifying technical vulnerabilities in the organization's network infrastructure.
b. Validating communication flows, roles, and decision-making processes during a breach.
c. Training IT personnel on how to use specific incident response tools and technologies.
d. Calculating the potential financial impact and reputation damage of various breach scenarios.

Answer: b. Validating communication flows, roles, and decision-making processes during a breach. Explanation: Tabletop exercises focus on the human process side of response. They are great for uncovering gaps in plans, coordination issues, and unclear responsibilities.

A red team exercise recently concluded. The results were eye-opening, exposing gaps your internal security team had missed. However, some stakeholders question the value, as no "real" damage occurred. How do you justify this exercise?
a. Emphasize that the red team simulated the attacker's mindset, revealing exploitable weaknesses.
b. Acknowledge it was a failure, as the whole point of security is to prevent any breach.
c. Downplay the findings, as red team exercises often use overly aggressive tactics.
d. Pivot to discussing how much it would have cost if a real attacker had these findings.

Answer: a. Emphasize that the red team simulated the attacker's mindset, revealing exploitable weaknesses. Explanation: Red teaming is proactive, uncovering issues BEFORE they're exploited for real. It's about finding what a determined adversary could do, not just what casually happens.

When planning a penetration test, what is the MOST important step to ensure its effectiveness and alignment with organizational goals?
a. Selecting automated penetration testing tools with the latest exploit databases.
b. Obtaining explicit written permission and defined scope from senior management.
c. Ensuring the penetration testing team has a diverse set of hacking skills.
d. Coordinating the penetration test timing to coincide with network low traffic periods.

Answer: b. Obtaining explicit written permission and defined scope from senior management. Explanation: The most critical step is to obtain explicit written permission and a well-defined scope from senior management to ensure the test is legally sanctioned, aligned with business objectives, and clearly outlines the boundaries to avoid unintended disruptions or legal issues.

In a security audit, which type of testing would provide the MOST in-depth understanding of an application's security posture without actual exploitation?
a. Vulnerability scanning using automated tools.
b. Manual penetration testing by ethical hackers.
c. Code review and static analysis.
d. Dynamic application security testing (DAST).

Answer: c. Code review and static analysis. Explanation: Code review and static analysis allow for an in-depth examination of the application's source code to identify potential security flaws, offering detailed insights into the security posture without the need for actual exploitation, which might be part of a penetration test.

What is the PRIMARY purpose of conducting a security control assessment on a new software deployment before going live?
a. To ensure that the software meets functional requirements and user expectations.

b. To identify any security vulnerabilities and weaknesses in the software.
c. To evaluate the software's performance and scalability under load.
d. To verify the compatibility of the software with existing systems.

Answer: b. To identify any security vulnerabilities and weaknesses in the software. Explanation: The primary purpose of conducting a security control assessment on new software before it goes live is to identify and mitigate any security vulnerabilities and weaknesses, ensuring that it does not introduce unacceptable risks to the organization.

During a security assessment, what indicates the need for a more detailed investigation when analyzing system logs?
a. Regular system updates and patch installations.
b. Consistent patterns of user access during business hours.
c. Unexplained discrepancies or gaps in log entries.
d. High volumes of data transfer during backup operations.

Answer: c. Unexplained discrepancies or gaps in log entries. Explanation: Unexplained discrepancies or gaps in system logs could indicate tampering, technical issues, or malicious activity, signaling the need for a more detailed investigation to understand the cause and potential security implications.

Why is it important to include both dynamic and static testing methods in a software security assessment strategy?
a. Dynamic testing validates the system's behavior during runtime, while static testing examines the codebase for vulnerabilities that might not be evident during operation.
b. Static testing can replace dynamic testing in agile development environments.
c. Dynamic testing alone is sufficient for uncovering all potential security issues.
d. Static testing is only useful for compliance purposes and not for security.

Answer: a. Dynamic testing validates the system's behavior during runtime, while static testing examines the codebase for vulnerabilities that might not be evident during operation. Explanation: Combining both dynamic and static testing methods provides a comprehensive view of a software's security posture by evaluating it both at rest (static) and in execution (dynamic), uncovering a wider range of potential vulnerabilities.

What is the key difference between a vulnerability assessment and a penetration test?
a. A vulnerability assessment identifies and quantifies vulnerabilities, while a penetration test attempts to exploit them to understand the actual impact.
b. Vulnerability assessments are performed by external parties, whereas penetration tests are only conducted internally.
c. Penetration testing is automated, while vulnerability assessments are manual processes.
d. Vulnerability assessments focus on system availability, whereas penetration tests focus on confidentiality and integrity.

Answer: a. A vulnerability assessment identifies and quantifies vulnerabilities, while a penetration test attempts to exploit them to understand the actual impact. Explanation: The key difference lies in the approach and objective; vulnerability assessments aim to identify and quantify vulnerabilities, while penetration tests go further by attempting to exploit these vulnerabilities to assess the real-world impact of an attack.

In security testing, why is it important to perform both authenticated and unauthenticated scans?
a. Authenticated scans assess external threats, while unauthenticated scans assess internal threats.
b. Unauthenticated scans are sufficient for systems exposed to the internet.
c. Authenticated scans provide a view of the system's vulnerabilities from a privileged user's perspective, while unauthenticated scans simulate an external attacker's perspective.
d. Authenticated scans are only necessary for compliance with regulatory standards.

Answer: c. Authenticated scans provide a view of the system's vulnerabilities from a privileged user's perspective, while unauthenticated scans simulate an external attacker's perspective. Explanation: Performing both types of scans offers a more complete assessment of the system's security by evaluating it from both an insider's view (authenticated) and an outsider's view (unauthenticated), identifying vulnerabilities that might be exploited in different attack scenarios.

What role does fuzz testing play in software security?
a. It checks for compliance with coding standards.
b. It verifies the accuracy of software documentation.
c. It identifies potential buffer overflows and input validation issues by inputting large amounts of random data.
d. It ensures that the user interface elements are aligned correctly.

Answer: c. It identifies potential buffer overflows and input validation issues by inputting large amounts of random data. Explanation: Fuzz testing, or fuzzing, involves inputting large amounts of random data ("fuzz") into software to identify potential vulnerabilities such as buffer overflows, crashes, and input validation issues, which could be exploited by attackers.

When establishing a program for regular security assessments, what is the MOST critical factor to ensure ongoing effectiveness?
a. Conducting assessments annually in compliance with industry standards.
b. Continuously updating the assessment tools and methodologies to adapt to new threats.
c. Limiting the assessments to high-risk areas to focus resources more efficiently.
d. Relying on automated tools to increase the frequency of assessments.

Answer: b. Continuously updating the assessment tools and methodologies to adapt to new threats. Explanation: The threat landscape is constantly evolving, so it's critical to continuously update tools and methodologies to ensure that security assessments remain effective in identifying new and emerging threats, ensuring the organization's defenses keep pace with the evolving threat environment.

A critical server crashes in the middle of the night. The on-call incident responder restores it from a recent backup, but the issue reoccurs hours later. What key step might they have missed in the response process?
a. Verifying the patch level of the server's operating system.
b. Implementing network segmentation to isolate the affected server.
c. Conducting root cause analysis to identify the underlying trigger of the crashes.
d. Escalating the incident to senior management for immediate approval of a replacement server.

Answer: c. Conducting root cause analysis to identify the underlying trigger of the crashes. Explanation: Restoration is just a temporary fix. Without determining why it crashed, it's likely to happen again. Patching, segmentation, etc., are good practices but might not address the immediate issue.

Your SIEM is constantly generating alerts about suspicious network traffic originating from IoT devices (security cameras, etc.). How should you best address this?
a. Disable the IoT devices as they are clearly compromised.
b. Update the SIEM rules to reduce the sensitivity for IoT-related alerts.
c. Investigate the traffic patterns to establish a baseline for normal IoT device behavior.
d. Implement stricter firewall rules to block all traffic from the IoT device subnet.

Answer: c. Investigate the traffic patterns to establish a baseline for normal IoT device behavior. Explanation: IoT devices can be noisy. You need a baseline to determine what's truly malicious vs. their typical chatter. Simply disabling, blocking, or ignoring alerts isn't good long-term practice.

Your new CISO wants to quantify how quickly the security operations team is detecting and responding to incidents. Which pair of metrics is MOST relevant?
a. Mean Time to Detect (MTTD) and Mean Time to Contain (MTTC)
b. Mean Time to Acknowledge (MTTA) and Mean Time to Resolve (MTTR)
c. Mean Time to Patch (MTP) and Mean Time to Fail (MTTF)
d. Return on Investment (ROI) and Total Cost of Ownership (TCO)

Answer: b. Mean Time to Acknowledge (MTTA) and Mean Time to Resolve (MTTR). Explanation: These focus on the operational efficiency of incident handling itself. MTTD/MTTC are important too, but the CISO's wording implies a focus on the team's process.

A new SIEM rule is generating too many false positive alerts, overwhelming analysts. What's the MOST appropriate course of action?
a. Disable the rule entirely to reduce the workload on the analysts.
b. Increase the severity threshold required for the rule to trigger an alert.
c. Refine the rule logic to focus on only the most critical attack indicators.
d. Invest in a more advanced SIEM solution with better machine-learning capabilities.

Answer: c. Refine the rule logic to focus on only the most critical attack indicators. Explanation: It's about striking a balance. Disabling the rule leaves a gap, buying a whole new SIEM is overkill, and simply raising the threshold might miss lower-level attacks.

In the event of a data breach, what is the FIRST action that should be taken according to best practices in incident response?
a. Notify affected customers and stakeholders immediately.
b. Isolate the affected systems to contain the breach.
c. Initiate a legal investigation to find the perpetrators.
d. Restore all systems from the most recent backup.

Answer: b. Isolate the affected systems to contain the breach. Explanation: The immediate isolation of affected systems is crucial to contain the breach and prevent further unauthorized access or data loss. This step is essential before proceeding with notifications, investigations, or system restorations to ensure that the breach does not spread.

When establishing a Security Operations Center (SOC), what is the MOST critical factor for its success?
a. State-of-the-art security monitoring tools and technologies.
b. A well-defined escalation process for incident management.
c. Continuous training and development programs for SOC staff.
d. Executive support and adequate funding.

Answer: d. Executive support and adequate funding. Explanation: While tools, processes, and training are important, the most critical factor for the success of a SOC is executive support and adequate funding. This ensures the SOC has the necessary resources and organizational backing to operate effectively and respond to threats.

In managing physical security, which of the following would be considered a deterrent control?
a. Surveillance cameras visibly placed around the perimeter.
b. Biometric access controls at entry points.
c. Motion detectors within secure areas.
d. Security personnel patrolling the facility.

Answer: a. Surveillance cameras visibly placed around the perimeter. Explanation: Visible surveillance cameras act as a deterrent control by making potential intruders aware they are being monitored, potentially preventing unauthorized access attempts before they occur.

For a global organization, what is the PRIMARY challenge in implementing a unified security policy across all locations?
a. Variations in local laws and regulations concerning data protection and privacy.
b. Differences in time zones affecting coordination and communication.
c. The need for translation services for non-English speaking employees.
d. Differing levels of technological advancement across offices.

Answer: a. Variations in local laws and regulations concerning data protection and privacy. Explanation: The primary challenge is accounting for variations in local laws and regulations, as these can significantly affect how data protection and privacy must be handled in different jurisdictions, requiring the policy to be adaptable while still maintaining a unified security posture.

When conducting a risk assessment, why is it important to consider the impact of identified risks in addition to their likelihood?
a. To prioritize risks based on their potential impact on organizational objectives.
b. Because low-likelihood risks can be ignored if their impact is minimal.
c. To ensure that all risks, regardless of impact, are mitigated.
d. Because the likelihood of risks is more challenging to determine than impact.

Answer: a. To prioritize risks based on their potential impact on organizational objectives. Explanation: Considering both the likelihood and impact of risks is essential to prioritize them effectively. This allows an organization to allocate resources to mitigate risks that pose the greatest threat to achieving its objectives.

In a digital forensics investigation, why is maintaining a chain of custody for evidence critical?
a. To ensure the evidence can be used in disciplinary actions against employees.
b. To prevent the legal team from accessing sensitive information.
c. To ensure the evidence remains uncontaminated and its integrity is preserved for legal proceedings.
d. To keep a record for potential insurance claims.

Answer: c. To ensure the evidence remains uncontaminated and its integrity is preserved for legal proceedings. Explanation: Maintaining a chain of custody is critical to ensure that digital evidence is not altered, tampered with, or contaminated, preserving its integrity and admissibility in legal proceedings.

What is the primary purpose of implementing a Security Information and Event Management (SIEM) system within an organization's security operations?
a. To automate the patch management process for all systems.
b. To provide real-time analysis and correlation of security alerts generated by applications and network hardware.
c. To enforce access controls and prevent unauthorized data access.
d. To serve as the primary data backup and recovery solution.

Answer: b. To provide real-time analysis and correlation of security alerts generated by applications and network hardware. Explanation: The primary purpose of a SIEM system is to offer real-time analysis, correlation, and aggregation of security alerts or events across various sources within the organization, enhancing the ability to detect and respond to threats.

During a business continuity planning process, what is the role of a Business Impact Analysis (BIA)?
a. To identify critical business processes and the impact of their disruption.
b. To outline the steps for recovering after a disaster.
c. To assign responsibilities for business continuity roles.
d. To document the backup procedures for all data.

Answer: a. To identify critical business processes and the impact of their disruption. Explanation: A BIA is conducted to identify and evaluate the impact of disruptions to business processes, determining which processes are critical to the organization's operational and financial stability, and thus guiding the prioritization in the business continuity plan.

In the context of incident response, what is the significance of the 'lessons learned' phase?
a. To provide a detailed account of the incident to law enforcement.
b. To review and analyze the incident response process for improvements.
c. To implement additional security controls to prevent future incidents.
d. To document the incident for insurance purposes.

Answer: b. To review and analyze the incident response process for improvements. Explanation: The 'lessons learned' phase is crucial for reviewing and analyzing the incident response process after an incident has been resolved to identify what was done well and what could be improved, enhancing the effectiveness of future responses.

When selecting a third-party vendor for processing sensitive data, what factor is MOST critical to ensure the protection of that data?
a. The cost-effectiveness of the vendor's services.
b. The physical security measures at the vendor's facilities.
c. The vendor's compliance with relevant security standards and regulations.
d. The geographic location of the vendor's data centers.

Answer: c. The vendor's compliance with relevant security standards and regulations. Explanation: Ensuring that the vendor complies with relevant security standards and regulations is critical to protect sensitive data. This compliance demonstrates that the vendor adheres to best practices and legal requirements for data security and privacy, reducing the risk to the data they process.

Your development team is resistant to integrating security practices, saying it slows them down. How can you best address this concern?
a. Mandate strict code reviews and security testing gates before any deployment can occur.
b. Focus on automated tools that integrate into the development pipeline, providing early feedback.
c. Hire dedicated security developers to embed within the development team.
d. Penalize developers with the most vulnerabilities found in their code after release.

Answer: b. Focus on automated tools that integrate into the development pipeline, providing early feedback. Explanation: Shifting left is about making security seamless. Automation is key, as it works at the pace of development. While the other options have merit, they're either less frictionless or focus on consequences rather than enablement.

You're conducting a threat modeling exercise for a web application that processes payments. Which methodology is best suited for this task?
a. PASTA (Process for Attack Simulation and Threat Analysis)
b. STRIDE (Spoofing, Tampering, Repudiation, Information Disclosure, Denial of Service, Elevation of Privilege)
c. CVSS (Common Vulnerability Scoring System)
d. OWASP Top 10 (List of most critical web application security risks)

Answer: b. STRIDE (Spoofing, Tampering, Repudiation, Information Disclosure, Denial of Service, Elevation of Privilege). Explanation: STRIDE is a threat modeling framework, designed to identify how an attacker could break the system. CVSS scores known vulns, OWASP is a high-level awareness list, and PASTA is broader risk-focused.

Your company needs its new patient portal app to be HIPAA compliant. When should security requirements be primarily addressed?
a. During the coding process to ensure secure coding practices are followed.
b. In the design phase, to architect privacy and security controls from the start.
c. After development, during a rigorous penetration testing phase.
d. By selecting a cloud vendor that can guarantee HIPAA compliance.

Answer: b. In the design phase, to architect privacy and security controls from the start. Explanation: Compliance is easiest when baked in, not bolted on. While testing, good coding, and vendor choice all matter, early design decisions have the broadest impact.

A critical vulnerability is found in a third-party library used heavily in your product. A patch isn't yet available, but the vendor provides a workaround. What's the MAIN risk to consider?
a. The workaround might introduce unexpected side effects or break existing functionality.
b. The workaround could increase the attack surface of your product.
c. Regulatory compliance might be jeopardized if you deviate from the vendor's standard.
d. The performance overhead of the workaround might negatively impact user experience.

Answer: a. The workaround might introduce unexpected side effects or break existing functionality. Explanation: While all the options are valid concerns, the immediate worry is if the 'fix' causes new problems. Workarounds often focus on the narrow vulnerability, not the holistic impact on your code.

In the Secure Software Development Life Cycle (SSDLC), at what stage should security requirements be defined to ensure the software is designed with security in mind from the beginning?
a. After the first prototype is developed.
b. During the testing phase, before deployment.
c. At the requirements gathering and analysis phase.
d. Post-deployment during the maintenance phase.

Answer: c. At the requirements gathering and analysis phase. Explanation: Security requirements should be defined at the very beginning of the SSDLC, specifically during the requirements gathering and analysis phase. This ensures that security is an integral part of the software design and development process, rather than an afterthought.

When implementing input validation in a web application to mitigate the risk of SQL injection attacks, which approach is MOST effective?
a. Relying on client-side validation using JavaScript.
b. Using prepared statements and parameterized queries.
c. Applying strict file permissions on the web server.
d. Encrypting data before inserting it into the database.

Answer: b. Using prepared statements and parameterized queries. Explanation: Prepared statements and parameterized queries are the most effective approach to prevent SQL injection, as they separate SQL logic from data. This method ensures that user input is treated as data, not executable code, regardless of any malicious content.

In the context of secure coding practices, why is it important to avoid hardcoding sensitive information, such as passwords or cryptographic keys, within the source code?
a. It can lead to performance degradation.
b. It complicates the code deployment process.
c. Such information can be easily extracted by decompiling the application.
d. It violates most coding style guidelines.

Answer: c. Such information can be easily extracted by decompiling the application. Explanation: Hardcoding sensitive information within the source code is insecure because if the application is decompiled, which can be relatively easy, especially for interpreted languages, this sensitive information can be extracted and exploited by attackers.

What principle underlies the concept of 'security by design' in software development?
a. Adding security features only when security issues are reported.
b. Considering security as an additional module rather than an integrated feature.
c. Incorporating security considerations throughout the software development process.
d. Focusing on usability and functionality before addressing security concerns.

Answer: c. Incorporating security considerations throughout the software development process. Explanation: 'Security by design' means that security is considered a fundamental component and is incorporated throughout the entire software development process, from the initial design to deployment, not treated as an add-on or afterthought.

In Agile software development, how should security-related tasks be integrated into the workflow?
a. By conducting a comprehensive security review at the end of each sprint.
b. Security tasks should be treated as separate from the development sprints.
c. Incorporating security user stories and tasks into each sprint backlog.
d. Only addressing security concerns during the final testing phase.

Answer: c. Incorporating security user stories and tasks into each sprint backlog. Explanation: In Agile development, security-related tasks should be integrated into the workflow by including security user stories and tasks in the sprint backlog, ensuring that security is addressed continuously and iteratively throughout the development process.

What is the primary purpose of conducting a threat modeling exercise during the design phase of software development?
a. To document the software's functionality and user interfaces.
b. To identify and assess potential threats and vulnerabilities within the system.
c. To establish a budget for the security components of the project.
d. To choose the programming languages and frameworks to be used.

Answer: b. To identify and assess potential threats and vulnerabilities within the system. Explanation: The primary purpose of threat modeling is to proactively identify and assess potential threats, vulnerabilities, and attack vectors within a system during the design phase, allowing for the implementation of appropriate security controls and mitigations.

When choosing a third-party library or framework for inclusion in a software project, what is the MOST critical security consideration?
a. The popularity and user community size of the library.
b. Compatibility with the current development environment.
c. The library's license and cost.
d. The security posture and history of vulnerabilities in the library.

Answer: d. The security posture and history of vulnerabilities in the library. Explanation: While compatibility, license, and community support are important, the most critical security consideration is the library's security posture, including its history of vulnerabilities and how they have been managed. This helps assess the potential risk the library might introduce into the project.

In ensuring the security of an application's session management, what is a key practice to prevent session hijacking?
a. Implementing long session timeout intervals.
b. Storing session identifiers in user-controllable locations like URLs.
c. Regenerating session IDs after a successful login.
d. Using predictable session IDs to improve user experience.

Answer: c. Regenerating session IDs after a successful login. Explanation: Regenerating session IDs after a successful login is a key practice to prevent session hijacking. This approach ensures that any session identifier a potential attacker may have obtained before login becomes invalid, thus protecting the session from being hijacked.

For a software project that processes sensitive personal data, what is an essential security feature to implement in order to ensure data privacy?
a. Mandatory password changes every 30 days.
b. Data encryption both at rest and in transit.
c. Captcha on every user input form.
d. Daily antivirus scans of the server.

Answer: b. Data encryption both at rest and in transit. Explanation: Encrypting sensitive personal data both at rest and in transit is essential to ensuring its privacy and protection from unauthorized access or interception, in line with best practices and regulatory requirements such as GDPR.

In the development of a mobile application, what security measure is crucial to protect sensitive information stored on the device?
a. Forcing the device to connect only to known Wi-Fi networks.
b. Implementing application sandboxing and least privilege access.
c. Requiring the user to update the application every two weeks.
d. Disabling screenshots within the application.

Answer: b. Implementing application sandboxing and least privilege access. Explanation: Application sandboxing and enforcing the principle of least privilege for the app's access to device resources and data are crucial measures to protect sensitive information stored on the device by isolating the application and minimizing its permissions to only what is necessary for its function.

Your organization is assessing the risk of a potential data breach due to unpatched systems. Subject matter experts disagree on the likelihood of an exploit. Which risk assessment method would be MOST suitable in this situation?
a. Quantitative analysis, assigning numerical values to asset value, threat frequency, and vulnerability severity.
b. The Delphi method, to anonymously gather expert opinions and reach a consensus.
c. Calculating the Annual Loss Expectancy (ALE), using industry-standard loss estimates.
d. A penetration test, to provide a concrete demonstration of the exploitability of the vulnerabilities.

Answer: b. The Delphi method, to anonymously gather expert opinions and reach a consensus. Explanation: Delphi is designed for handling subjective judgments. Quantitative analysis and ALE rely on hard numbers, which aren't available, and a pen test, while valuable, doesn't directly resolve the likelihood disagreement.

You're calculating the ALE for a customer database breach. The asset value is $1 million, an exploit is deemed to have a 20% chance of success each year, and security controls would reduce the loss by 50%. What is the ALE?
a. $100,000
b. $200,000
c. $500,000
d. $1,000,000

Answer: a. $100,000. Explanation: ALE = Asset Value ($1M) * Exploit Likelihood (0.2) * Exposure Factor (0.5) = $100,000

A quantitative risk assessment report highlights a specific server compromise could cost the company $250,000. However, this doesn't consider the harder-to-measure reputational damage. What key concept does this illustrate?
a. Quantitative analysis fails to account for intangible risks, and should be supplemented.
b. The chosen exposure factor (EF) for the calculation likely was too low.
c. Single loss expectancy (SLE) should have been calculated alongside ALE.
d. Risk assessments are inherently unreliable and should not be used for decision-making.

Answer: a. Quantitative analysis fails to account for intangible risks, and should be supplemented. Explanation: It demonstrates the strength and weakness of quantitative methods – they're precise on what you CAN quantify, but not the whole picture.

During a risk assessment brainstorming session, a few senior leaders dominate the conversation, silencing others. Which risk is MOST likely to be underestimated as a result?
a. Risks associated with commonly exploited vulnerabilities.
b. Risks arising from zero-day attacks.
c. Risks that the senior leadership is personally familiar with.
d. Risks stemming from newer technologies that the senior leaders may be less familiar with.

Answer: d. Risks stemming from newer technologies that the senior leaders may be less familiar with. Explanation: This highlights the danger of groupthink and lack of diverse perspectives. Senior leadership might miss 'blind spots' due to their experience not aligning with the latest threats.

In the context of risk assessment, which method is best suited for identifying and prioritizing risks when expert opinions are diverse and a consensus needs to be reached without direct confrontation?
a. Monte Carlo simulation
b. Checklist
c. Delphi technique
d. Brainstorming

Answer: c. Delphi technique. Explanation: The Delphi technique is designed to gather expert opinions independently and anonymously, with subsequent rounds of questioning leading to a convergence of opinions, making it ideal for reaching a consensus on risk priorities without direct confrontation among experts.

When calculating the Annual Loss Expectancy (ALE) for a specific asset, which two components are essential to determine its value?
a. Asset value (AV) and Exposure Factor (EF)
b. Single Loss Expectancy (SLE) and Annual Rate of Occurrence (ARO)
c. Total Cost of Ownership (TCO) and Residual Risk (RR)
d. Risk Appetite (RA) and Inherent Risk (IR)

Answer: b. Single Loss Expectancy (SLE) and Annual Rate of Occurrence (ARO). Explanation: ALE is calculated by multiplying the Single Loss Expectancy (SLE), which is the financial loss expected from a single risk event, by the Annual Rate of Occurrence (ARO), which is the expected frequency of that event per year.

During a qualitative risk assessment, which technique involves team members adding risks to a list without criticism or immediate discussion, fostering a creative and uninhibited flow of ideas?
a. SWOT analysis
b. Brainstorming
c. Fault tree analysis
d. Root cause analysis

Answer: b. Brainstorming. Explanation: Brainstorming is a technique used in qualitative risk assessments to generate a broad list of risks by encouraging free thinking and creativity among team members, without immediate critique or evaluation, to ensure a comprehensive identification of potential risks.

In a quantitative risk assessment, what is the primary purpose of using the Monte Carlo simulation?
a. To determine the exact financial impact of each identified risk
b. To provide a detailed analysis of root causes for potential risks
c. To simulate a range of possible outcomes to understand the potential variability in risk impact
d. To calculate the minimum investment needed to mitigate all identified risks

Answer: c. To simulate a range of possible outcomes to understand the potential variability in risk impact. Explanation: The Monte Carlo simulation is used in quantitative risk assessments to model the probability of different outcomes in a process that cannot easily be predicted due to the intervention of random variables, helping to understand the variability and uncertainty in risk impact.

When employing the checklist method in a risk assessment, what is a key limitation that should be considered?
a. The high cost and time required to develop comprehensive checklists
b. The potential for overlooking risks not included in the checklist
c. The difficulty in engaging all stakeholders in the checklist creation process
d. The complexity of analyzing results obtained from the checklist

Answer: b. The potential for overlooking risks not included in the checklist. Explanation: While checklists can be an efficient way to identify common and known risks, a key limitation is the potential to overlook or miss out on identifying new, emerging, or unique risks that are not included in the checklist.

In calculating the Single Loss Expectancy (SLE) for a given risk, which two factors must be combined?
a. Threat frequency and control effectiveness
b. Vulnerability and threat capability
c. Asset value (AV) and Exposure Factor (EF)
d. Risk tolerance and impact scale

Answer: c. Asset value (AV) and Exposure Factor (EF). Explanation: SLE is calculated by multiplying the asset value (AV), representing the value of the asset at risk, by the Exposure Factor (EF), which represents the percentage of loss a realized threat could have on the asset.

For an organization with limited resources for conducting a risk assessment, which approach provides a balance between detail and resource expenditure?
a. Full quantitative risk assessment for all assets
b. Qualitative risk assessment using the Delphi technique
c. Semi-quantitative risk assessment
d. Outsourcing the risk assessment to a third party

Answer: c. Semi-quantitative risk assessment. Explanation: A semi-quantitative risk assessment offers a balance by combining elements of both qualitative and quantitative assessments, providing more detail than a purely qualitative approach without the extensive resource requirements of a full quantitative analysis.

When considering the implementation of a new security control, which metric is crucial for determining its cost-effectiveness in terms of risk reduction?
a. Return on Investment (ROI)
b. Total Cost of Ownership (TCO)
c. Risk Acceptance Threshold (RAT)
d. Cost-Benefit Analysis (CBA)

Answer: d. Cost-Benefit Analysis (CBA). Explanation: Cost-Benefit Analysis (CBA) is crucial for evaluating the cost-effectiveness of a new security control by comparing the expected reduction in risk (benefit) against the cost of implementing the control, providing a basis for informed decision-making.

In a risk assessment scenario, how can the "4-2-1" rule be applied to prioritize risks?
a. Assigning a probability score based on a scale of 4, a severity score on a scale of 2, and then adding 1 as a constant to ensure no risk is undervalued.
b. For every 4 qualitative risks identified, quantitatively analyze 2, and ensure at least 1 has a mitigation plan developed.
c. Categorizing risks into 4 major groups, identifying 2 primary risks within each group, and focusing on 1 with the highest impact.
d. Applying the rule to resource allocation: 4 parts to high risks, 2 parts to medium risks, and 1 part to low risks.

Answer: b. For every 4 qualitative risks identified, quantitatively analyze 2, and ensure at least 1 has a mitigation plan developed. Explanation: While the "4-2-1" rule is not a standard risk management principle, in a hypothetical application, it could suggest a prioritization strategy where for every 4 qualitative risks identified, 2 should be analyzed quantitatively, and at least 1 of these should have a detailed mitigation plan, ensuring a focused approach to risk management.

What is the main advantage of using the brainstorming method in qualitative risk assessment over the Delphi technique?
a. Brainstorming allows for immediate discussion and development of ideas, which can lead to faster consensus and idea generation.
b. Brainstorming requires less time and fewer resources to implement compared to the Delphi technique.
c. The Delphi technique is more suited for quantitative risk assessments, making brainstorming the preferred method for qualitative analysis.
d. Brainstorming provides a more structured approach to risk identification, which results in a more comprehensive risk list.

Answer: a. Brainstorming allows for immediate discussion and development of ideas, which can lead to faster consensus and idea generation. Explanation: The main advantage of brainstorming in qualitative risk assessment is the dynamic and interactive nature of the method, which encourages immediate discussion, rapid idea generation, and can often lead to faster consensus among participants, unlike the Delphi technique, which is more structured and iterative, involving rounds of independent and anonymous input.

Your company relies on a sole-source supplier for a crucial component. A risk assessment shows their facility is in an area prone to devastating floods, potentially disrupting your supply chain. Which response is MOST appropriate?
a. Risk transference: Purchase insurance to offset the potential financial losses from a disruption.
b. Risk mitigation: Help the vendor relocate their facility to a less flood-prone area.
c. Risk avoidance: Immediately redesign your product to eliminate the need for this component.
d. Risk acceptance: Accept the risk, as redesign and supplier changes would be too costly.

Answer: a. Risk transference: Purchase insurance to offset the potential financial losses from a disruption. Explanation: You can't control the flood risk itself. Insurance shifts the financial burden, while the other options are overly drastic or don't address the core issue.

An audit reveals that several employees have been repeatedly violating your organization's acceptable use policy for internet browsing. Which risk response strategy is likely the BEST fit?
a. Risk avoidance: Block all non-essential internet access for general employees.
b. Risk mitigation: Implement stricter technical controls and enhance security awareness training.
c. Risk transference: Update your cybersecurity insurance policy to cover losses from employee actions.
d. Risk acceptance: Accept the risk, as occasional policy violations are expected.

Answer: b. Risk mitigation: Implement stricter technical controls and enhance security awareness training. Explanation: Mitigation aims to reduce the likelihood/impact. Avoidance is too extreme, insurance doesn't fix the root cause, and acceptance ignores a fixable problem.

A zero-day vulnerability is discovered in your core business system. Exploitation would cause catastrophic downtime, but the vendor patch has significant known side effects. What's the MAIN factor in deciding whether to accept the risk?
a. The potential financial losses weighed against the cost of patching side-effects.
b. The documented process for obtaining senior management sign-off on risk acceptance.
c. The likelihood of an exploit being developed before the next scheduled patch cycle.
d. The availability of compensating controls that could partially reduce the risk.

Answer: a. The potential financial losses weighed against the cost of patching side-effects. Explanation: This is a classic cost-benefit analysis at the heart of risk decisions. The other factors are important, but secondary to the core business impact.

A moderately severe vulnerability is found in a non-critical internal system. The vendor states they won't provide a patch due to the product being end-of-life. Which risk response might be MOST reasonable?
a. Risk transference: Reassess if it's worth keeping your current cybersecurity insurance coverage.
b. Risk mitigation: Immediately retire the system and migrate to a supported alternative.
c. Risk avoidance: Segment the system from the rest of the network to limit potential impact.
d. Risk acceptance: Document the risk and continue using the system as-is.

Answer: d. Risk acceptance: Document the risk and continue using the system as-is. Explanation: The severity, lack of patch, and non-criticality make this a candidate for acceptance. The other options incur cost or disruption that might be disproportionate to the risk.

When a company decides not to enter a potentially lucrative market due to high cybersecurity risks, which risk response strategy is it employing?
a. Mitigation
b. Acceptance
c. Transference
d. Avoidance

Answer: d. Avoidance. Explanation: Avoidance involves deciding not to proceed with an activity or action that presents too high a risk. In this scenario, the company is avoiding the risk associated with entering a new market that has high cybersecurity threats, thereby eliminating the risk altogether.

A software development firm uses third-party code libraries to speed up its development process. To manage the risk of potential vulnerabilities in these libraries, the firm regularly updates them to the latest versions. This approach is an example of which risk response strategy?
a. Avoidance
b. Mitigation
c. Acceptance
d. Transference

Answer: b. Mitigation. Explanation: Mitigation involves taking steps to reduce the severity or likelihood of a risk. By regularly updating third-party code libraries to address known vulnerabilities, the firm is mitigating the risk of security breaches, not eliminating, transferring, or accepting it.

An organization decides to purchase cyber insurance to cover potential financial losses from data breaches. This approach to managing risk is known as:
a. Avoidance
b. Mitigation
c. Transference
d. Acceptance

Answer: c. Transference. Explanation: Transference involves shifting the potential impact of a risk to a third party. By purchasing cyber insurance, the organization transfers the financial risk associated with data breaches to the insurance company.

After conducting a risk assessment, a small business owner decides that the cost of mitigating the risk of theft of physical assets is higher than the potential loss. The owner decides to take no specific action. This decision is an example of which risk response strategy?
a. Avoidance
b. Mitigation
c. Transference
d. Acceptance

Answer: d. Acceptance. Explanation: Acceptance involves acknowledging a risk and deciding to take no action to mitigate it, often due to cost-benefit analysis results. In this case, the business owner accepts the risk of theft because the cost of mitigation exceeds the potential loss.

When a company outsources its customer service to a third-party provider to manage the risk associated with maintaining a large in-house team, it is applying which risk response strategy?
a. Avoidance
b. Mitigation
c. Transference
d. Acceptance

Answer: c. Transference. Explanation: Outsourcing functions to third-party providers is a form of risk transference where the company shifts some of the operational risks to the outsourcing provider, in this case, the risks associated with maintaining a large customer service team.

In response to the risk of losing critical data due to hardware failure, a company implements a robust backup and recovery plan. This plan includes regular backups and drills to ensure data can be restored quickly. This approach is best described as:
a. Avoidance
b. Mitigation

c. Transference
d. Acceptance

Answer: b. Mitigation. Explanation: Implementing a backup and recovery plan to reduce the impact of data loss due to hardware failure is a mitigation strategy. It doesn't eliminate the risk of hardware failure but reduces the impact on the business by ensuring that data can be restored.

A multinational corporation identifies political instability in a region where it has significant investments. After evaluation, it decides to continue operations but closely monitors the situation. This approach to risk is known as:
a. Avoidance
b. Mitigation
c. Transference
d. Acceptance

Answer: d. Acceptance. Explanation: The decision to continue operations despite recognizing the risk of political instability, with plans to monitor the situation, is an example of risk acceptance. The company is willing to accept the risk without taking significant steps to avoid, mitigate, or transfer it.

For a company operating in an area prone to natural disasters, building a secondary data center in a geographically diverse location is an example of which risk response strategy?
a. Avoidance
b. Mitigation
c. Transference
d. Acceptance

Answer: b. Mitigation. Explanation: Building a secondary data center in a different location to ensure business continuity in the event of a natural disaster is a mitigation strategy. It reduces the risk of total data loss and service disruption by spreading out the risk geographically.

When a technology firm identifies a new regulatory requirement that imposes severe penalties for non-compliance, but chooses to adhere strictly to the regulations to avoid potential fines, this approach is an example of:
a. Avoidance
b. Mitigation
c. Transference
d. Acceptance

Answer: a. Avoidance. Explanation: Strictly adhering to new regulatory requirements to avoid penalties is an avoidance strategy. The firm is avoiding the risk of non-compliance and its associated costs by ensuring that it meets all regulatory standards.

In the context of managing risks associated with a new IT project, the project manager decides to conduct a series of expert interviews to gain insights into potential risks and their impacts. This technique is most closely associated with which risk assessment method?
a. Quantitative analysis
b. Qualitative analysis
c. Delphi method
d. Brainstorming

Answer: c. Delphi method. Explanation: The Delphi method involves soliciting the opinions of experts through a series of rounds of questioning, with the aim of reaching a consensus on a particular issue, in this case, potential project risks. This method is particularly useful for gathering expert insights in a structured manner.

In threat modeling, which STRIDE component is primarily concerned with the unauthorized modification of data?
a. Spoofing
b. Tampering

c. Repudiation
d. Information disclosure

Answer: b. Tampering. Explanation: Tampering refers to the unauthorized alteration of data. It involves modifying data, code, or an application in an unauthorized manner to compromise a system. Threat modeling identifies tampering as a key concern where data integrity is a primary security objective.

Which STRIDE threat is directly addressed by implementing non-repudiation controls in a system?
a. Spoofing
b. Tampering
c. Repudiation
d. Denial of Service

Answer: c. Repudiation. Explanation: Repudiation threats involve denying the performance of an action or transaction. Non-repudiation controls, such as digital signatures, ensure that actions or transactions cannot be denied by the parties involved, thus addressing the repudiation threat.

When evaluating a system for STRIDE threats, which component would be MOST concerned with an attacker gaining higher-level permissions than allocated?
a. Information disclosure
b. Denial of Service
c. Elevation of privilege
d. Spoofing

Answer: c. Elevation of privilege. Explanation: Elevation of privilege occurs when an attacker gains higher-level permissions than intended, allowing unauthorized access to restricted data or functions. This threat is critical in systems where hierarchical access levels are enforced.

In the context of DREAD risk assessment model, which factor assesses the potential total harm that a successful exploit might cause?
a. Damage
b. Reproducibility
c. Exploitability
d. Affected users

Answer: a. Damage. Explanation: The Damage component of the DREAD model assesses the potential impact or harm that could result from a successful exploit, ranging from data loss to system downtime or other operational impacts.

Which STRIDE element focuses on the unauthorized access to confidential information?
a. Tampering
b. Repudiation
c. Information disclosure
d. Elevation of privilege

Answer: c. Information disclosure. Explanation: Information disclosure involves the exposure of information to individuals who are not supposed to have access to it. This STRIDE component is concerned with protecting the confidentiality of data against unauthorized access.

In threat modeling, how does the "Spoofing" threat relate to identity security?
a. It involves altering data to corrupt its integrity.
b. It pertains to unauthorized elevation of user privileges.
c. It concerns the denial of service to legitimate users.
d. It involves impersonating another user or device.

Answer: d. It involves impersonating another user or device. Explanation: Spoofing is a threat where an attacker impersonates another user, device, or resource to gain unauthorized access to systems or information, directly impacting identity security and authentication mechanisms.

When utilizing the DREAD model to assess a vulnerability, which aspect evaluates how easy it is for an attacker to exploit the vulnerability?
a. Damage
b. Reproducibility
c. Exploitability
d. Discoverability

Answer: c. Exploitability. Explanation: Exploitability in the DREAD model assesses the ease with which a vulnerability can be exploited by an attacker. A high exploitability score indicates that an attacker can easily exploit the vulnerability, increasing the risk associated with it.

Considering the Denial of Service (DoS) threat in STRIDE, what is the primary security goal compromised by this threat?
a. Confidentiality
b. Integrity
c. Availability
d. Authenticity

Answer: c. Availability. Explanation: Denial of Service (DoS) attacks primarily compromise the availability of services or resources, making them inaccessible to legitimate users. This threat targets the availability aspect of the CIA (Confidentiality, Integrity, Availability) triad.

In the STRIDE model, which threat is specifically addressed by implementing strong authentication mechanisms and secure communication channels?
a. Spoofing
b. Tampering
c. Repudiation
d. Information disclosure

Answer: a. Spoofing. Explanation: Spoofing threats involve pretending to be something or someone else to gain unauthorized access. Strong authentication mechanisms and secure communication channels help prevent spoofing by ensuring that identities are correctly verified and data in transit is protected.

Within the DREAD model, how is "Affected users" evaluated in terms of risk assessment?
a. By the number of users who have the necessary privileges to exploit the vulnerability.
b. By the total number of users who could potentially be harmed by an exploit.
c. By the percentage of time that the system is vulnerable to attack.
d. By the number of users required to reproduce the vulnerability.

Answer: b. By the total number of users who could potentially be harmed by an exploit. Explanation: "Affected users" in the DREAD model refers to the scope or number of users who could be impacted by an exploit of the vulnerability, assessing the potential breadth of impact on the user base.

An organization is planning to implement a new IT service management system. To ensure alignment with best practices for service management, which framework should be primarily consulted?
a) ISO 27001
b) NIST CSF
c) COBIT
d) ITIL

Answer: d) ITIL. Explanation: ITIL (Information Technology Infrastructure Library) is specifically designed to standardize the selection, planning, delivery, and support of IT services within a business. The goal is to improve efficiency and achieve predictable service levels. The other frameworks, while valuable, have different primary focuses: ISO 27001 focuses on information security management, NIST CSF on cybersecurity, and COBIT on governance and management of enterprise IT.

When assessing the cybersecurity posture of an organization, which framework provides a comprehensive, flexible, and cost-effective approach to managing cybersecurity risk?
a) ISO 27001
b) NIST CSF
c) COBIT
d) ITIL

Answer: b) NIST CSF. Explanation: The NIST Cybersecurity Framework (CSF) is designed to help organizations manage and reduce cybersecurity risk in a comprehensive manner. It provides a flexible and cost-effective approach through its core functions: Identify, Protect, Detect, Respond, and Recover. ISO 27001, while also focused on security, is more prescriptive and focused on establishing and maintaining an ISMS (Information Security Management System). COBIT and ITIL are more focused on governance, management, and service delivery, respectively.

An enterprise wants to establish a governance and management framework for IT that aligns IT goals with strategic business goals. Which framework should they adopt?
a) ISO 27001
b) NIST CSF
c) COBIT
d) ITIL

Answer: c) COBIT. Explanation: COBIT (Control Objectives for Information and Related Technologies) is designed for the governance and management of enterprise IT. It aims to align IT services and technology with business goals, manage risks, and ensure resource optimization. Unlike COBIT, ISO 27001 is focused on information security, NIST CSF on cybersecurity risk management, and ITIL on IT service management.

A company is seeking to certify its Information Security Management System (ISMS) to demonstrate its commitment to information security. Which standard should the company seek certification against?
a) ISO 27001
b) NIST CSF
c) COBIT
d) ITIL

Answer: a) ISO 27001. Explanation: ISO 27001 is the international standard for Information Security Management Systems (ISMS) and provides requirements for establishing, implementing, maintaining, and continuously improving an ISMS. It is the only standard in this list that offers certification to demonstrate compliance and commitment to information security. NIST CSF, COBIT, and ITIL provide guidelines and best practices but do not offer certification for organizations.

In the context of IT service delivery, which framework focuses on best practices for creating value with IT services?
a) ISO 27001
b) NIST CSF
c) COBIT
d) ITIL

Answer: d) ITIL. Explanation: ITIL is focused on IT service management (ITSM) and aims to provide best practices for creating value through IT services. It emphasizes aligning IT services with the needs of businesses and customers, covering service strategy, design, transition, operation, and continual improvement. The other frameworks have different focuses: ISO 27001 on information security management, NIST CSF on cybersecurity risk management, and COBIT on governance of enterprise IT.

For an organization aiming to enhance its cybersecurity defenses by adopting a framework that allows for prioritization and flexibility based on business needs, which would be most appropriate?
a) ISO 27001
b) NIST CSF
c) COBIT
d) ITIL

Answer: b) NIST CSF. Explanation: The NIST Cybersecurity Framework is designed to be flexible and prioritizable, allowing organizations to adapt its practices based on their specific business needs, risk tolerance, and the current cyber threat landscape. It provides a prioritized, flexible, repeatable, and cost-effective approach, unlike the other frameworks which have more rigid guidelines and different primary focuses.

When a multinational corporation seeks to ensure its IT processes and controls are aligned with global standards for effective governance, which framework should it consider?
a) ISO 27001
b) NIST CSF
c) COBIT
d) ITIL

Answer: c) COBIT. Explanation: COBIT is widely recognized for its comprehensive framework for the governance and management of enterprise IT. It provides globally accepted principles, practices, analytical tools, and models designed to help business leaders ensure their IT processes and controls are effective and aligned with their business objectives. COBIT is particularly well-suited for multinational corporations looking for a governance framework that is accepted worldwide.

For a company looking to implement a robust process for continuous improvement in its IT services, which framework offers a structured approach for this purpose?
a) ISO 27001
b) NIST CSF
c) COBIT
d) ITIL

Answer: d) ITIL. Explanation: ITIL includes a continuous service improvement (CSI) stage within its lifecycle approach to IT service management. This stage is dedicated to evaluating and improving IT services and processes over time. ITIL's structured approach to CSI involves the use of methods and metrics to identify and implement improvements in efficiency, effectiveness, and can adapt to changing business needs.

In the event of a cyber incident, which framework provides structured response and recovery guidelines to minimize impact and restore services quickly?
a) ISO 27001
b) NIST CSF
c) COBIT
d) ITIL

Answer: b) NIST CSF. Explanation: The NIST Cybersecurity Framework includes specific categories and guidelines for Respond and Recover functions, which are critical during and after a cybersecurity incident. These guidelines help organizations to have a structured approach to detect incidents, respond effectively, and restore any impaired services or capabilities, thereby minimizing the impact of the incident.

A company focused on aligning its IT investments with business objectives would benefit most from which framework's guidelines?
a) ISO 27001
b) NIST CSF
c) COBIT

d) ITIL

Answer: c) COBIT. Explanation: COBIT is specifically designed to bridge the gap between business risks, control needs, and technical issues. It provides a set of best practices for IT management and governance, focusing on aligning IT investments and activities with business objectives and delivering value to the business. COBIT's framework helps in understanding and managing the risks associated with IT and ensuring that IT supports the overall business goals.

In a large organization, who is primarily responsible for defining the overall direction and strategy of information security?
a. Chief Information Officer (CIO)
b. Chief Information Security Officer (CISO)
c. Data Owners
d. Security Steering Committee

Answer: b. Chief Information Security Officer (CISO). Explanation: The CISO is primarily responsible for defining the overall direction and strategy of information security within an organization. They focus on protecting information assets and managing risks related to information security.

When a new regulatory requirement mandates enhanced data protection measures, who would typically be responsible for ensuring that the IT infrastructure and applications comply with these requirements?
a. Chief Information Officer (CIO)
b. Chief Information Security Officer (CISO)
c. Data Owners
d. Security Steering Committee

Answer: a. Chief Information Officer (CIO). Explanation: The CIO is typically responsible for the overall IT infrastructure and applications and would ensure that they comply with new regulatory requirements, including enhanced data protection measures.

In the event of a data breach involving customer information, who would be responsible for determining the impact of the breach on the data and deciding on the necessary response actions?
a. Chief Information Officer (CIO)
b. Chief Information Security Officer (CISO)
c. Data Owners
d. Security Steering Committee

Answer: c. Data Owners. Explanation: Data Owners are responsible for the data's integrity, confidentiality, and availability. In the event of a data breach, they would determine the impact on the data they own and decide on the necessary response actions.

How does the Security Steering Committee support the CISO in an organization?
a. By providing technical solutions to security problems
b. By offering legal advice on compliance issues
c. By aligning information security strategies with business objectives
d. By directly managing security operations

Answer: c. By aligning information security strategies with business objectives. Explanation: The Security Steering Committee supports the CISO by ensuring that the information security strategies are aligned with the wider business objectives, providing guidance, and facilitating cross-departmental collaboration.

Who is typically responsible for approving access to sensitive systems and ensuring that access rights are reviewed periodically?
a. Chief Information Officer (CIO)
b. Chief Information Security Officer (CISO)

c. Data Owners
d. Security Steering Committee

Answer: c. Data Owners. Explanation: Data Owners are typically responsible for approving access to the systems and data they own and ensuring that access rights are reviewed periodically to maintain security and compliance.

In a scenario where a new technology introduces potential security vulnerabilities, who would be responsible for evaluating the risks and proposing mitigation strategies?
a. Chief Information Officer (CIO)
b. Chief Information Security Officer (CISO)
c. Data Owners
d. Security Steering Committee

Answer: b. Chief Information Security Officer (CISO). Explanation: The CISO is responsible for evaluating risks associated with new technologies and proposing appropriate mitigation strategies to manage those risks effectively.

When developing a new business application, who ensures that security requirements are integrated into the project from the initial phase?
a. Chief Information Officer (CIO)
b. Chief Information Security Officer (CISO)
c. Data Owners
d. Security Steering Committee

Answer: b. Chief Information Security Officer (CISO). Explanation: The CISO ensures that security requirements are integrated into the development of new business applications from the start, in line with the principle of 'security by design.'

Who plays a key role in communicating and enforcing IT security policies across different departments within an organization?
a. Chief Information Officer (CIO)
b. Chief Information Security Officer (CISO)
c. Data Owners
d. Security Steering Committee

Answer: a. Chief Information Officer (CIO). Explanation: The CIO plays a key role in communicating and enforcing IT security policies across different departments, ensuring that the organization's IT strategy aligns with its security policies.

In the process of risk assessment for a new IT project, who would typically be responsible for identifying and classifying the information assets involved?
a. Chief Information Officer (CIO)
b. Chief Information Security Officer (CISO)
c. Data Owners
d. Security Steering Committee

Answer: c. Data Owners. Explanation: Data Owners are responsible for identifying and classifying the information assets involved in a new IT project, as they have the best understanding of the data's value and sensitivity.

If an organization decides to adopt a cloud-based service, who would be responsible for ensuring that the cloud service provider meets the organization's security standards and compliance requirements?
a. Chief Information Officer (CIO)
b. Chief Information Security Officer (CISO)
c. Data Owners
d. Security Steering Committee

Answer: b. Chief Information Security Officer (CISO). Explanation: The CISO is responsible for ensuring that any third-party service providers, including cloud services, meet the organization's security standards and compliance requirements, safeguarding the organization's information assets.

A multinational corporation experiences a significant cyber-attack, leading to a prolonged IT systems outage. To maintain critical business functions during this period, the company should primarily rely on which plan?
a) Business Continuity Plan (BCP)
b) Disaster Recovery Plan (DRP)
c) Continuity of Operations Plan (COOP)
d) Crisis Management Plan (CMP)

Answer: a) Business Continuity Plan (BCP). Explanation: A Business Continuity Plan (BCP) focuses on maintaining and quickly resuming business functions in the event of major disruptions. While a DRP focuses specifically on IT restoration and a COOP on maintaining essential functions, a BCP provides a comprehensive approach to ensuring the entire business can continue to operate during crises such as significant cyber-attacks.

During an unexpected natural disaster, the IT infrastructure of a company is severely damaged. Which plan is primarily concerned with the restoration of IT systems and data?
a) Business Continuity Plan (BCP)
b) Disaster Recovery Plan (DRP)
c) Continuity of Operations Plan (COOP)
d) Incident Response Plan (IRP)

Answer: b) Disaster Recovery Plan (DRP). Explanation: The Disaster Recovery Plan (DRP) is specifically designed to restore IT systems, applications, and data after a disaster. Unlike BCP, which covers the whole business, or COOP, which focuses on essential functions, DRP is centered on IT restoration to support business recovery efforts.

In the event of a government agency being temporarily unable to access its primary facilities, which plan ensures that critical government functions continue?
a) Business Continuity Plan (BCP)
b) Disaster Recovery Plan (DRP)
c) Continuity of Operations Plan (COOP)
d) Emergency Response Plan (ERP)

Answer: c) Continuity of Operations Plan (COOP). Explanation: The Continuity of Operations Plan (COOP) is specifically designed for government agencies and organizations to ensure that essential functions continue during a wide range of emergencies, including situations where primary facilities are inaccessible.

A large financial institution has identified several critical business processes that must be maintained in the event of any disruption. The plan that outlines strategies for risk mitigation and recovery for these processes is known as:
a) Business Continuity Plan (BCP)
b) Disaster Recovery Plan (DRP)
c) Continuity of Operations Plan (COOP)
d) Operational Resilience Plan (ORP)

Answer: a) Business Continuity Plan (BCP). Explanation: The Business Continuity Plan (BCP) identifies critical business processes and outlines strategies for risk mitigation and rapid recovery in the event of a disruption, ensuring the business can maintain essential functions.

Following a data breach, a company needs to recover lost data and restore system functions. Which plan provides detailed instructions for this recovery process?
a) Business Continuity Plan (BCP)
b) Disaster Recovery Plan (DRP)
c) Continuity of Operations Plan (COOP)
d) Cyber Incident Response Plan (CIRP)

Answer: b) Disaster Recovery Plan (DRP). Explanation: The Disaster Recovery Plan (DRP) includes detailed instructions for recovering lost data and restoring system functions after an IT-specific disruption, such as a data breach, focusing on IT infrastructure and applications.

An organization's critical operations are identified to ensure they can be sustained during a wide range of emergencies. This identification is most likely part of which planning process?
a) Business Continuity Planning
b) Disaster Recovery Planning
c) Continuity of Operations Planning
d) Crisis Communications Planning

Answer: c) Continuity of Operations Planning. Explanation: Continuity of Operations Planning (COOP) involves identifying critical operations and functions that must be sustained during various emergencies, ensuring that essential services and functions remain available.

In the wake of a disruptive event that halts business operations, the immediate plan that encompasses management decision-making, communication, and stakeholder coordination is known as:
a) Business Continuity Plan (BCP)
b) Disaster Recovery Plan (DRP)
c) Continuity of Operations Plan (COOP)
d) Crisis Management Plan (CMP)

Answer: d) Crisis Management Plan (CMP). Explanation: The Crisis Management Plan (CMP) is designed for immediate response to a disruptive event, focusing on management decision-making, communication, and coordination among stakeholders to manage the situation effectively.

For an online retailer, ensuring the availability of its e-commerce platform is paramount. In the event of a system failure, which plan details the technical steps to restore online services?
a) Business Continuity Plan (BCP)
b) Disaster Recovery Plan (DRP)
c) Continuity of Operations Plan (COOP)
d) Technology Recovery Plan (TRP)

Answer: b) Disaster Recovery Plan (DRP). Explanation: The Disaster Recovery Plan (DRP) for an online retailer would detail the technical steps necessary to restore online services, focusing on IT systems and data essential to resuming e-commerce operations.

A healthcare provider must maintain patient care services even when its main facility is unavailable. This requirement would be addressed in a:
a) Business Continuity Plan (BCP)
b) Disaster Recovery Plan (DRP)
c) Continuity of Operations Plan (COOP)
d) Health and Safety Plan (HSP)

Answer: c) Continuity of Operations Plan (COOP). Explanation: For a healthcare provider, a Continuity of Operations Plan (COOP) would outline how to maintain patient care services in the event the main facility is unavailable, ensuring that essential healthcare functions continue.

In planning for potential disruptions, a company outlines the roles and responsibilities of its employees to ensure business operations can continue with minimal impact. This planning is part of the:
a) Business Continuity Plan (BCP)
b) Disaster Recovery Plan (DRP)
c) Continuity of Operations Plan (COOP)
d) Human Resources Continuity Plan (HRCP)

Answer: a) Business Continuity Plan (BCP). Explanation: The Business Continuity Plan (BCP) includes outlining roles and responsibilities of employees to ensure that business operations can continue with minimal impact during and after a disruption, focusing on the resilience of the entire business.

In a Business Impact Analysis (BIA), why is it crucial to identify critical assets?
a. To ensure all assets receive equal investment in security controls.
b. To prioritize recovery efforts based on the assets' importance to business operations.
c. To calculate the total asset value for insurance purposes.
d. To facilitate the disposal of redundant assets.

Answer: b. To prioritize recovery efforts based on the assets' importance to business operations. Explanation: Identifying critical assets in a BIA is essential for prioritizing recovery efforts, ensuring that the most important assets for maintaining key business operations are restored first in the event of a disruption, thereby minimizing the impact on the business.

What does the Recovery Time Objective (RTO) represent in a business continuity plan?
a. The maximum duration for which data might be lost from an IT service due to a major incident.
b. The total length of time a service interruption is acceptable, including recovery operations.
c. The targeted duration within which business functions or IT services must be restored after a disruption.
d. The window of time available for conducting post-incident reviews after recovery.

Answer: c. The targeted duration within which business functions or IT services must be restored after a disruption. Explanation: RTO is a key goal in business continuity and disaster recovery planning, representing the targeted duration within which a business process must be restored after a disruption to avoid unacceptable consequences.

How does the Recovery Point Objective (RPO) differ from the Recovery Time Objective (RTO)?
a. RPO focuses on the age of files that must be recovered from backup storage for normal operations to resume.
b. RPO is the maximum tolerable duration that a system can be offline, whereas RTO is the maximum allowable data loss.
c. RPO describes the minimum frequency of backups, while RTO outlines the timeline for disaster recovery activities.
d. RPO and RTO are interchangeable terms used in disaster recovery planning.

Answer: a. RPO focuses on the age of files that must be recovered from backup storage for normal operations to resume. Explanation: The Recovery Point Objective (RPO) defines the maximum age of files that must be recovered from backup storage for normal operations to resume without significant loss. Unlike RTO, which focuses on time to recover, RPO focuses on the point in time to which data must be restored.

What is the Maximum Tolerable Period of Disruption (MTPD) in the context of a Business Impact Analysis?
a. The maximum time that an organization can sustain financial losses before it must declare bankruptcy.
b. The longest period a business process can be disrupted without causing irreparable harm to the business.
c. The duration for which IT systems can be down before a backup site must be activated.
d. The time it takes for a business to fully recover to its pre-disruption state and operational capacity.

Answer: b. The longest period a business process can be disrupted without causing irreparable harm to the business. Explanation: MTPD represents the maximum duration that a business process can be disrupted before causing unacceptable damage or consequences to the organization, beyond which the viability of the business might be at risk.

During a BIA, what is the primary reason for assessing the impact of disruptions over different time intervals?
a. To determine the most cost-effective time to purchase insurance.
b. To identify the best times of the year to schedule maintenance and avoid disruptions.
c. To understand how the severity of impact might escalate as the duration of the disruption increases.
d. To schedule staff vacations around the periods of least impact.

Answer: c. To understand how the severity of impact might escalate as the duration of the disruption increases. Explanation: Assessing the impact over different time intervals is crucial to understand how the impact severity might change over time, allowing for more effective prioritization and resource allocation in recovery efforts.

In the context of disaster recovery, why might an organization with a high RTO choose to use a cold site instead of a hot site?
a. Cold sites provide faster recovery speeds.
b. Cold sites are more cost-effective for longer RTOs.
c. Hot sites do not offer customizable recovery options.
d. Cold sites typically have more advanced technology and resources.

Answer: b. Cold sites are more cost-effective for longer RTOs. Explanation: Cold sites, which are essentially office spaces equipped for IT operations but without active hardware and data, offer a more cost-effective recovery solution for organizations with longer RTOs, where immediate resumption of IT services is not critical.

When conducting a BIA, how should the interdependencies between business processes be handled?
a. By assuming all processes are independent to simplify the analysis.
b. By documenting and analyzing the interdependencies to understand cascading effects of disruptions.
c. By focusing only on the processes with the highest revenue impact.
d. By outsourcing the management of interdependent processes to third-party vendors.

Answer: b. By documenting and analyzing the interdependencies to understand cascading effects of disruptions. Explanation: Understanding the interdependencies between business processes is crucial in a BIA to identify cascading effects of disruptions, ensuring that recovery strategies address not just individual processes but also the impact on interconnected processes.

For an online retailer, which factor would be MOST critical when determining the RPO for its transaction database?
a. The average number of transactions per hour.
b. The time it takes to switch to a backup power supply.
c. The frequency of changes to the product catalog.
d. The marketing team's content update schedule.

Answer: a. The average number of transactions per hour. Explanation: For an online retailer, the RPO for the transaction database would be most influenced by the average number of transactions per hour, as this reflects the criticality of data and the potential for loss, guiding how often data backups should occur.

If an organization's MTPD for its customer service operations is 24 hours, what does this imply about its tolerance for disruptions in this area?
a. Customer service operations can be offline for up to a week without significant impact.
b. The organization must be able to resume customer service operations within 24 hours of a disruption to avoid unacceptable consequences.
c. Customer service is not a critical function for the organization.
d. The organization has overestimated the importance of customer service operations.

Answer: b. The organization must be able to resume customer service operations within 24 hours of a disruption to avoid unacceptable consequences. Explanation: An MTPD of 24 hours for customer service operations implies that the organization considers this function critical and must restore it within 24 hours post-disruption to prevent unacceptable business impact or loss.

A cybersecurity analyst is classifying data for a multinational corporation. Which classification should be applied to employee personal information, such as home addresses and social security numbers?
a) Public
b) Internal
c) Confidential
d) Restricted

Answer: c) Confidential. Explanation: Employee personal information, such as home addresses and social security numbers, should be classified as Confidential due to its sensitive nature. This classification helps in applying appropriate safeguards to protect the information from unauthorized access and ensure privacy.

A government document outlines detailed operational capabilities of a newly developed military technology. This document is most likely classified as:
a) Unclassified
b) Confidential
c) Secret
d) Top Secret

Answer: d) Top Secret. Explanation: Documents detailing advanced military technology and operational capabilities are typically classified as Top Secret, the highest level of classification, due to the significant impact on national security if such information were disclosed.

During a security audit, an auditor finds a document on an open file share that details a company's merger plans with financial forecasts. This document should have been classified as:
a) Public
b) Internal Use Only
c) Confidential
d) Restricted

Answer: c) Confidential. Explanation: Documents containing sensitive business information, such as merger plans and financial forecasts, should be classified as Confidential. This classification helps in implementing controls to prevent unauthorized access and protect the company's interests.

An organization's IT department is distributing a new security policy update via email to all employees. This document is best classified as:
a) Public
b) Internal Use Only
c) Confidential
d) Restricted

Answer: b) Internal Use Only. Explanation: Documents like security policy updates that are intended for all employees but not for external distribution should be classified as Internal Use Only. This classification ensures that the information remains within the organization while not being as tightly controlled as Confidential or Restricted information.

A defense contractor is working on a project that involves national security information. The contractor receives a document labeled "NOFORN". This indicates that the document:
a) Can be shared with foreign nationals
b) Cannot be shared with foreign nationals
c) Is declassified and public
d) Requires a specific clearance level to view

Answer: b) Cannot be shared with foreign nationals. Explanation: The "NOFORN" (No Foreign Nationals) marking indicates that the document cannot be shared with foreign nationals. This restriction is applied to sensitive information that could harm national security if disclosed to non-U.S. persons.

A research institution publishes a paper on the findings of a publicly funded study on climate change. This paper is most appropriately classified as:
a) Internal Use Only
b) Confidential
c) Restricted

d) Public

Answer: d) Public. Explanation: Research findings from publicly funded studies, especially on topics like climate change, are typically intended for wide dissemination to advance knowledge and public awareness, making the Public classification most appropriate.

A company's internal newsletter, which includes employee birthdays, work anniversaries, and corporate event photos, should be classified as:
a) Public
b) Internal Use Only
c) Confidential
d) Restricted

Answer: b) Internal Use Only. Explanation: An internal newsletter containing employee-related information and corporate events is intended for company employees and should be classified as Internal Use Only to restrict access to company personnel.

A document containing the minutes of a strategic meeting discussing potential acquisition targets should be classified as:
a) Internal Use Only
b) Confidential
c) Restricted
d) Public

Answer: c) Restricted. Explanation: Minutes from meetings discussing sensitive topics such as potential acquisitions should be classified as Restricted, a higher level of classification than Confidential, to limit access to a select group within the organization who are directly involved in the matter.

An IT administrator is instructed to release a set of anonymized data for academic research. This data set should be classified as:
a) Confidential
b) Restricted
c) Public
d) Internal Use Only

Answer: c) Public. Explanation: Anonymized data sets intended for academic research are often made publicly available to contribute to scientific knowledge and innovation. The Public classification facilitates access by the broader academic and research community.

A project manager is documenting the detailed specifications of a proprietary product development process. This document should be classified as:
a) Public
b) Internal Use Only
c) Confidential
d) Restricted

Answer: d) Restricted. Explanation: Detailed specifications of a proprietary product development process are highly sensitive and should be restricted to prevent competitive disadvantage, making the Restricted classification most appropriate to ensure that access is tightly controlled.

When conducting a tabletop exercise for a disaster recovery plan, what is the PRIMARY objective?
a. To physically test the functionality of backup systems.
b. To validate communication plans among the response team.
c. To discuss hypothetical disaster scenarios and review response procedures.
d. To conduct a full-scale evacuation drill.

Answer: c. To discuss hypothetical disaster scenarios and review response procedures. Explanation: The primary objective of a tabletop exercise is to bring together key personnel to discuss hypothetical disaster scenarios in a structured but informal setting, allowing for the review and validation of response procedures, roles, and responsibilities without the stress of an actual emergency.

In a full-scale simulation of a cyber-attack, what aspect is MOST critical to evaluate?
a. The time it takes for the IT department to become aware of the attack.
b. The effectiveness of the incident response team in isolating the breach.
c. The ability to recover lost data from backups.
d. The impact on customer service during the simulation.

Answer: b. The effectiveness of the incident response team in isolating the breach. Explanation: While all aspects are important, the effectiveness of the incident response team in quickly and effectively isolating the breach is crucial to minimize damage and restore normal operations, making it a critical focus of a full-scale cyber-attack simulation.

During a security walkthrough, what is the main focus?
a. Identifying physical security breaches in the facility.
b. Ensuring all employees understand their roles in the security plan.
c. Verifying the alignment of security practices with policy.
d. Testing the response time of external emergency services.

Answer: c. Verifying the alignment of security practices with policy. Explanation: A security walkthrough primarily focuses on verifying that actual security practices within the organization align with the established policies and procedures, identifying any discrepancies that need to be addressed.

What is the advantage of conducting a tabletop exercise over a full-scale simulation for a business continuity plan?
a. Tabletop exercises are more cost-effective and less disruptive to daily operations.
b. Tabletop exercises provide a more realistic experience of an actual disaster.
c. Tabletop exercises are mandatory for compliance with certain regulations.
d. Tabletop exercises allow for actual testing of physical security controls.

Answer: a. Tabletop exercises are more cost-effective and less disruptive to daily operations. Explanation: Tabletop exercises offer a cost-effective way to review and test business continuity plans without the expense and disruption of a full-scale simulation, making them an efficient tool for identifying issues and improving plans.

In preparing for a tabletop exercise involving a data breach scenario, which material is MOST important to provide to participants beforehand?
a. A detailed script of how the exercise will unfold.
b. Background information on recent real-world data breaches.
c. The current incident response plan and relevant policies.
d. A list of emergency contacts and external resources.

Answer: c. The current incident response plan and relevant policies. Explanation: Providing participants with the current incident response plan and relevant policies beforehand is crucial to ensure they understand the procedures and protocols they will be expected to follow during the exercise, making the discussion more focused and productive.

When evaluating the effectiveness of a disaster recovery plan through a full-scale simulation, which metric is MOST important to measure?
a. The number of participants who attend the simulation.
b. The total cost of conducting the simulation.
c. The recovery time objective (RTO) compliance.
d. The satisfaction level of the simulation participants.

Answer: c. The recovery time objective (RTO) compliance. Explanation: The most important metric to measure during a full-scale simulation of a disaster recovery plan is compliance with the recovery time objective (RTO), as it directly relates to the plan's effectiveness in restoring critical operations within the targeted time frame.

During a security walkthrough, what activity is typically performed?
a. Detailed inspection of firewall and intrusion detection system configurations.
b. Physical inspection of facilities to identify potential security weaknesses.
c. Review of employee web browsing history for policy violations.
d. Simulation of social engineering attacks on staff.

Answer: b. Physical inspection of facilities to identify potential security weaknesses. Explanation: A security walkthrough typically involves a physical inspection of the facilities to visually identify potential security weaknesses, such as unsecured entry points, inadequate lighting, or improper handling of sensitive materials.

How does a tabletop exercise for incident response differ from a technical drill?
a. Tabletop exercises require the actual use of technical tools and systems.
b. Tabletop exercises focus on discussion and decision-making processes rather than technical execution.
c. Technical drills are less formal and require no documentation.
d. Technical drills involve the entire organization, while tabletop exercises are limited to IT staff.

Answer: b. Tabletop exercises focus on discussion and decision-making processes rather than technical execution. Explanation: Tabletop exercises for incident response focus on the discussion of hypothetical scenarios and decision-making processes to evaluate the incident response plan, as opposed to technical drills, which involve the actual use of technical tools and systems to validate specific capabilities.

What is a key objective when conducting a walkthrough of the incident response process with business unit leaders?
a. To test the technical skills of the IT security team.
b. To ensure business unit leaders are familiar with their specific roles and responsibilities in the incident response plan.
c. To evaluate the physical security measures of each business unit's workspace.
d. To assess the business unit leaders' ability to technically mitigate security threats.

Answer: b. To ensure business unit leaders are familiar with their specific roles and responsibilities in the incident response plan. Explanation: The key objective of conducting a walkthrough of the incident response process with business unit leaders is to ensure they understand their specific roles and responsibilities within the plan, fostering a coordinated and effective response to incidents.

In a full-scale simulation of a business continuity plan, what element is essential to test the plan's effectiveness?
a. The ability to conduct business as usual during the simulation.
b. The participation of local emergency services in the exercise.
c. The activation of backup sites and failover systems.
d. The performance of the company's stock during the simulation.

Answer: c. The activation of backup sites and failover systems. Explanation: An essential element to test during a full-scale simulation of a business continuity plan is the activation of backup sites and failover systems, as this directly tests the plan's effectiveness in maintaining critical operations during a disruption.

To secure data at rest in a highly regulated industry, which encryption standard is most appropriate due to its strong security and widespread acceptance?
a) DES
b) 3DES
c) AES
d) Blowfish

Answer: c) AES. Explanation: AES (Advanced Encryption Standard) is widely recognized and accepted in highly regulated industries for its strong security features and efficiency in protecting data at rest. Unlike DES and 3DES, which are older and have known vulnerabilities, AES offers a robust security framework that is more resistant to cryptographic attacks.

For securing data transmissions over the Internet, which encryption method is commonly used due to its ability to support digital signatures and public key infrastructure?
a) AES
b) RSA
c) DES
d) RC4

Answer: b) RSA. Explanation: RSA is widely used for securing data transmissions over the Internet, especially because it supports digital signatures and public key infrastructure (PKI), facilitating secure data exchange and authentication. AES is primarily used for encrypting data at rest, while DES and RC4 are considered less secure for Internet transmissions.

When implementing a Data Loss Prevention (DLP) solution, which feature is crucial for identifying sensitive data across various endpoints in an organization?
a) Content discovery
b) Encryption enforcement
c) Bandwidth management
d) Application whitelisting

Answer: a) Content discovery. Explanation: Content discovery is a crucial feature of DLP solutions, enabling the identification and classification of sensitive data across various endpoints within an organization. This feature helps in effectively applying data protection policies and preventing unauthorized access or data leaks.

In the context of tokenization, which scenario best illustrates its use?
a) Encrypting data stored on a laptop
b) Substituting a credit card number with a unique identifier in a database
c) Securing email communications
d) Authenticating users accessing a secure system

Answer: b) Substituting a credit card number with a unique identifier in a database. Explanation: Tokenization involves substituting sensitive data, such as a credit card number, with a unique identifier or token that has no exploitable value. This process is commonly used in payment systems to enhance security while maintaining the utility of the database for transactions and processing.

For a multinational corporation seeking to protect intellectual property in collaborative environments, which data security control is most effective?
a) Full disk encryption
b) Digital rights management (DRM)
c) Transport Layer Security (TLS)
d) Virtual private network (VPN)

Answer: b) Digital rights management (DRM). Explanation: Digital rights management (DRM) is particularly effective for protecting intellectual property in collaborative environments. DRM controls access to and usage of documents and files, ensuring that intellectual property is only used in authorized manners, even within distributed teams.

In a scenario where legacy systems must be secured, which encryption algorithm presents a viable option due to its compatibility and lower resource requirements?
a) AES
b) RSA
c) 3DES
d) ECC

Answer: c) 3DES. Explanation: 3DES (Triple Data Encryption Standard) offers a balance between maintaining compatibility with legacy systems and providing an acceptable level of security, despite being less efficient than newer algorithms. Its lower resource requirements make it suitable for older hardware or systems that cannot support the computational demands of more advanced encryption methods like AES or RSA.

For ensuring the confidentiality of messages in a peer-to-peer communication system, which of the following is the most suitable option?
a) Symmetric encryption with a shared secret key
b) Asymmetric encryption using a public key infrastructure
c) Hashing with SHA-256
d) Tokenization of message content

Answer: a) Symmetric encryption with a shared secret key. Explanation: In a peer-to-peer communication system, symmetric encryption with a shared secret key is most suitable for ensuring message confidentiality. This method allows for efficient encryption and decryption of messages between peers, assuming the secure exchange of the secret key.

When dealing with high volumes of transaction data requiring quick, reversible obfuscation, which data security method is preferred?
a) Tokenization
b) AES encryption
c) Data masking
d) RSA encryption

Answer: a) Tokenization. Explanation: Tokenization is preferred for high-volume transactions that require quick, reversible obfuscation, as it replaces sensitive data elements with non-sensitive equivalents or tokens. This method is particularly useful in payment processing systems, where it secures data without significantly affecting processing speed.

In the implementation of a secure email system for an organization, which protocol ensures the confidentiality and integrity of the messages?
a) SMTP
b) FTP
c) S/MIME
d) HTTP

Answer: c) S/MIME. Explanation: S/MIME (Secure/Multipurpose Internet Mail Extensions) is specifically designed to ensure the confidentiality and integrity of email messages through encryption and digital signatures. Unlike SMTP, FTP, or HTTP, S/MIME provides the necessary security features for protecting email communications within an organization.

To meet compliance requirements for securing customer data in a cloud storage solution, which approach provides robust encryption for data in transit and at rest?
a) Implementing TLS for data in transit and AES for data at rest
b) Using VPNs for data in transit and disk encryption for data at rest
c) Applying RSA encryption for both data in transit and at rest
d) Utilizing ECC for data in transit and tokenization for data at rest

Answer: a) Implementing TLS for data in transit and AES for data at rest. Explanation: To secure customer data in a cloud storage solution, implementing TLS (Transport Layer Security) for data in transit and AES (Advanced Encryption Standard) for data at rest provides a robust encryption approach. TLS ensures secure data transmission, while AES offers strong encryption for stored data, meeting compliance requirements for data security.

Who is primarily responsible for defining the classifications of data within an organization?
a) Data Custodian

b) Data Owner
c) IT Support Staff
d) End User

Answer: b) Data Owner. Explanation: Data Owners are typically senior members of an organization who have overarching responsibility for specific sets of data. Their responsibilities include defining the classifications of data, which involves determining the sensitivity and criticality of the data they oversee, to ensure appropriate handling and protection measures are implemented.

In a scenario where a new security policy requires the encryption of all stored customer data, who would be responsible for implementing the encryption mechanisms?
a) Data Custodian
b) Data Owner
c) IT Support Staff
d) End User

Answer: a) Data Custodian. Explanation: Data Custodians are responsible for the technical implementation of data protection measures as directed by the data policies and standards defined by Data Owners. In this scenario, implementing encryption mechanisms for stored customer data would fall under the custodian's responsibilities, ensuring the security policy is technically enforced.

When deciding who should have access to a sensitive project's documents, the decision is typically made by the:
a) Data Custodian
b) Data Owner
c) IT Support Staff
d) End User

Answer: b) Data Owner. Explanation: Data Owners have the authority to decide on access rights to the data they oversee, based on the business purpose and the sensitivity of the information. They are responsible for determining who should have access to sensitive project documents, ensuring that access is aligned with organizational policies and the data's classification.

If a database requires a patch to address a new security vulnerability, who would typically ensure that the patch is applied?
a) Data Custodian
b) Data Owner
c) IT Support Staff
d) End User

Answer: a) Data Custodian. Explanation: Data Custodians are tasked with the operational maintenance and protection of data assets, which includes applying necessary patches to databases and other systems to address security vulnerabilities. This role involves the technical management of data to ensure its integrity, confidentiality, and availability.

In the event of a data breach involving customer information, who is primarily responsible for communicating with stakeholders and the public?
a) Data Custodian
b) Data Owner
c) IT Support Staff
d) End User

Answer: b) Data Owner. Explanation: Data Owners, given their overarching responsibility for specific data sets, are typically the ones responsible for communicating with stakeholders and the public in the event of a data breach. Their role involves managing the business aspects of data, including stakeholder communication and compliance with legal and regulatory requirements.

Who would be most directly involved in the routine backup of an organization's data as part of a disaster recovery plan?
a) Data Custodian
b) Data Owner
c) IT Support Staff
d) End User

Answer: a) Data Custodian. Explanation: Data Custodians are responsible for the technical aspects of data management, which includes performing routine backups as part of a disaster recovery plan. This ensures that data can be restored in the event of loss or corruption, maintaining the availability and integrity of the organization's data assets.

In a situation where a new type of sensitive data is being introduced into the organization, who would be responsible for determining the appropriate data handling and protection policies?
a) Data Custodian
b) Data Owner
c) IT Support Staff
d) End User

Answer: b) Data Owner. Explanation: Data Owners are responsible for determining the appropriate handling and protection policies for the data under their purview, based on its classification, sensitivity, and business requirements. This includes new types of sensitive data being introduced into the organization.

Who is typically responsible for conducting regular audits of data access logs to ensure compliance with data access policies?
a) Data Custodian
b) Data Owner
c) IT Support Staff
d) End User

Answer: a) Data Custodian. Explanation: Data Custodians are involved in the operational aspects of data security, which includes conducting regular audits of data access logs to ensure that access to data is in compliance with established policies and procedures. This helps in identifying any unauthorized access or anomalies in data usage.

When a business unit requires additional data analytics capabilities, who would typically authorize the procurement of the necessary software and tools?
a) Data Custodian
b) Data Owner
c) IT Support Staff
d) End User

Answer: b) Data Owner. Explanation: Data Owners have the authority to make business decisions related to the data they oversee, including authorizing the procurement of additional software and tools for data analytics. This decision is based on the business needs and strategic objectives related to the data.

If an organization is transitioning to a new data storage platform, who would be responsible for ensuring that the data is migrated accurately and securely?
a) Data Custodian
b) Data Owner
c) IT Support Staff
d) End User

Answer: a) Data Custodian. Explanation: Data Custodians are responsible for the technical aspects of data management, including the accurate and secure migration of data to new storage platforms. This involves ensuring that data integrity is maintained and that security measures are in place during the migration process.

What is the MOST critical factor to consider when securing data during its creation phase?
a. Implementing access controls to restrict who can create data.
b. Ensuring data is encrypted as soon as it is created.
c. Classifying data based on its sensitivity and value to the organization.
d. Establishing a strong password policy for users who create data.

Answer: c. Classifying data based on its sensitivity and value to the organization. Explanation: Data classification at the point of creation is critical as it dictates the level of protection and handling procedures required based on the data's sensitivity and value. This foundational step ensures appropriate security measures are applied throughout the data's lifecycle.

When considering secure data storage, what is a key practice to protect data at rest?
a. Regularly changing storage access passwords.
b. Implementing strict physical security controls at data storage sites.
c. Encrypting data stored on any physical or virtual storage medium.
d. Using off-site storage facilities for all backup data.

Answer: c. Encrypting data stored on any physical or virtual storage medium. Explanation: Encrypting data at rest is a fundamental security practice that protects the confidentiality and integrity of data stored on any physical or virtual medium, ensuring that even if unauthorized access occurs, the data remains unintelligible and secure.

What is the PRIMARY security concern when handling data in use?
a. Ensuring that data is not accidentally deleted by the user.
b. Protecting data against unauthorized access and modification.
c. Backing up data frequently to prevent loss.
d. Maintaining the physical security of devices accessing the data.

Answer: b. Protecting data against unauthorized access and modification. Explanation: When data is in use, the primary security concern is protecting it from unauthorized access and modification, ensuring that only authorized users can interact with the data and that it remains accurate and reliable.

In the context of secure data disposal, what is the MOST effective method for preventing data recovery from decommissioned hard drives?
a. Deleting all files and emptying the recycle bin.
b. Performing a high-level format of the drive.
c. Degaussing the hard drive to disrupt the magnetic domains.
d. Overwriting the drive multiple times with random data.

Answer: d. Overwriting the drive multiple times with random data. Explanation: Overwriting a hard drive multiple times with random data is a highly effective method of secure disposal, as it ensures that residual data cannot be recovered, even with advanced forensic techniques.

When developing software, what practice ensures security considerations are integrated from the beginning of the software development lifecycle?
a. Conducting a security review only at the final stage of development.
b. Incorporating security requirements in the initial functional specifications.
c. Relying solely on automated security scanning tools during development.
d. Outsourcing security testing to third-party vendors after development.

Answer: b. Incorporating security requirements in the initial functional specifications. Explanation: Integrating security requirements into the initial functional specifications ensures that security considerations are embedded within the software from the outset, promoting a 'security by design' approach throughout the development lifecycle.

For secure data handling, why is it important to establish a clear data retention policy?
a. To ensure sufficient storage is available for new data.

b. To comply with legal and regulatory requirements and minimize risk by not retaining data longer than necessary.
c. To make data recovery easier in case of accidental deletion.
d. To keep track of which employees have accessed the data.

Answer: b. To comply with legal and regulatory requirements and minimize risk by not retaining data longer than necessary. Explanation: A clear data retention policy helps an organization comply with legal and regulatory requirements and minimizes security risks by ensuring that data is not kept longer than necessary, reducing the volume of data susceptible to potential breaches.

What is a key consideration when implementing encryption for data in transit?
a. The physical security of the network infrastructure.
b. The strength of the encryption algorithm and key management practices.
c. The cost of the encryption software.
d. The geographical location of the data endpoints.

Answer: b. The strength of the encryption algorithm and key management practices. Explanation: When securing data in transit, the strength of the encryption algorithm used and the effectiveness of key management practices are crucial to ensuring that data remains confidential and secure as it moves across networks.

During the disposal phase of the data lifecycle, what step is essential to prevent unauthorized disclosure of paper records?
a. Storing paper records in a locked cabinet until disposal.
b. Shredding paper records to a fine, unreadable consistency.
c. Digitizing paper records and deleting the digital copies.
d. Marking paper records as "confidential" before disposal.

Answer: b. Shredding paper records to a fine, unreadable consistency. Explanation: Shredding paper records to a fine, unreadable consistency is essential during the disposal phase to ensure that sensitive information cannot be reconstructed or disclosed unauthorizedly.

When transitioning data from the 'in-use' phase back to storage, what practice ensures its continued protection?
a. Logging off the system where the data was accessed.
b. Re-encrypting the data before it is stored.
c. Performing a virus scan on the data.
d. Updating the data's classification level.

Answer: b. Re-encrypting the data before it is stored. Explanation: Re-encrypting data before returning it to storage ensures that its confidentiality and integrity are maintained, protecting the data against unauthorized access or alteration while at rest.

In the secure handling of data creation within cloud environments, what is a fundamental security control?
a. Ensuring physical access to cloud data centers is restricted.
b. Utilizing cloud services with built-in encryption for data at rest and in transit.
c. Keeping all data on-premises and using the cloud only for processing.
d. Using only private clouds to prevent data from being stored on shared infrastructure.

Answer: b. Utilizing cloud services with built-in encryption for data at rest and in transit. Explanation: Employing cloud services that offer built-in encryption for data at rest and in transit is fundamental to securing data created within cloud environments, ensuring that data is protected throughout its lifecycle, regardless of the underlying infrastructure.

To securely erase data from an SSD, which method is most effective?
a) Degaussing
b) Physical destruction
c) Overwriting with software tools
d) Formatting

Answer: c) Overwriting with software tools. Explanation: For SSDs, overwriting with specialized software tools that are designed for SSD secure erasure is most effective. Degaussing is ineffective on SSDs because they do not store data magnetically. Physical destruction is also effective but may not be practical for all organizational needs. Formatting is not sufficient for secure data erasure as it does not actually overwrite the data.

When decommissioning a magnetic hard drive, which method ensures that the data cannot be reconstructed?
a) Standard formatting
b) Degaussing
c) High-level formatting
d) Simple deletion of files

Answer: b) Degaussing. Explanation: Degaussing is a process that demagnetizes the entire hard drive, effectively destroying all data and making it unrecoverable. Standard formatting, high-level formatting, and simple deletion do not securely erase the data and can potentially be recovered with specialized tools.

For a company needing to dispose of a large volume of magnetic tapes containing sensitive information, the most secure method of ensuring data is irrecoverable is:
a) Overwriting each tape multiple times
b) Degaussing the tapes
c) Deleting files from the tapes
d) Encrypting data on the tapes

Answer: b) Degaussing the tapes. Explanation: Degaussing magnetic tapes will disrupt the magnetic field and destroy the data stored on them, ensuring it is irrecoverable. Overwriting and deleting files may not be as secure because of the potential for data remnants. Encrypting data does not remove existing data but rather secures it, which does not meet the disposal requirement.

In an environment where data security is paramount, and HDDs are being decommissioned, the preferred method to ensure data remanence is not a risk is:
a) Encryption
b) Degaussing followed by physical destruction
c) Low-level formatting
d) Disk defragmentation

Answer: b) Degaussing followed by physical destruction. Explanation: Degaussing will erase the data on HDDs by disrupting the magnetic field, and physical destruction will ensure that the drives are physically incapable of being used again. This combination is the most secure method to prevent data remanence. Encryption, low-level formatting, and disk defragmentation do not guarantee that data is irrecoverable.

For encrypted data on a flash drive, which method is most appropriate to ensure data is securely erased?
a) Degaussing
b) Overwriting with zeros
c) Physical destruction
d) High-level formatting

Answer: c) Physical destruction. Explanation: For flash drives, physical destruction (such as shredding or incineration) is the most reliable method to ensure that encrypted data cannot be recovered, as it completely destroys the physical medium. Degaussing is ineffective on flash drives, overwriting may not reach all areas due to wear leveling, and formatting does not securely erase data.

When dealing with the secure erasure of data from an organization's smartphones, the most effective method is:
a) Factory reset
b) Degaussing
c) Encryption followed by deletion

d) Physical destruction

Answer: a) Factory reset. Explanation: A factory reset on smartphones is designed to erase all data and restore the device to its original state, making it an effective method for secure data erasure. Degaussing is not applicable to smartphones, encryption followed by deletion does not guarantee data erasure, and physical destruction, while effective, is not always practical or necessary.

In a scenario where a company must ensure that backup tapes are securely erased before disposal, the most fail-safe method would be:
a) Overwriting data with multiple passes
b) Degaussing
c) Encrypting data on the tapes
d) Storing the tapes in a secure location

Answer: b) Degaussing. Explanation: Degaussing backup tapes is the most fail-safe method for ensuring that the data is securely erased, as it disrupts the magnetic fields that store the data, making it unrecoverable. Overwriting may not be as secure due to potential data remnants, encrypting does not erase existing data, and storing tapes does not address the need for disposal.

To mitigate the risk of data remanence on a decommissioned server's RAID array, the best approach is:
a) Degaussing the entire array
b) Overwriting all drives individually
c) Physical destruction of the array
d) Removing and storing the drives securely

Answer: b) Overwriting all drives individually. Explanation: Overwriting all drives individually in a RAID array is the best approach to securely erase data and mitigate remanence risks, as it ensures that all areas of each drive are covered. Degaussing may not be effective on all RAID configurations, physical destruction is definitive but not always practical, and simply removing and storing the drives does not erase the data.

For an organization looking to repurpose laptops for new employees, ensuring no data remanence from previous users involves:
a) Performing a standard OS reinstall
b) Using a certified data wiping tool to overwrite the drives
c) Encrypting the hard drive
d) Conducting a quick format

Answer: b) Using a certified data wiping tool to overwrite the drives. Explanation: Using a certified data wiping tool to securely overwrite the drives is the most effective way to ensure no data remanence on laptops being repurposed, as it ensures all previously stored data is irrecoverably removed. A standard OS reinstall, encrypting the hard drive, or conducting a quick format may not be sufficient to remove all data traces.

When ensuring the secure disposal of old photocopier machines with internal storage, the most comprehensive method is:
a) Conducting a factory reset
b) Overwriting the internal storage multiple times
c) Degaussing the internal storage unit
d) Physical destruction of the internal storage unit

Answer: d) Physical destruction of the internal storage unit. Explanation: Physical destruction of the internal storage unit in old photocopier machines is the most comprehensive method to ensure secure disposal and prevent data remanence, as it completely destroys the medium, making data recovery impossible. Factory resets, overwriting, or degaussing may not be as effective due to the complexity and accessibility of internal storage in such devices.

How do Information Rights Management (IRM) solutions primarily protect sensitive documents within an organization?
a. By encrypting the network traffic between document repositories and end-users.
b. By enforcing digital rights and permissions directly embedded within the documents themselves.
c. By using antivirus scanning to prevent unauthorized access to documents.
d. By creating secure backups of all documents at regular intervals.

Answer: b. By enforcing digital rights and permissions directly embedded within the documents themselves. Explanation: IRM solutions protect sensitive information by embedding digital rights and permissions directly into the documents, controlling actions such as viewing, editing, printing, and forwarding, regardless of where the document is stored or who possesses it.

What is a common limitation of Rights Management Technologies when considering collaborative work environments?
a. They can significantly enhance collaboration by streamlining document sharing.
b. They may introduce compatibility issues with common file formats.
c. They can restrict document access too tightly, hindering collaboration and sharing.
d. They are often too complex for end-users to understand and use effectively.

Answer: c. They can restrict document access too tightly, hindering collaboration and sharing. Explanation: While Rights Management Technologies are designed to secure sensitive information, a common limitation is that they can impose overly restrictive controls on documents, potentially hindering collaboration and sharing in environments where teamwork and information exchange are crucial.

In the context of IRM, what does "persistent protection" refer to?
a. The continuous updating of antivirus definitions to protect documents.
b. The protection that remains with the document, no matter where it is stored or sent.
c. The uninterrupted power supply systems that prevent data loss.
d. The regular audits performed to ensure document integrity.

Answer: b. The protection that remains with the document, no matter where it is stored or sent. Explanation: Persistent protection in IRM means that the security measures, such as encryption and access controls, are embedded within the document itself and remain effective regardless of the document's location or how it is transmitted, ensuring continuous protection.

Which scenario best illustrates the use of Rights Management Technologies in a corporate setting?
a. A company encrypts its entire database to prevent data breaches.
b. An employee uses a VPN to access the corporate network remotely.
c. A document containing sensitive project details is set to expire and become unreadable after a project's completion date.
d. A firewall is configured to block all incoming traffic from unknown IP addresses.

Answer: c. A document containing sensitive project details is set to expire and become unreadable after a project's completion date. Explanation: This scenario exemplifies the use of Rights Management Technologies by embedding a policy within a document that controls its lifecycle, such as making it expire and become unreadable after a certain date, which is a key feature of IRM solutions.

What challenge do organizations face when implementing IRM solutions across diverse platforms and devices?
a. Ensuring that all platforms support the encryption algorithms used by the IRM solution.
b. Standardizing document formats across all platforms to ensure compatibility.
c. Managing the different user interfaces presented by the IRM solution on various devices.
d. Ensuring the IRM solution is compatible and functions consistently across all platforms and devices.

Answer: d. Ensuring the IRM solution is compatible and functions consistently across all platforms and devices. Explanation: A significant challenge with IRM implementation is ensuring compatibility and consistent functionality across diverse platforms and devices, which may have different capabilities, operating systems, and application environments.

How can Rights Management Technologies impact an organization's compliance with data protection regulations?
a. By making it difficult to delete data permanently, potentially violating regulations that require data erasure.
b. By automatically generating compliance reports for regulatory bodies.
c. By providing controls that help ensure data is handled in accordance with regulatory requirements.
d. By encrypting data, which is prohibited under certain regulations.

Answer: c. By providing controls that help ensure data is handled in accordance with regulatory requirements. Explanation: Rights Management Technologies can positively impact compliance by offering granular controls over how sensitive data is accessed, used, and shared, helping organizations adhere to data protection regulations by enforcing policies that align with regulatory standards.

When considering IRM for protecting video content, what is a key consideration to ensure effective management and protection?
a. The impact of IRM on video streaming performance and latency.
b. The ability of IRM solutions to generate subtitles automatically.
c. The compatibility of IRM with virtual reality platforms.
d. The ease of integrating IRM with social media platforms for content sharing.

Answer: a. The impact of IRM on video streaming performance and latency. Explanation: A key consideration when applying IRM to video content is its impact on streaming performance and latency. Protecting video content requires balancing security measures with the need for smooth playback and minimal buffering, ensuring that DRM controls do not degrade the user experience.

In deploying an IRM solution, what is a critical factor to ensure user adoption and compliance within an organization?
a. Implementing the most restrictive access controls to ensure maximum security.
b. Ensuring the IRM solution is transparent and minimally intrusive to users' workflows.
c. Mandating extensive training programs for all employees on the technical aspects of IRM.
d. Requiring all documents, regardless of sensitivity, to be protected by the IRM solution.

Answer: b. Ensuring the IRM solution is transparent and minimally intrusive to users' workflows. Explanation: For successful IRM implementation and user compliance, it's crucial that the solution integrates seamlessly with existing workflows, being as transparent and minimally intrusive as possible, to avoid disrupting users' normal activities and productivity.

What is the primary advantage of using IRM solutions for remote work environments?
a. They provide a secure VPN connection for remote workers.
b. They ensure that sensitive information remains protected, regardless of the employee's location.
c. They monitor remote employees' productivity and screen activity.
d. They offer a unified communication platform for all remote workers.

Answer: b. They ensure that sensitive information remains protected, regardless of the employee's location. Explanation: The primary advantage of IRM solutions in remote work environments is their ability to protect sensitive information persistently, ensuring that data remains secure regardless of where employees are working or the devices they use, by enforcing predefined access and usage policies.

Which hashing algorithm is considered broken and unsuitable for secure applications due to its vulnerability to collision attacks?
a) SHA-256
b) SHA-1
c) MD5
d) HMAC-SHA1

Answer: c) MD5. Explanation: MD5 is considered broken and unsuitable for secure applications due to its vulnerability to collision attacks, where two different inputs produce the same hash output. This vulnerability undermines the integrity of the hash function, making it possible to forge documents or data with the same hash value.

When storing passwords in a database, what technique should be used in conjunction with hashing to make it more resistant to rainbow table attacks?
a) Encryption
b) Compression
c) Salting
d) Tokenization

Answer: c) Salting. Explanation: Salting involves adding a unique, random value to each password before it is hashed. This technique prevents attackers from effectively using precomputed tables (rainbow tables) for cracking passwords, as it ensures that even identical passwords will have different hash values due to the unique salt.

Which of the following hashing algorithms offers the highest level of security?
a) MD5
b) SHA-1
c) SHA-256
d) HMAC-MD5

Answer: c) SHA-256. Explanation: SHA-256, part of the SHA-2 family, offers a higher level of security compared to MD5 and SHA-1, due to its larger bit size and resistance to known vulnerabilities such as collision attacks. HMAC-MD5 uses MD5, which is considered insecure.

In the context of digital signatures, what role does hashing play?
a) Encrypts the message to ensure confidentiality
b) Compresses the message to reduce its size
c) Ensures the integrity and authenticity of the message
d) Facilitates the routing of the message over the network

Answer: c) Ensures the integrity and authenticity of the message. Explanation: In digital signatures, hashing is used to create a digest of the message, which is then encrypted with the sender's private key. This ensures the integrity (that the message has not been altered) and authenticity (verification of the sender's identity) of the message.

For a system that requires fast and secure generation of hash values for large data sets, which of the following algorithms would be most suitable?
a) MD5
b) SHA-1
c) SHA-256
d) SHA-3

Answer: d) SHA-3. Explanation: SHA-3 is designed for both speed and security, making it suitable for systems that require fast hash generation for large data sets. It is the latest member of the Secure Hash Algorithm family, providing enhanced security features and efficiency over its predecessors.

When implementing HMAC for message authentication, what key information is required?
a) The public key of the sender
b) The private key of the sender
c) A shared secret key known to both sender and receiver
d) The receiver's public key

Answer: c) A shared secret key known to both sender and receiver. Explanation: HMAC (Hash-based Message Authentication Code) requires a shared secret key known to both the sender and receiver. This key is combined with the message hash to provide authentication and integrity, ensuring that both parties have the same shared secret.

In which scenario is salting ineffective in enhancing security?
a) Storing user passwords in a database
b) Transmitting data over an insecure network
c) Protecting against brute force attacks
d) Defending against dictionary attacks

Answer: b) Transmitting data over an insecure network. Explanation: Salting enhances security by adding a unique value to passwords before hashing, making precomputed attack methods like dictionary attacks less effective. However, salting does not encrypt or protect data during transmission over networks, which requires encryption methods.

What is a primary advantage of using SHA-256 over SHA-1 for hashing sensitive information?
a) Faster computation speeds
b) Reduced storage requirements for hash values
c) Enhanced resistance to collision attacks
d) Improved compression of original data

Answer: c) Enhanced resistance to collision attacks. Explanation: SHA-256, part of the SHA-2 family, provides enhanced resistance to collision attacks compared to SHA-1. Its larger bit size (256 bits versus 160 bits for SHA-1) makes it more computationally difficult for attackers to find two different inputs that produce the same hash output.

For an application requiring both data integrity verification and authentication, which of the following is the most appropriate choice?
a) SHA-256
b) HMAC-SHA256
c) MD5
d) Salting with SHA-1

Answer: b) HMAC-SHA256. Explanation: HMAC-SHA256 provides both data integrity verification and authentication by combining the SHA-256 hash function with a secret key. This makes it suitable for applications requiring assurance that the message has not been altered and is from a verified sender.

In the process of digitally signing a document, where does hashing occur?
a) After the document is encrypted with the sender's private key
b) Before the document is encrypted with the sender's private key
c) After the document is encrypted with the receiver's public key
d) Before the document is encrypted with the receiver's public key

Answer: b) Before the document is encrypted with the sender's private key. Explanation: In digital signing, the document is first hashed, and then the hash is encrypted with the sender's private key. This process ensures that any alteration to the document after signing can be detected, as the hash value will change and not match the decrypted hash value from the digital signature.

What mechanism does the Diffie-Hellman key exchange use to securely establish a symmetric key over an insecure channel?
a. Digital signatures
b. Public-key encryption
c. Hash functions
d. Mathematical properties of discrete logarithms

Answer: d. Mathematical properties of discrete logarithms. Explanation: The Diffie-Hellman key exchange leverages the mathematical properties of discrete logarithms to allow two parties to securely establish a shared symmetric key over an insecure channel, without the need to exchange any secret information directly.

In what scenario is symmetric encryption preferred over asymmetric encryption?
a. When encrypting data stored on a device that remains at rest
b. For digital signature verification
c. During the initial key exchange process in a secure communication session
d. For encrypting large volumes of data due to its faster processing speeds

Answer: d. For encrypting large volumes of data due to its faster processing speeds. Explanation: Symmetric encryption is preferred for encrypting large volumes of data because it generally offers faster processing speeds than asymmetric encryption, making it more suitable for situations where performance and efficiency are critical.

Which of the following is a primary strength of asymmetric encryption?
a. Lower computational overhead compared to symmetric encryption
b. The ability to securely exchange keys over an insecure channel
c. More efficient for encrypting large volumes of data
d. Simpler key management for large networks

Answer: b. The ability to securely exchange keys over an insecure channel. Explanation: A primary strength of asymmetric encryption is its ability to securely exchange keys over an insecure channel, utilizing a pair of keys (public and private) for encryption and decryption, which facilitates secure communications without needing a pre-shared secret.

Why might an organization use symmetric encryption for internal communications but asymmetric encryption for external communications?
a. Symmetric encryption cannot be used for external communications.
b. Asymmetric encryption provides non-repudiation, which is often required for external communications.
c. Symmetric encryption keys are too difficult to distribute to external parties securely.
d. Asymmetric encryption is legally mandated for external communications.

Answer: c. Symmetric encryption keys are too difficult to distribute to external parties securely. Explanation: Organizations might prefer symmetric encryption for its efficiency in internal communications but opt for asymmetric encryption for external communications due to the challenge of securely distributing symmetric keys to external parties.

What is the primary use-case for employing asymmetric encryption in digital signatures?
a. To ensure the confidentiality of the message being signed
b. To verify the integrity and authenticity of the message
c. To encrypt the entire message for secure transmission
d. To increase the speed of the signature verification process

Answer: b. To verify the integrity and authenticity of the message. Explanation: Asymmetric encryption is used in digital signatures primarily to verify the integrity and authenticity of a message. The sender signs the message with their private key, and the recipient uses the sender's public key to verify that the message was indeed sent by the purported sender and has not been altered.

In the context of secure email communication, what role does asymmetric encryption play?
a. It is used to encrypt the body of the email for confidentiality.
b. It facilitates the secure exchange of symmetric keys for encrypting email content.
c. It compresses the email content for faster transmission.
d. It serves as a spam filter by verifying the sender's identity.

Answer: b. It facilitates the secure exchange of symmetric keys for encrypting email content. Explanation: In secure email communication, asymmetric encryption is often used to facilitate the secure exchange of symmetric keys, which are then used to encrypt the actual content of the emails for confidentiality.

How does symmetric encryption differ from asymmetric encryption in terms of key management for large-scale systems?
a. Symmetric encryption requires a public key infrastructure (PKI) for key management.
b. Asymmetric encryption uses a single key for both encryption and decryption, simplifying management.
c. Symmetric encryption requires each pair of communicating parties to share a unique key, complicating key management.
d. Asymmetric encryption requires manual exchange of keys, making it less practical for large systems.

Answer: c. Symmetric encryption requires each pair of communicating parties to share a unique key, complicating key management. Explanation: In symmetric encryption, the need for each pair of communicating parties to have a shared unique key significantly complicates key management in large-scale systems, as the number of required keys grows exponentially with the number of participants.

What advantage does the Diffie-Hellman key exchange offer in a public network environment?
a. It eliminates the need for encryption.
b. It allows for the anonymous exchange of messages.
c. It enables the secure establishment of a shared secret key without prior direct contact.
d. It provides a method for digital signatures and non-repudiation.

Answer: c. It enables the secure establishment of a shared secret key without prior direct contact. Explanation: The Diffie-Hellman key exchange offers the advantage of allowing two parties to securely establish a shared secret key over a public network without needing to have had any prior secure communication or direct contact, facilitating encrypted communications.

When considering the secure transfer of data between a client and a server, why might a hybrid encryption approach be used?
a. To comply with international encryption standards
b. To benefit from the strengths of both symmetric and asymmetric encryption
c. Because symmetric encryption cannot be used over the internet
d. To reduce the dependency on third-party encryption tools

Answer: b. To benefit from the strengths of both symmetric and asymmetric encryption. Explanation: A hybrid encryption approach combines the strengths of both symmetric (speed and efficiency for large data volumes) and asymmetric encryption (secure key exchange over insecure channels), providing a balanced solution for secure data transfer between a client and a server.

Which entity in a PKI infrastructure is responsible for issuing digital certificates to subjects?
a) Registration Authority (RA)
b) Certificate Authority (CA)
c) Online Certificate Status Protocol (OCSP) responder
d) Subscriber

Answer: b) Certificate Authority (CA). Explanation: In a Public Key Infrastructure (PKI) system, the Certificate Authority (CA) is responsible for issuing digital certificates to subjects. The CA verifies the identity of the subject and then issues a certificate that associates the subject's identity with their public key.

When a digital certificate needs to be checked for its current status, which protocol provides a real-time verification response?
a) LDAP
b) HTTPS
c) OCSP
d) S/MIME

Answer: c) OCSP (Online Certificate Status Protocol). Explanation: OCSP provides real-time verification of a digital certificate's status, allowing entities to query an OCSP responder for the current status of a specific certificate, making it a more efficient alternative to downloading and checking entire Certificate Revocation Lists (CRLs).

What is the primary purpose of a Certificate Revocation List (CRL)?
a) To list all active digital certificates
b) To provide encryption keys for certificate holders
c) To list digital certificates that have been revoked
d) To validate the identity of certificate holders

Answer: c) To list digital certificates that have been revoked. Explanation: A Certificate Revocation List (CRL) is used to list digital certificates that have been revoked by the Certificate Authority (CA) before their scheduled expiration date, due to reasons such as compromise, loss of private keys, or change in affiliation.

In PKI, what role does the Registration Authority (RA) play?
a) Issues digital certificates directly to users
b) Validates user identities before certificate issuance
c) Revokes digital certificates when necessary
d) Manages the distribution of public keys

Answer: b) Validates user identities before certificate issuance. Explanation: The Registration Authority (RA) is responsible for validating the identities of entities (users or devices) before a Certificate Authority (CA) issues a digital certificate. The RA does not issue certificates directly but acts as a verifier of the subjects' identities.

Which of the following best describes a self-signed certificate?
a) A certificate signed by a trusted CA
b) A certificate signed by the subject's private key
c) A certificate validated by OCSP
d) A certificate listed in a CRL

Answer: b) A certificate signed by the subject's private key. Explanation: A self-signed certificate is one in which the digital signature is created using the subject's own private key. This means that the subject is vouching for its own identity, which is common in scenarios where a trusted CA is not required or available.

When implementing OCSP in a PKI system, what is a common method to address the privacy concern of the OCSP responder knowing the entire certificate check history?
a) Using OCSP stapling
b) Encrypting OCSP requests
c) Utilizing a CRL instead of OCSP
d) Employing anonymous authentication methods

Answer: a) Using OCSP stapling. Explanation: OCSP stapling is a method where the server obtains a current OCSP response for its own certificate and "staples" it to the TLS/SSL handshake, thus preventing the OCSP responder from seeing individual client requests, enhancing privacy.

Which component of PKI is specifically designed to handle the revocation of digital certificates?
a) Key Recovery Agent
b) OCSP responder
c) CA
d) CRL

Answer: d) CRL (Certificate Revocation List). Explanation: The CRL is a specific component within PKI designed to handle the revocation of digital certificates. It is a list published by the CA that contains all the certificates that have been revoked before their expiration.

In a scenario where immediate revocation of a digital certificate is necessary, which of the following would provide the fastest notification of revocation?
a) Publication of a new CRL
b) An updated entry in an OCSP responder
c) Notification via email to relevant parties
d) Manual distribution of a new certificate

Answer: b) An updated entry in an OCSP responder. Explanation: An OCSP responder provides real-time status checks of digital certificates. Updating the revocation status in an OCSP responder would be the fastest way to notify systems and users of the revocation, compared to the periodic nature of CRL updates or manual communication methods.

In the context of PKI, what is the main advantage of using OCSP over CRLs?
a) OCSP provides encryption while CRLs do not
b) OCSP can revoke certificates, while CRLs cannot
c) OCSP offers real-time certificate status checks
d) OCSP requires less computational power than CRLs

Answer: c) OCSP offers real-time certificate status checks. Explanation: The main advantage of OCSP over CRLs is the ability to offer real-time checks of a certificate's revocation status. This is more efficient and timely compared to CRLs, which may not be updated immediately and can be cumbersome to download and parse for large lists.

For a high-security environment requiring two-factor authentication, what role can digital certificates play?
a) Serve as the sole authentication method
b) Act as one of the two required factors, typically the "something you have" factor
c) Replace the need for a password or PIN
d) Function as a backup authentication method in case primary methods fail

Answer: b) Act as one of the two required factors, typically the "something you have" factor. Explanation: In a two-factor authentication setup, digital certificates stored on a device like a smart card or a mobile device can serve as the "something you have" factor. This, combined with a second factor like a password ("something you know") or biometrics ("something you are"), enhances security by requiring two distinct forms of verification.

In the context of Defense in Depth, which of the following is considered a physical control?
a) Antivirus software
b) Biometric access controls
c) Firewalls
d) Intrusion Detection Systems (IDS)

Answer: b) Biometric access controls. Explanation: Biometric access controls are physical controls that restrict access to facilities and resources based on physical characteristics, such as fingerprints or retina scans. This type of control is a key component of a layered security approach, providing a tangible barrier to unauthorized physical access.

Which layer in a Defense in Depth strategy involves the implementation of network segmentation?
a) Physical layer
b) Application layer
c) Network layer
d) Data layer

Answer: c) Network layer. Explanation: Network segmentation is implemented at the network layer as part of a Defense in Depth strategy. It involves dividing the network into smaller, manageable segments or subnetworks to control traffic flow, reduce the attack surface, and limit the potential impact of security breaches.

For securing sensitive data at rest, which of the following is an effective data layer control in a Defense in Depth approach?
a) Encryption

b) Subnetting
c) VLANs
d) NAC (Network Access Control)

Answer: a) Encryption. Explanation: Encryption is an effective data layer control for securing sensitive data at rest within a Defense in Depth strategy. It ensures that even if data is accessed unauthorizedly, it remains unintelligible without the corresponding decryption key, thus protecting the confidentiality and integrity of the data.

In a Defense in Depth strategy, which control is primarily used to monitor and analyze network traffic for signs of unauthorized or malicious activity?
a) Firewalls
b) Antivirus software
c) Intrusion Detection Systems (IDS)
d) Physical security measures

Answer: c) Intrusion Detection Systems (IDS). Explanation: Intrusion Detection Systems (IDS) are used within a Defense in Depth strategy to monitor and analyze network traffic for signs of unauthorized or malicious activity. IDS can alert administrators to potential security incidents, helping to detect and respond to threats in real-time.

What is the role of security awareness training in a Defense in Depth strategy?
a) To ensure all physical security controls are bypass-proof
b) To enhance the effectiveness of technical controls through educated user behavior
c) To replace the need for application security measures
d) To solely focus on improving network security protocols

Answer: b) To enhance the effectiveness of technical controls through educated user behavior. Explanation: Security awareness training plays a critical role in a Defense in Depth strategy by ensuring that users are informed about security policies, potential threats, and best practices. This education enhances the effectiveness of technical controls by promoting secure user behavior, reducing the risk of security breaches caused by human error.

Which of the following best describes the purpose of implementing redundant security controls in a Defense in Depth strategy?
a) To ensure that each control serves a unique security function with no overlap
b) To reduce the overall cost of the security infrastructure
c) To provide backup controls in case the primary control fails
d) To focus solely on strengthening the perimeter defense

Answer: c) To provide backup controls in case the primary control fails. Explanation: Redundant security controls in a Defense in Depth strategy ensure that if one control fails or is bypassed, other controls are in place to provide continued protection. This redundancy enhances overall security by creating multiple layers of defense against threats.

How does the principle of least privilege support a Defense in Depth strategy?
a) By giving users full access to all system resources by default
b) By restricting user access rights to only those necessary for their roles
c) By implementing a single, strong access control at the network perimeter
d) By focusing exclusively on external threats

Answer: b) By restricting user access rights to only those necessary for their roles. Explanation: The principle of least privilege supports a Defense in Depth strategy by minimizing each user's access rights to only those resources absolutely necessary for their job functions. This reduces the potential impact of a breach by limiting the access an attacker can gain through compromised credentials.

In the context of application security within a Defense in Depth approach, which of the following is a key consideration?
a) Ensuring all users have admin privileges to report security flaws

b) Developing applications without input validation to speed up performance
c) Using secure coding practices to prevent vulnerabilities
d) Relying solely on network security to protect applications

Answer: c) Using secure coding practices to prevent vulnerabilities. Explanation: Secure coding practices are a key consideration for application security within a Defense in Depth approach. These practices help prevent vulnerabilities that could be exploited by attackers, such as SQL injection or cross-site scripting (XSS), ensuring that applications have their own layer of defense in addition to network and other security measures.

When implementing a Defense in Depth strategy, what is the significance of having diverse security technologies?
a) It ensures compliance with international security standards
b) It provides a single point of failure for simplicity
c) It reduces the likelihood that one exploit can bypass multiple layers of defense
d) It guarantees the elimination of all security risks

Answer: c) It reduces the likelihood that one exploit can bypass multiple layers of defense. Explanation: The use of diverse security technologies in a Defense in Depth strategy reduces the likelihood that a single exploit or vulnerability can compromise multiple layers of defense. This diversity ensures that even if one technology is compromised, others will still function effectively to protect the system.

What is the primary benefit of incorporating incident response planning into a Defense in Depth strategy?
a) To eliminate the need for proactive security measures
b) To ensure that incidents are never experienced
c) To provide a structured approach for managing and mitigating incidents
d) To focus solely on recovery of financial losses post-incident

Answer: c) To provide a structured approach for managing and mitigating incidents. Explanation: Incorporating incident response planning into a Defense in Depth strategy provides a structured approach to managing and mitigating security incidents when they occur. This ensures that the organization can quickly respond to, contain, and recover from incidents, minimizing potential damage and disruption.

When implementing the principle of least privilege in a software development environment, what is the BEST practice regarding developer access to production data?
a. Developers should have unrestricted access to production data for debugging purposes.
b. Developers should have read-only access to production data to understand real-world application usage.
c. Developers should not have direct access to production data, and anonymized data sets should be used for testing.
d. Developers should have access to production data but only during off-peak hours to minimize risk.

Answer: c. Developers should not have direct access to production data, and anonymized data sets should be used for testing. Explanation: To adhere to the principle of least privilege and protect sensitive information, developers should not have direct access to production data. Instead, anonymized or sanitized data sets that mimic real-world data should be used for development and testing purposes, minimizing potential data exposure risks.

In a secure network architecture, how should access to sensitive system logs be managed to comply with the principle of least privilege?
a. All IT staff should have access to system logs to ensure transparency and collective responsibility.
b. Only network administrators and security personnel should have access to sensitive system logs based on their specific job requirements.
c. Access to system logs should be open to all employees to facilitate self-auditing and personal data tracking.
d. System logs should be automatically emailed to all stakeholders to ensure immediate action on anomalies.

Answer: b. Only network administrators and security personnel should have access to sensitive system logs based on their specific job requirements. Explanation: Access to sensitive system logs should be strictly controlled and limited to individuals whose job roles specifically require it, such as network administrators and security personnel. This approach is in line with the principle of least privilege, ensuring that only those with a legitimate need can access potentially sensitive or critical information contained within the logs.

How should a financial organization implement the principle of least privilege in relation to its customer service representatives' access to client accounts?
a. Representatives should have full access to all client accounts to efficiently resolve any issues that may arise.
b. Representatives should have access limited to the specific functions and data necessary to service the client's immediate needs.
c. Representatives should only be able to access client accounts after receiving explicit permission from the client for each session.
d. Representatives should have access to client accounts only through a supervisor's oversight and approval for each transaction.

Answer: b. Representatives should have access limited to the specific functions and data necessary to service the client's immediate needs. Explanation: Customer service representatives should have their access rights limited to only those specific functions and pieces of information necessary to assist clients with their immediate needs. This approach minimizes the risk of unauthorized access to sensitive client information and aligns with the principle of least privilege by ensuring representatives can only access information pertinent to their job tasks.

In the context of a hospital information system, how should the principle of least privilege be applied to a nurse's access to patient records?
a. Nurses should have access to all patient records to ensure comprehensive care.
b. Nurses should only have access to the records of patients currently under their care.
c. Nurses should request access from a doctor each time they need to view a patient's record.
d. Nurses should have view-only access to a patient's record, with edits requiring approval from senior medical staff.

Answer: b. Nurses should only have access to the records of patients currently under their care. Explanation: In a hospital information system, applying the principle of least privilege means that nurses should have access only to the records of patients who are currently under their care. This approach ensures that patient data remains confidential and is only accessible by healthcare professionals directly involved in the patient's treatment, thereby minimizing the risk of unauthorized access or data breaches.

For a cloud storage solution used by a corporation, what access control model aligns BEST with the principle of least privilege for managing employee access to files?
a. All employees have full access by default, with restrictions applied as needed.
b. Employees are granted access on a need-to-know basis, with permissions tailored to job functions.
c. Access is granted only to department heads, who then delegate file access to their team members.
d. Employees request access to each file on an ad-hoc basis, with each request reviewed by IT security.

Answer: b. Employees are granted access on a need-to-know basis, with permissions tailored to job functions. Explanation: The most effective way to manage employee access to files in a cloud storage solution, in line with the principle of least privilege, is to grant access based on a need-to-know basis. This means customizing permissions so that they align with each employee's specific job functions, thereby ensuring that individuals can only access information that is necessary for their work.

When configuring a role-based access control (RBAC) system for a new enterprise resource planning (ERP) software, how should roles be defined to adhere to the principle of least privilege?
a. By assigning all users to a general role with broad permissions to ensure operational efficiency.
b. By creating specific roles that correspond to job functions, with permissions limited to the necessary resources for each role.
c. By allowing users to self-assign roles based on their personal assessment of their job needs.
d. By defaulting all users to the most restrictive role and requiring a formal review process for additional access.

Answer: b. By creating specific roles that correspond to job functions, with permissions limited to the necessary resources for each role. Explanation: To adhere to the principle of least privilege in an RBAC system, roles should be defined based on specific job functions, with each role having access only to the resources necessary to perform associated tasks. This approach ensures that users have access only to what they need to do their jobs, minimizing the potential for unauthorized access to sensitive information.

In a scenario where an employee transitions from one department to another, how should the principle of least privilege be maintained regarding system access?
a. The employee retains access to systems from the previous department to maintain institutional knowledge.
b. The employee's access rights should be immediately revised to reflect only those necessary for the new role.
c. Access to new systems should be granted on a provisional basis, with permanent access dependent on performance.
d. System access should be suspended until the employee demonstrates proficiency in the new role.

Answer: b. The employee's access rights should be immediately revised to reflect only those necessary for the new role. Explanation: When an employee transitions to a new department, maintaining the principle of least privilege requires that their access rights be promptly reviewed and adjusted to align with the needs of the new role. This ensures that the employee only has access to systems and information pertinent to their current job responsibilities, minimizing the risk associated with unnecessary access to sensitive or irrelevant systems.

How should a company's IT department apply the principle of least privilege to third-party vendors requiring temporary access to the network for maintenance purposes?
a. Granting vendors unrestricted access to ensure they can resolve any issues without delay.
b. Providing vendors with limited access that is strictly necessary for the maintenance task, with a defined expiration date.
c. Allowing vendors access only under the direct supervision of an IT staff member.
d. Requiring vendors to submit detailed access requests for each system or resource needed, approved on a case-by-case basis.

Answer: b. Providing vendors with limited access that is strictly necessary for the maintenance task, with a defined expiration date.

In an emergency, a fail-safe door lock system will:
a) Remain locked to secure the area.
b) Unlock to allow egress.
c) Lock all exits to prevent entry or exit.
d) Require manual intervention to determine the state.

Answer: b) Unlock to allow egress. Explanation: A fail-safe door lock system is designed to unlock in the event of an emergency, such as a power failure or fire alarm, to ensure that individuals can exit the area safely. This approach prioritizes the safety of individuals by allowing egress.

A data center door equipped with a fail-secure locking mechanism will, in the event of a power outage:
a) Unlock, allowing unrestricted access.
b) Remain locked, securing the data center.
c) Alternate between locked and unlocked states.
d) Unlock only for individuals with override keys.

Answer: b) Remain locked, securing the data center. Explanation: A fail-secure locking mechanism remains locked in the event of a power outage, ensuring that the data center remains secure and access is restricted, even when electrical systems fail.

In a high-security government building, which access control philosophy would likely be applied to the storage room containing classified documents?
a) Fail-safe, to ensure evacuation routes are accessible.
b) Fail-secure, to prevent unauthorized access at all times.
c) Fail-safe, to allow easy access during emergencies.
d) Fail-secure, to allow free egress during power outages.

Answer: b) Fail-secure, to prevent unauthorized access at all times. Explanation: In high-security areas, especially those containing classified documents, a fail-secure system would be employed to ensure that the storage room remains locked and secure, even in the event of a power outage, preventing unauthorized access.

For a hospital's fire exit doors, which access control philosophy is most appropriate?
a) Fail-secure, to control access during emergencies.
b) Fail-safe, to ensure doors unlock for evacuation.
c) Fail-secure, to maintain security at all exit points.
d) Fail-safe, to lock doors and prevent entry from outside.

Answer: b) Fail-safe, to ensure doors unlock for evacuation. Explanation: Hospital fire exit doors would use a fail-safe system to ensure that in the event of an emergency, such as a fire, the doors automatically unlock, allowing for the safe and rapid evacuation of patients, staff, and visitors.

An electronic gate controlling vehicle access to a secure facility fails. Following the fail-secure principle, the gate will:
a) Open automatically to prevent vehicles from being trapped.
b) Remain closed, securing the facility from vehicle access.
c) Cycle through opening and closing to reset itself.
d) Open for emergency vehicles only.

Answer: b) Remain closed, securing the facility from vehicle access. Explanation: Following the fail-secure principle, an electronic gate would remain closed in the event of a failure, ensuring that the facility remains secure by preventing unauthorized vehicle access.

In a smart home security system, a fail-safe mechanism would ensure that in case of a system failure:
a) The alarm remains activated to deter intruders.
b) Doors automatically lock to secure the premises.
c) Doors unlock to allow occupants to exit safely.
d) Security cameras switch to a backup power source.

Answer: c) Doors unlock to allow occupants to exit safely. Explanation: A smart home security system with a fail-safe mechanism would ensure that doors unlock in case of a system failure, prioritizing the occupants' safety by allowing them to exit the premises safely.

A museum implementing a fail-secure system for its artifact display cases would result in the cases:
a) Unlocking in the event of a power failure for easy removal of artifacts.
b) Remaining locked to protect the artifacts, even during a power failure.
c) Cycling through security modes to find the most secure setting.
d) Unlocking only for authorized personnel with manual keys.

Answer: b) Remaining locked to protect the artifacts, even during a power failure. Explanation: A fail-secure system for museum artifact display cases would ensure that the cases remain locked in the event of a power failure, protecting the artifacts from theft or unauthorized access.

For an office building's main entry, employing a fail-safe access control system would mean that during a power outage, the doors would:
a) Remain securely locked to prevent unauthorized entry.
b) Unlock, allowing both entry and exit for safety reasons.
c) Lock, permitting only exit to prevent unauthorized access.
d) Require manual unlocking by security personnel.

Answer: b) Unlock, allowing both entry and exit for safety reasons. Explanation: A fail-safe access control system for an office building's main entry would unlock the doors during a power outage, allowing both entry and exit to ensure the safety of the occupants and facilitate emergency response if needed.

In a scenario where a server room's cooling system is connected to an access control system, a fail-safe approach would ensure that in case of an access control system failure:
a) The server room remains locked to protect sensitive equipment.
b) The cooling system shuts down to prevent damage.
c) The server room door unlocks to allow maintenance access.
d) The cooling system switches to a high-security mode.

Answer: c) The server room door unlocks to allow maintenance access. Explanation: A fail-safe approach in this scenario would unlock the server room door in case of an access control system failure, ensuring that maintenance personnel can access the room to address any issues with the cooling system or servers, preventing overheating and potential damage to sensitive equipment.

For a bank's safe deposit box room, which access control philosophy would be most suitable to ensure the security of customer valuables?
a) Fail-safe, to allow customer access during emergencies.
b) Fail-secure, to keep the room locked if the system fails.
c) Fail-safe, to ensure all boxes unlock during a system failure.
d) Fail-secure, to allow timed access based on banking hours.

Answer: b) Fail-secure, to keep the room locked if the system fails. Explanation: A fail-secure access control philosophy would be most suitable for a bank's safe deposit box room to ensure the security of customer valuables. In the event of a system failure, the room would remain locked, preventing unauthorized access and protecting the valuables.

Which fire suppression system is most suitable for data centers where water damage to equipment must be avoided?
a) Wet pipe sprinkler system
b) Dry pipe sprinkler system
c) Gas-based suppression system (e.g., FM200, Inergen)
d) Water mist system

Answer: c) Gas-based suppression system (e.g., FM200, Inergen). Explanation: Gas-based suppression systems, such as those using FM200 or Inergen, are most suitable for data centers as they extinguish fires without causing damage to electronic equipment, unlike water-based systems that can lead to significant water damage.

For areas containing high-voltage equipment, which fire suppression method is preferred to ensure safety and equipment integrity?
a) Carbon dioxide system
b) Foam system
c) Wet chemical system
d) Portable fire extinguishers

Answer: a) Carbon dioxide system. Explanation: Carbon dioxide fire suppression systems are preferred for areas with high-voltage equipment because CO2 efficiently extinguishes fires by displacing oxygen, without leaving residue or conducting electricity, thus ensuring both safety and equipment integrity.

In a healthcare facility, which type of fire suppression system would be least likely to pose a risk to patients and staff during deployment?
a) Halon system
b) Dry chemical system
c) Clean agent system (e.g., FM200)
d) Water sprinkler system

Answer: c) Clean agent system (e.g., FM200). Explanation: Clean agent systems, like FM200, are designed to suppress fires without posing a significant risk to humans and without leaving a residue. They are suitable for occupied spaces like healthcare facilities, where safety of patients and staff is a priority.

What is the primary purpose of HVAC systems in maintaining data center environmental controls?
a) To reduce the risk of fire by lowering oxygen levels
b) To maintain optimal humidity and temperature for equipment operation
c) To filter out contaminants from the air
d) To provide a comfortable working environment for staff

Answer: b) To maintain optimal humidity and temperature for equipment operation. Explanation: The primary purpose of HVAC systems in data centers is to maintain optimal humidity and temperature levels to ensure the reliable operation of equipment, prevent overheating, and minimize the risk of static electricity buildup.

Which environmental control is essential for preventing EMI/RFI interference in server rooms?
a) Conductive flooring
b) Faraday cage or shielding
c) Ultrasonic humidifiers
d) Raised flooring systems

Answer: b) Faraday cage or shielding. Explanation: Faraday cages or EMI/RFI shielding materials are essential for preventing electromagnetic interference (EMI) or radio frequency interference (RFI) in server rooms. These controls protect sensitive equipment from external electromagnetic fields that could disrupt operations.

When is a water mist system preferred over traditional sprinkler systems for fire suppression?
a) In historic buildings where water damage to artifacts must be minimized
b) In outdoor storage areas with heavy equipment
c) In settings where rapid fire spread is a significant risk
d) In areas with flammable liquid storage

Answer: a) In historic buildings where water damage to artifacts must be minimized. Explanation: Water mist systems are preferred in environments like historic buildings, where it's crucial to minimize water damage to valuable artifacts. These systems use fine water mist to suppress fires effectively while reducing water-related damage.

In designing a server room, why is raised flooring commonly used?
a) To facilitate underfloor cable management and airflow
b) To increase the structural strength of the floor
c) To provide additional insulation and reduce heating costs
d) To prevent pest infestations

Answer: a) To facilitate underfloor cable management and airflow. Explanation: Raised flooring is commonly used in server rooms to facilitate underfloor cable management and airflow. This design allows for efficient cooling of equipment by distributing conditioned air through the raised floor and provides a tidy solution for managing the extensive cabling associated with server equipment.

For a manufacturing facility with high levels of airborne particulates, which HVAC system feature is most critical?
a) High-capacity cooling towers
b) Advanced air filtration systems
c) Humidity control units
d) Energy recovery ventilators

Answer: b) Advanced air filtration systems. Explanation: In manufacturing facilities with high levels of airborne particulates, advanced air filtration systems are critical to remove contaminants from the air, protecting both the equipment and the health of the workers.

In a scenario where a fire suppression system needs to be quickly deactivated to prevent accidental discharge during maintenance, which feature is most important?
a) Manual pull stations
b) Automatic detection sensors
c) Isolation valves
d) Smoke detectors

Answer: c) Isolation valves. Explanation: Isolation valves are crucial in a fire suppression system for quickly deactivating the system to prevent accidental discharge during maintenance. These valves can isolate sections of the system, allowing for safe maintenance without compromising the overall protection.

What is the main advantage of using an inert gas (e.g., Nitrogen, Argon) suppression system in occupied spaces compared to chemical-based systems?
a) Inert gases are more effective in suppressing chemical fires.
b) They do not deplete oxygen levels to dangerously low levels.
c) They leave no residue and are non-toxic to humans.
d) They are less expensive and easier to maintain.

Answer: c) They leave no residue and are non-toxic to humans. Explanation: The main advantage of using inert gas suppression systems in occupied spaces is that they leave no residue and are non-toxic to humans, making them safe for use in areas where people are present. Unlike chemical-based systems, inert gases suppress fires by reducing oxygen levels without the risk of toxic exposure or residue.

When assessing a potential site for a new data center, which factor should be considered to mitigate the risk of natural disasters?
a. Proximity to the city center for ease of access
b. Availability of high-bandwidth network infrastructure
c. Historical data on earthquakes, floods, and hurricanes in the area
d. Local tax incentives for technology companies

Answer: c. Historical data on earthquakes, floods, and hurricanes in the area. Explanation: Assessing historical data on natural disasters such as earthquakes, floods, and hurricanes is crucial in site selection for a data center. This information helps in evaluating the natural disaster risk of the location, which is essential for ensuring the resilience and continuity of operations.

In site selection for a critical infrastructure facility, why is it important to consider crime rates in the potential area?
a. Higher crime rates may indicate a well-connected transportation network
b. Areas with high crime rates often have lower property costs
c. High crime rates can pose a risk to physical security and increase the threat of unauthorized access
d. Crime rates are directly correlated with the availability of skilled labor

Answer: c. High crime rates can pose a risk to physical security and increase the threat of unauthorized access. Explanation: Considering crime rates is important because high crime rates can compromise the physical security of the facility, increasing the risk of theft, vandalism, and unauthorized access, which can threaten the integrity and availability of critical infrastructure.

What is a primary consideration when selecting a site for a new server farm in terms of natural disaster risk?
a. The aesthetic appeal of the surrounding area
b. The site's elevation and floodplain status
c. The proximity to recreational facilities
d. The cultural significance of the location

Answer: b. The site's elevation and floodplain status. Explanation: The site's elevation and floodplain status are critical considerations for a server farm to mitigate the risk of flooding. A location outside of high-risk flood zones reduces the potential for water damage, which is crucial for the reliability and uptime of server operations.

How does the proximity to law enforcement and emergency services impact site selection for high-security facilities?
a. It ensures faster response times in case of emergencies, enhancing overall security
b. It generally increases the cost of utilities and services due to higher taxes
c. It is only a minor consideration compared to aesthetic factors
d. It decreases the site's desirability due to potential noise and disruption

Answer: a. It ensures faster response times in case of emergencies, enhancing overall security. Explanation: Proximity to law enforcement and emergency services is a significant factor in site selection for high-security facilities because it ensures quicker response times during security incidents or emergencies, which can critically enhance the facility's overall security posture.

When evaluating a location for its susceptibility to natural disasters, which resource provides the most comprehensive information for decision-making?
a. Local real estate trends
b. Government-issued natural disaster risk maps
c. Tourist attraction ratings
d. Public transportation schedules

Answer: b. Government-issued natural disaster risk maps. Explanation: Government-issued natural disaster risk maps are invaluable resources for evaluating a location's susceptibility to natural disasters. These maps offer detailed information on historical and potential future risks, aiding in informed site selection to minimize disaster impact.

What role does a site's geological stability play in the selection process for critical operational facilities?
a. It is crucial for determining the aesthetic value of the site
b. Geological stability is important for minimizing the risk of disruptions due to seismic activity
c. It primarily affects the insurance premiums for the property
d. It determines the site's eligibility for certain types of construction permits

Answer: b. Geological stability is important for minimizing the risk of disruptions due to seismic activity. Explanation: Geological stability is a critical factor in selecting sites for critical operational facilities to minimize the risk of disruptions and damage due to seismic activities, such as earthquakes, ensuring the continuity and reliability of operations.

In site selection, why might a lower crime rate area be preferred for data-sensitive operations?
a. Lower crime rates correlate with higher property values, ensuring a prestigious location
b. Areas with lower crime rates typically offer better infrastructure
c. Lower crime rates reduce the risk of physical security breaches and theft of sensitive information
d. They are often closer to technological hubs and innovation centers

Answer: c. Lower crime rates reduce the risk of physical security breaches and theft of sensitive information. Explanation: Choosing an area with lower crime rates for data-sensitive operations is crucial because it reduces the risk of physical security breaches, vandalism, and theft, protecting sensitive information and ensuring operational integrity.

During the site selection process for a new IT hub, what is the significance of evaluating the local infrastructure's resilience to extreme weather events?
a. It is essential for planning outdoor recreational activities for employees
b. Infrastructure resilience is key to ensuring uninterrupted power and network services during extreme weather
c. It primarily affects the architectural design of the building
d. It determines the availability of local talent and skilled workforce

Answer: b. Infrastructure resilience is key to ensuring uninterrupted power and network services during extreme weather. Explanation: Evaluating the resilience of local infrastructure to extreme weather is critical in site selection for an IT hub to ensure that essential services like power and network connectivity remain uninterrupted during adverse weather conditions, maintaining operational continuity.

What impact does a high-risk flood zone designation have on the site selection for a mission-critical data center?
a. It increases the likelihood of receiving government subsidies for development
b. Sites in high-risk flood zones are typically avoided to reduce the risk of water damage and service disruption
c. It is preferred for its cooling potential for data center equipment
d. It has no significant impact as long as insurance is obtainable

Answer: b. Sites in high-risk flood zones are typically avoided to reduce the risk of water damage and service disruption. Explanation: Choosing a site for a mission-critical data center in a high-risk flood zone is generally avoided due to the elevated risk of water damage and potential service disruptions, which can have severe consequences for data integrity and availability.

In Software-Defined Networking (SDN), what is the primary function of the control plane?
a) To forward network traffic based on ACLs
b) To make decisions about where traffic is sent
c) To encrypt data packets for secure transmission
d) To physically transmit data packets over the network

Answer: b) To make decisions about where traffic is sent. Explanation: In SDN, the control plane is responsible for making decisions about where traffic should be sent within the network. It determines the most efficient path for data packets based on various factors, such as network policies and current network conditions, separating the network's brain (control plane) from the muscle (data plane) that forwards the traffic.

Which protocol is commonly used in SDN architectures to communicate between the control plane and the data plane?
a) HTTPS
b) SSH
c) OpenFlow
d) SNMP

Answer: c) OpenFlow. Explanation: OpenFlow is a widely adopted protocol in SDN architectures that provides the interface for communication between the control plane and the data plane. It allows the control plane to direct network devices on the forwarding of packets, enabling more dynamic and programmable network management.

What is a key advantage of separating the control and data planes in an SDN environment?
a) Reduced network latency
b) Enhanced physical security of network devices
c) Centralized network management and control
d) Increased bandwidth availability

Answer: c) Centralized network management and control. Explanation: The separation of control and data planes in SDN allows for centralized network management and control, enabling administrators to programmatically configure, manage, and optimize network resources dynamically. This centralization simplifies network design and operation, providing greater flexibility and efficiency.

In the context of SDN, how does the data plane function?
a) It analyzes network policies to make routing decisions.
b) It serves as the interface for network configuration and management.
c) It forwards data packets based on instructions from the control plane.
d) It encrypts and decrypts traffic for secure data transmission.

Answer: c) It forwards data packets based on instructions from the control plane. Explanation: In SDN, the data plane is responsible for the actual forwarding of data packets across the network based on instructions received from the control plane. Its primary function is to handle the packet forwarding (switching/routing) process, without being involved in the decision-making process.

What role does the SDN controller play in a software-defined network?
a) Acts as the main routing device within the network
b) Functions as the central point of control, making decisions for the data plane
c) Serves as the primary firewall and security device
d) Operates as the main data storage unit for network configurations

Answer: b) Functions as the central point of control, making decisions for the data plane. Explanation: The SDN controller is a critical component in a software-defined network, acting as the central point of control that communicates with the network devices (data plane) to make routing decisions, enforce policies, and manage traffic flow based on a comprehensive view of the network.

How does SDN improve network flexibility and scalability compared to traditional networks?
a) By using physical switches for direct packet routing
b) By enabling manual configuration of each network device
c) By allowing network behavior to be controlled by software
d) By increasing the number of hardware devices in the network

Answer: c) By allowing network behavior to be controlled by software. Explanation: SDN improves network flexibility and scalability by abstracting the control logic to software that can run on various hardware, allowing network behavior to be dynamically adjusted through software controls. This software-centric approach enables rapid adaptation to changing network requirements and easier scalability.

In an SDN architecture, which component is responsible for translating high-level network policies into device configurations?
a) Network switch
b) SDN controller
c) Physical router
d) Application layer

Answer: b) SDN controller. Explanation: The SDN controller is responsible for translating high-level network policies into specific device configurations. It acts as an intermediary between the applications that define network policies and the network devices that execute them, ensuring that the network behaves in accordance with the defined policies.

What is the impact of implementing SDN on network security?
a) Decreases network visibility and control
b) Limits the network's ability to respond to security threats
c) Enhances security through centralized control and dynamic policy enforcement
d) Requires additional physical security measures for network devices

Answer: c) Enhances security through centralized control and dynamic policy enforcement. Explanation: Implementing SDN can enhance network security by providing centralized control over the network, which improves visibility and allows for dynamic policy enforcement. The centralized nature of SDN enables more coherent and agile responses to security threats and anomalies.

When considering disaster recovery in an SDN environment, what capability does the separation of control and data planes offer?
a) Faster physical replacement of damaged hardware
b) Immediate failover to backup data centers without reconfiguration
c) Automatic physical repairs to damaged network infrastructure
d) Enhanced manual control over individual packet routing

Answer: b) Immediate failover to backup data centers without reconfiguration. Explanation: The separation of control and data planes in an SDN environment allows for immediate failover to backup data centers without the need for manual reconfiguration of network devices. The centralized control plane can quickly redirect traffic to alternative paths or sites, enhancing disaster recovery capabilities.

How does SDN address network complexity and operational challenges?
a) By increasing the number of network protocols required
b) Through decentralized management and configuration of network devices
c) By simplifying network management through abstraction and programmability
d) By relying on proprietary hardware solutions for network optimization

Answer: c) By simplifying network management through abstraction and programmability. Explanation: SDN addresses network complexity and operational challenges by abstracting the control logic from the underlying hardware, allowing for simplified, centralized, and programmable network management. This reduces the complexity associated with traditional network configurations and adjustments, making the network more adaptable and easier to manage.

What is a foundational principle of Zero Trust security architecture?
a) Trust but verify all network traffic
b) Never trust, always verify all network traffic
c) Trust internal network traffic by default
d) Only verify external network traffic

Answer: b) Never trust, always verify all network traffic. Explanation: Zero Trust security architecture operates on the principle of "Never trust, always verify." This means that no entity, whether inside or outside the network, is trusted by default, and all access requests must be verified before granting access.

Which component is crucial in a Zero Trust model to enforce dynamic access control policies?
a) Firewall
b) Intrusion Detection System (IDS)
c) Network Access Control (NAC)
d) Policy Engine

Answer: d) Policy Engine. Explanation: In Zero Trust architecture, a Policy Engine is crucial for making real-time decisions on access requests based on the Zero Trust principles. It evaluates the context of each request and enforces dynamic access control policies.

In implementing Zero Trust, what role does microsegmentation play?
a) Decreases network complexity
b) Increases the attack surface
c) Divides the network into smaller, more secure zones
d) Consolidates network resources for easier management

Answer: c) Divides the network into smaller, more secure zones. Explanation: Microsegmentation is a key component in Zero Trust implementation, as it divides the network into smaller, more secure zones. This limits lateral movement within the network, enhancing security by isolating workloads from one another.

Which technology is essential in Zero Trust architecture for verifying user identity and granting access to resources?
a) Virtual Private Network (VPN)
b) Antivirus software
c) Multi-factor Authentication (MFA)
d) Deep Packet Inspection (DPI)

Answer: c) Multi-factor Authentication (MFA). Explanation: Multi-factor Authentication (MFA) is essential in Zero Trust architecture as it provides a method to verify the user's identity beyond just a username and password. MFA requires one or more additional verification factors, which significantly enhances security by ensuring that only authorized users gain access to resources.

In Zero Trust, how is the concept of "least privilege access" applied?

a) By giving users full access to network resources
b) By restricting user access to only the resources necessary for their role
c) By allowing unrestricted access to external networks
d) By providing temporary admin privileges to all users

Answer: b) By restricting user access to only the resources necessary for their role. Explanation: The concept of "least privilege access" in Zero Trust means restricting users' access rights to only the minimum resources necessary for their specific roles. This limits potential damage from breaches or insider threats by reducing the number of resources an attacker can access.

What is a key advantage of implementing Zero Trust in an organization's network?
a) Reduced need for security monitoring
b) Increased trust in internal network traffic
c) Enhanced security posture through continuous verification
d) Simplification of network infrastructure

Answer: c) Enhanced security posture through continuous verification. Explanation: A key advantage of implementing Zero Trust is the enhanced security posture achieved through continuous verification of all access requests, regardless of their origin. This approach ensures that security is maintained continuously and is not reliant on the perimeter defense alone.

In a Zero Trust model, which element is considered a potential threat vector that must be continuously monitored and verified?
a) Only external devices attempting to access the network
b) Only user access requests from within the network
c) Both internal and external access requests and devices
d) Only devices that have previously shown suspicious behavior

Answer: c) Both internal and external access requests and devices. Explanation: In a Zero Trust model, both internal and external access requests and devices are considered potential threat vectors. Continuous monitoring and verification are applied to all access requests, regardless of their origin, to maintain security.

How does Zero Trust architecture typically handle data access and protection?
a) By encrypting all data by default
b) By using a single encryption key for simplicity
c) By applying data access policies based on user role and context
d) By storing all data in a centralized, unencrypted database

Answer: c) By applying data access policies based on user role and context. Explanation: Zero Trust architecture handles data access and protection by applying data access policies that consider the user's role and the context of the access request. This ensures that users can only access the data necessary for their job functions, and sensitive data is protected according to its classification.

What is an essential step in transitioning to a Zero Trust network?
a) Disabling all existing security controls
b) Implementing a perimeter-based security model
c) Conducting a comprehensive inventory of all assets and their flows
d) Centralizing all data processing to reduce complexity

Answer: c) Conducting a comprehensive inventory of all assets and their flows. Explanation: An essential step in transitioning to a Zero Trust network is conducting a comprehensive inventory of all assets and understanding their flows within the network. This knowledge is crucial for implementing effective microsegmentation and access control policies based on the Zero Trust principles.

Which principle underlies the Zero Trust strategy regarding network traffic?

a) Assume breach
b) Trust but verify
c) Open by default
d) Closed unless proven open

Answer: a) Assume breach. Explanation: The Zero Trust strategy is underpinned by the principle of "assume breach," which implies that organizations should operate under the assumption that their defenses have been or could be compromised. This approach emphasizes the need for rigorous verification of all access requests to prevent unauthorized access.

What is the primary purpose of a Demilitarized Zone (DMZ) in network security?
a. To serve as a secure area for the organization's most sensitive data.
b. To act as a buffer zone between the internal network and untrusted external networks, like the internet.
c. To monitor and filter all outgoing organizational traffic.
d. To provide a public Wi-Fi network for guests and visitors.

Answer: b. To act as a buffer zone between the internal network and untrusted external networks, like the internet. Explanation: The primary purpose of a DMZ is to improve security by segregating an area of the network that is exposed to an untrusted network, typically the internet, from the internal, trusted network. It contains external-facing services to reduce the risk of an attacker gaining access to the internal network.

Where is a DMZ commonly placed within a network architecture?
a. Between two firewall devices, with one facing the internal network and the other facing the external network.
b. Directly within the internal network to protect the most critical servers.
c. Outside the external firewall, exposed directly to the internet.
d. Parallel to the internal network, without any firewall protection.

Answer: a. Between two firewall devices, with one facing the internal network and the other facing the external network. Explanation: A DMZ is typically placed between two firewalls: one that separates the DMZ from the internal network and one that separates it from the external network. This placement ensures that traffic to and from the DMZ can be carefully controlled and monitored, providing an additional layer of security.

In a DMZ configuration, which type of server is MOST likely to be placed within the DMZ?
a. A database server containing sensitive customer information.
b. A web server hosting the organization's public website.
c. An internal mail server used exclusively by company employees.
d. A file server containing confidential corporate documents.

Answer: b. A web server hosting the organization's public website. Explanation: Web servers hosting public websites are commonly placed within the DMZ. This allows public access to the website while isolating the server from the internal network, thereby reducing the risk of an external attack reaching the internal, more sensitive areas of the network.

What is a key security feature of a DMZ that helps protect internal networks?
a. The ability to serve as a collection point for internal logs and audit data.
b. The use of strong encryption for data stored within the DMZ.
c. The implementation of strict access controls and monitoring of traffic passing through it.
d. The deployment of antivirus software on all devices within the DMZ.

Answer: c. The implementation of strict access controls and monitoring of traffic passing through it. Explanation: A key security feature of a DMZ is the implementation of strict access controls and the monitoring of all traffic to and from the DMZ. This helps to ensure that only authorized traffic can enter the internal network from the DMZ, and suspicious activity can be detected and responded to promptly.

How does placing servers in a DMZ enhance security for services exposed to the internet?

a. By encrypting all data stored on DMZ servers.
b. By physically isolating these servers from the main organizational network.
c. By ensuring all internet traffic is scanned for malware before reaching the DMZ servers.
d. By limiting the exposure of the internal network to external threats and providing a controlled interface.

Answer: d. By limiting the exposure of the internal network to external threats and providing a controlled interface. Explanation: Placing servers in a DMZ enhances security for internet-exposed services by limiting the exposure of the internal network to external threats. The DMZ acts as a controlled interface for public access, ensuring that direct access to the internal network is not possible from the internet.

What role does a DMZ play in mitigating the risks associated with external web traffic?
a. It completely blocks all external web traffic.
b. It allows unrestricted access to internal resources from the web.
c. It provides a secure area for external web traffic to access hosted services without reaching the internal network.
d. It encrypts external web traffic to prevent eavesdropping.

Answer: c. It provides a secure area for external web traffic to access hosted services without reaching the internal network. Explanation: A DMZ mitigates risks by providing a secure, isolated area where external web traffic can access certain services (like a web server) hosted within the DMZ without having direct access to the internal network. This reduces the risk of external threats reaching the internal network.

When designing a network with a DMZ, what is a common practice for handling incoming internet traffic?
a. Directing all traffic directly to the internal network for efficiency.
b. Filtering traffic through an external firewall before it reaches the DMZ, and then through an internal firewall before it reaches the internal network.
c. Allowing internet traffic to bypass the DMZ to reduce latency.
d. Using a single firewall with no DMZ for simplicity.

Answer: b. Filtering traffic through an external firewall before it reaches the DMZ, and then through an internal firewall before it reaches the internal network. Explanation: A common practice in DMZ design is to filter incoming internet traffic through an external firewall before it enters the DMZ and then through a separate internal firewall before it can access the internal network. This dual-layer firewall configuration adds an extra layer of security.

In the event of a compromise, how does a DMZ help in containing the attack?
a. By automatically redirecting the attack to law enforcement agencies.
b. By ensuring the attack is confined to the DMZ, preventing direct access to the internal network.
c. By shutting down all network operations to stop the attack.
d. By broadcasting the attack details to all users for heightened awareness.

Answer: b. By ensuring the attack is confined to the DMZ, preventing direct access to the internal network. Explanation: In the event of a compromise, a DMZ helps contain the attack within the DMZ itself, due to the isolation and additional firewall protections in place. This prevents the attacker from gaining direct access to the internal network, thereby limiting the potential damage.

For an organization that requires high availability of its public-facing web applications, how might a DMZ be configured?
a. By including redundant DMZ segments with load balancing.
b. By excluding web applications from the DMZ to reduce load.
c. By connecting the web servers directly to the internet without a DMZ.
d. By using a single server in the DMZ to simplify management.

Answer: a. By including redundant DMZ segments with load balancing. Explanation: For high availability of public-facing web applications, a DMZ might be configured with redundant segments and load balancing. This setup ensures that if one server or segment fails, others can take over, maintaining the availability of services without exposing the internal network.

What consideration is important when placing a mail server in a DMZ?
a. Ensuring the mail server has a direct, unfiltered connection to the internet for maximum speed.
b. Configuring the mail server to store only non-sensitive information to mitigate potential data breaches.
c. Isolating the mail server from the rest of the DMZ to prevent access to other services.
d. Allowing all internal users unrestricted access to the mail server in the DMZ.

Answer: b. Configuring the mail server to store only non-sensitive information to mitigate potential data breaches. Explanation: When placing a mail server in a DMZ, it's important to configure it so that it stores only non-sensitive information or minimizes the amount of sensitive information it holds. This approach helps mitigate the impact of potential data breaches, as the server is more exposed to external threats than servers within the internal network.

What is the primary function of the Transport layer in the TCP/IP suite?
a) Routing data packets between networks
b) Providing end-to-end communication services for applications
c) Assigning IP addresses to devices
d) Encrypting data for secure transmission

Answer: b) Providing end-to-end communication services for applications. Explanation: The Transport layer in the TCP/IP suite, which includes protocols like TCP and UDP, is responsible for providing end-to-end communication services for applications. It ensures that data is delivered across the network reliably (in the case of TCP) or quickly (in the case of UDP), handling error correction, flow control, and segmentation or reassembly of data.

In the TCP/IP model, which layer is responsible for determining the best path for data to reach its destination?
a) Application
b) Transport
c) Internet
d) Link

Answer: c) Internet. Explanation: The Internet layer of the TCP/IP suite, primarily through the Internet Protocol (IP), is responsible for routing packets across network boundaries. It determines the best path for data to reach its destination by using IP addresses and routing protocols.

Which field in the TCP header is crucial for ensuring data is reassembled in the correct order at the destination?
a) Checksum
b) Source Port
c) Sequence Number
d) Window Size

Answer: c) Sequence Number. Explanation: The Sequence Number field in the TCP header is crucial for ensuring that data segments are reassembled in the correct order at the destination. Each byte of data in a TCP segment is sequentially numbered, allowing the receiver to reorder segments that may have arrived out of order.

What role does the Application layer in the TCP/IP model play?
a) It provides data encoding and serialization.
b) It offers routing protocols for inter-networking.
c) It facilitates user or software interactions with the network.
d) It manages physical network connections and hardware.

Answer: c) It facilitates user or software interactions with the network. Explanation: The Application layer in the TCP/IP model provides protocols that facilitate user or software interactions with the network, such as HTTP for web browsing, SMTP for email, and FTP for file transfers. This layer is where communication services are provided directly to the application processes.

In the context of the TCP/IP suite, what is the significance of port numbers in the Transport layer?

a) They specify the physical path for data transmission.
b) They identify the sending and receiving applications.
c) They determine the data packet size.
d) They encrypt data for secure transmission.

Answer: b) They identify the sending and receiving applications. Explanation: In the Transport layer of the TCP/IP suite, port numbers are used to identify sending and receiving application processes. This allows multiple applications to use the network simultaneously on a single device.

How does the Link layer of the TCP/IP suite differ from the Network layer?
a) The Link layer controls physical hardware devices, while the Network layer manages logical addressing.
b) The Link layer encrypts data, while the Network layer ensures data integrity.
c) The Link layer is responsible for routing, while the Network layer manages data framing.
d) The Link layer handles application requests, while the Network layer provides user interfaces.

Answer: a) The Link layer controls physical hardware devices, while the Network layer manages logical addressing. Explanation: The Link layer of the TCP/IP suite is responsible for controlling physical hardware devices and media that make up the network, such as Ethernet. In contrast, the Network layer (Internet layer in TCP/IP) manages logical addressing (IP addresses) and routing of packets across different networks.

What is the primary purpose of the Window Size field in the TCP header?
a) To specify the number of bytes that the sender is willing to receive
b) To indicate the sequence number of the next expected byte
c) To verify the integrity of the header and data
d) To identify the maximum transmission unit size

Answer: a) To specify the number of bytes that the sender is willing to receive. Explanation: The Window Size field in the TCP header specifies the size of the sender's receive window, or the number of bytes it is willing to receive and buffer at one time. This is crucial for flow control, allowing the sender to manage the rate of data transmission based on available buffer space.

In a secure web transaction, which Application layer protocol works directly with TCP to provide encrypted communication?
a) HTTP
b) DNS
c) HTTPS
d) FTP

Answer: c) HTTPS. Explanation: HTTPS (Hypertext Transfer Protocol Secure) works directly with TCP at the Transport layer to provide encrypted communication for secure web transactions. It uses SSL/TLS protocols to encrypt the data, ensuring secure transmission between client and server.

For real-time video streaming, which protocol in the Transport layer is preferred for its speed and efficiency?
a) TCP
b) UDP
c) FTP
d) SMTP

Answer: b) UDP. Explanation: For real-time video streaming, the User Datagram Protocol (UDP) in the Transport layer is preferred for its speed and efficiency. Unlike TCP, UDP does not provide mechanisms for error correction, making it faster and more suitable for applications where timely delivery is more critical than perfect accuracy, such as streaming media.

What mechanism is used by the IP protocol in the Internet layer to handle data packets that exceed the maximum transmission unit (MTU) size?

a) Encryption
b) Compression
c) Fragmentation
d) Multiplexing

Answer: c) Fragmentation. Explanation: The IP protocol in the Internet layer uses fragmentation to handle data packets that exceed the maximum transmission unit (MTU) size of the network path. Fragmentation divides the packets into smaller units that can be transmitted independently and then reassembled at the destination.

What is the primary function of Network Access Control (NAC)?
a. To increase the speed of the network
b. To monitor network traffic and optimize performance
c. To enforce security policies on devices attempting to access network resources
d. To provide encryption for data transmitted across the network

Answer: c. To enforce security policies on devices attempting to access network resources. Explanation: NAC's primary function is to enforce security policies on all devices attempting to access network resources. It determines whether a device complies with a set of predefined security rules before allowing access, thereby protecting the network from potentially insecure or compromised devices.

Which NAC deployment method involves assessing a device's security posture before granting network access and continually after access is granted?
a. Agent-less NAC
b. Agent-based NAC
c. 802.1X-based NAC
d. VPN-based NAC

Answer: b. Agent-based NAC. Explanation: Agent-based NAC involves installing software on client devices to assess their security posture both before granting network access and continually thereafter. This allows for real-time compliance checks and ensures devices maintain required security standards to remain on the network.

How does 802.1X NAC enforce access controls at the data link layer?
a. By encrypting data packets transmitted over the network
b. By requiring a firewall at each endpoint
c. By utilizing a RADIUS server to authenticate devices based on credentials or certificates before allowing access to the network
d. By scanning devices for malware before granting network access

Answer: c. By utilizing a RADIUS server to authenticate devices based on credentials or certificates before allowing access to the network. Explanation: 802.1X NAC works at the data link layer by using a RADIUS server to authenticate devices attempting to connect to the network. It requires valid credentials or certificates for the device to be granted access, ensuring only authorized devices can connect.

In what scenario is an agent-less NAC solution most appropriate?
a. In environments with high-speed network requirements
b. When devices on the network cannot have software installed, such as IoT devices
c. For networks that solely rely on virtual private networks (VPNs) for access
d. When all devices on the network are company-owned and managed

Answer: b. When devices on the network cannot have software installed, such as IoT devices. Explanation: Agent-less NAC solutions are most appropriate in environments where devices cannot have software agents installed, such as IoT devices, guest devices, or other non-traditional network endpoints. These solutions assess and control access without the need for client-side software.

What is the key benefit of implementing NAC in a Bring Your Own Device (BYOD) environment?

a. It allows for unrestricted access to network resources for all devices
b. It enables the network to support more devices without upgrading infrastructure
c. It ensures that only devices complying with the organization's security policies can access the network
d. It provides a faster Internet connection for personal devices

Answer: c. It ensures that only devices complying with the organization's security policies can access the network. Explanation: In a BYOD environment, NAC is crucial for ensuring that personal devices comply with the organization's security policies before they can access network resources. This helps to maintain network security despite the diversity and variability of personal devices.

Which feature of NAC allows for the automatic remediation of non-compliant devices?
a. Quarantine network
b. Virtual LAN (VLAN) assignment
c. Guest networking
d. Continuous monitoring

Answer: a. Quarantine network. Explanation: NAC can automatically place non-compliant devices into a quarantine network, where they have limited access and can be remediated, such as by updating software or removing malware, before being allowed onto the main network.

How does NAC contribute to regulatory compliance within an organization?
a. By ensuring all network traffic is encrypted
b. By providing detailed usage statistics for billing purposes
c. By enforcing access policies that align with regulatory requirements for data protection
d. By increasing network bandwidth to support compliance-related tasks

Answer: c. By enforcing access policies that align with regulatory requirements for data protection. Explanation: NAC helps organizations comply with regulatory requirements by enforcing access policies that ensure only devices meeting specific security standards can access sensitive data, thereby contributing to the protection of regulated information.

What role does a RADIUS server play in a NAC implementation utilizing 802.1X for access control?
a. It acts as a firewall to block unauthorized access
b. It serves as the authentication server, verifying user credentials before granting network access
c. It encrypts network traffic between the device and the access point
d. It distributes IP addresses to devices on the network

Answer: b. It serves as the authentication server, verifying user credentials before granting network access. Explanation: In an 802.1X NAC implementation, the RADIUS server functions as the authentication server. It verifies the credentials of devices and users attempting to access the network and communicates with the access control equipment to grant or deny access based on authentication results.

In a network with NAC implemented, what happens when a guest device attempts to connect to the network for the first time?
a. The device is granted immediate full access to the network
b. The device is redirected to a captive portal for registration or compliance checks
c. The device's traffic is automatically encrypted and tunneled through the network
d. The device is assigned a static IP address for monitoring

Answer: b. The device is redirected to a captive portal for registration or compliance checks. Explanation: In networks with NAC, guest devices attempting to connect for the first time are typically redirected to a captive portal. This portal may require the user to register the device, agree to terms of service, or undergo compliance checks before being granted restricted or full access to the network.

What mechanism does BGP (Border Gateway Protocol) use to establish secure and reliable sessions between peers?
a) SSL/TLS encryption

b) MD5 hashing for session authentication
c) SSH tunneling for all BGP sessions
d) IPsec for encrypting BGP messages

Answer: b) MD5 hashing for session authentication. Explanation: BGP uses MD5 hashing for session authentication to establish secure and reliable sessions between peers. This mechanism involves configuring a shared secret key on both ends of the BGP session, which is then used to generate an MD5 hash of BGP messages, ensuring the authenticity and integrity of the communication.

In OSPF (Open Shortest Path First), what feature can be implemented to prevent unauthorized routers from participating in the OSPF domain?
a) Route redistribution
b) Area Border Routers (ABRs)
c) OSPF authentication
d) Link-state advertisements (LSAs) throttling

Answer: c) OSPF authentication. Explanation: OSPF authentication is used to prevent unauthorized routers from participating in the OSPF domain. It can be configured for an OSPF area to ensure that only routers with matching authentication credentials (password or cryptographic key) can exchange OSPF routing information, thereby enhancing the security of the OSPF domain.

How can route hijacking be mitigated in BGP?
a) Implementing route summarization
b) Using Route Origin Authorizations (ROAs) with RPKI
c) Increasing the BGP keepalive timer
d) Deploying additional BGP peers

Answer: b) Using Route Origin Authorizations (ROAs) with RPKI. Explanation: Route hijacking in BGP can be mitigated by using Route Origin Authorizations (ROAs) with the Resource Public Key Infrastructure (RPKI). ROAs are cryptographically signed objects that state which ASes (Autonomous Systems) are authorized to originate certain IP prefixes. This helps to ensure that only legitimate ASes can announce routes for specific IP address blocks, thereby preventing unauthorized route announcements.

Which feature in OSPF ensures that routing information is not accepted from unknown sources?
a) Stub areas
b) Demand circuits
c) OSPF flooding mechanism
d) Neighbor authentication

Answer: d) Neighbor authentication. Explanation: Neighbor authentication in OSPF ensures that routing information is not accepted from unknown sources. By configuring OSPF neighbor authentication, routers must authenticate each other before exchanging routing information, which helps to secure the OSPF domain against unauthorized access and route information tampering.

In the context of secure routing, what is the primary purpose of implementing BGP session hijacking protection mechanisms?
a) To reduce the convergence time of BGP
b) To prevent unauthorized ASes from establishing BGP sessions
c) To ensure optimal routing paths
d) To minimize the impact of DDoS attacks

Answer: b) To prevent unauthorized ASes from establishing BGP sessions. Explanation: The primary purpose of implementing BGP session hijacking protection mechanisms is to prevent unauthorized Autonomous Systems (ASes) from establishing BGP sessions with legitimate ASes. This is crucial for maintaining the integrity of BGP routing tables and ensuring that traffic is not maliciously rerouted or intercepted.

What OSPF security feature can protect against malicious or accidental flooding of LSAs (Link State Advertisements)?
a) LSA throttling
b) Area partitioning
c) SPF algorithm tuning
d) Virtual links

Answer: a) LSA throttling. Explanation: LSA throttling in OSPF is a security feature that can protect against malicious or accidental flooding of Link State Advertisements. By controlling the generation, reception, and processing rates of LSAs, OSPF routers can prevent excessive LSA traffic from overwhelming the network or being used as a vector for attacks.

How does implementing BGP prefix filtering contribute to secure routing?
a) By accelerating BGP convergence
b) By preventing the propagation of incorrect routing information
c) By encrypting BGP update messages
d) By reducing the size of the BGP table

Answer: b) By preventing the propagation of incorrect routing information. Explanation: Implementing BGP prefix filtering contributes to secure routing by preventing the propagation of incorrect or unauthorized routing information. Prefix filters are configured on BGP routers to allow only legitimate prefixes to be advertised or received from peers, thus helping to prevent route hijacking and leakage.

In secure routing practices, what is the role of AS_PATH filtering in BGP?
a) To optimize the selection of routing paths
b) To authenticate BGP peers
c) To prevent routing loops and route leaks
d) To compress routing update messages

Answer: c) To prevent routing loops and route leaks. Explanation: AS_PATH filtering in BGP plays a crucial role in preventing routing loops and route leaks. By inspecting the AS_PATH attribute of BGP routes, routers can apply policies to deny routes that contain certain ASes in their path, or that have been artificially extended, thus enhancing the security and integrity of BGP routing.

What technique can be used in OSPF to control which routes are imported into an OSPF area from external sources?
a) Route tagging
b) Cost manipulation
c) Redistribution filters
d) Passive interfaces

Answer: c) Redistribution filters. Explanation: Redistribution filters can be used in OSPF to control which routes are imported into an OSPF area from external sources. These filters are applied during the route redistribution process to selectively permit or deny specific routes from being introduced into the OSPF domain, allowing for more granular control over the routing table and enhancing security.

In securing BGP against route hijacking, what is the function of a Max Prefix Limit?
a) To specify the maximum number of prefixes an AS can advertise
b) To encrypt BGP update messages
c) To set the maximum number of hops in the AS_PATH
d) To limit the rate of BGP update messages

Answer: a) To specify the maximum number of prefixes an AS can advertise. Explanation: The Max Prefix Limit in BGP is used to specify the maximum number of prefixes that an AS (Autonomous System) can advertise to a BGP peer. This limit helps to prevent route hijacking and accidental misconfigurations by detecting and potentially shutting down sessions that exceed the expected number of prefixes, thereby maintaining the integrity of the routing table.

What is the first step in the SSL/TLS handshake process?
a. The client sends a list of supported cipher suites.
b. The server sends its digital certificate to the client.
c. The client sends a pre-master secret to the server.
d. The server sends a list of supported SSL/TLS versions.

Answer: a. The client sends a list of supported cipher suites. Explanation: The first step in the SSL/TLS handshake process involves the client sending a "ClientHello" message to the server, which includes a list of supported cipher suites, SSL/TLS versions, and other session data necessary to begin the secure communication.

Which known vulnerability affects SSL 3.0 and led to its deprecation?
a. Heartbleed
b. POODLE
c. BEAST
d. FREAK

Answer: b. POODLE. Explanation: The POODLE (Padding Oracle On Downgraded Legacy Encryption) vulnerability specifically affects SSL 3.0. It allows an attacker to decrypt the contents of encrypted messages using a padding oracle attack, leading to the deprecation of SSL 3.0 in favor of more secure versions of TLS.

During the SSL/TLS handshake, when is the symmetric session key first used?
a. Immediately after the "ClientHello" message
b. When the server sends its digital certificate
c. After the client and server exchange "Finished" messages
d. Upon the client's receipt and validation of the server's digital certificate

Answer: c. After the client and server exchange "Finished" messages. Explanation: The symmetric session key, derived from the pre-master secret, is first used after the client and server have exchanged "Finished" messages. This marks the completion of the handshake, and subsequent communication is encrypted using this symmetric key.

What purpose does the "ServerHello" message serve in the SSL/TLS handshake?
a. To provide the client with the server's public key
b. To confirm the symmetric cipher suite that will be used for the session
c. To request the client's digital certificate for mutual authentication
d. To initiate the handshake termination process

Answer: b. To confirm the symmetric cipher suite that will be used for the session. Explanation: The "ServerHello" message serves to confirm the selection of the symmetric cipher suite that will be used for the session, based on the list provided by the client in the "ClientHello" message. It also includes the server's chosen SSL/TLS version and other session-specific details.

In the context of SSL/TLS, what is a cipher suite?
a. A collection of algorithms used for securing network communications
b. A database of all possible encryption keys
c. A protocol for exchanging cryptographic keys
d. A software suite for managing digital certificates

Answer: a. A collection of algorithms used for securing network communications. Explanation: A cipher suite in SSL/TLS is a combination of cryptographic algorithms used to secure network communications. It includes key exchange algorithms, encryption algorithms, and message authentication codes (MACs).

Which vulnerability is specifically related to the use of weak cryptographic algorithms in older versions of SSL/TLS?
a. CRIME
b. Logjam

c. Sweet32
d. DROWN

Answer: b. Logjam. Explanation: The Logjam vulnerability affects SSL/TLS protocols that support weak Diffie-Hellman key exchange parameters, allowing attackers to downgrade vulnerable TLS connections to 512-bit export-grade cryptography, which can then be decrypted or altered.

What is the role of the "ChangeCipherSpec" message in the SSL/TLS handshake process?
a. To notify that subsequent messages will be encrypted
b. To change the cryptographic algorithms used in the session
c. To update the digital certificates used for the session
d. To renegotiate the session keys

Answer: a. To notify that subsequent messages will be encrypted. Explanation: The "ChangeCipherSpec" message is sent by both the client and server to signal that subsequent messages will be encrypted using the session keys established during the handshake. It marks the transition from unencrypted to encrypted communication.

How does the SSL/TLS handshake facilitate forward secrecy?
a. By using the server's private key to encrypt all session data
b. By encrypting the "ClientHello" message with the server's public key
c. Through the use of ephemeral Diffie-Hellman key exchange mechanisms
d. By requiring client-side certificates for all connections

Answer: c. Through the use of ephemeral Diffie-Hellman key exchange mechanisms. Explanation: Forward secrecy is achieved in SSL/TLS through the use of ephemeral Diffie-Hellman key exchange, where temporary, per-session keys are generated and used for encryption, ensuring that the compromise of long-term keys does not compromise past session keys.

When implementing TLS 1.2, which algorithm set is considered best practice for secure communications?
a. RSA for key exchange, AES for encryption, and SHA-1 for MAC
b. Diffie-Hellman for key exchange, 3DES for encryption, and MD5 for MAC
c. ECDHE for key exchange, AES-GCM for encryption, and SHA-256 for MAC
d. PGP for key exchange, Blowfish for encryption, and SHA-512 for MAC

Answer: c. ECDHE for key exchange, AES-GCM for encryption, and SHA-256 for MAC. Explanation: For TLS 1.2, using Elliptic Curve Diffie-Hellman Ephemeral (ECDHE) for key exchange, AES in Galois/Counter Mode (AES-GCM) for encryption, and SHA-256 for the Message Authentication Code (MAC) is considered best practice, offering a strong combination of security and performance.

What mechanism does TLS 1.3 use to reduce the number of round-trips required in the handshake process compared to TLS 1.2?
a. Pre-shared keys
b. Session resumption
c. Zero Round-Trip Time (0-RTT) resumption
d. Asynchronous key exchange

Answer: c. Zero Round-Trip Time (0-RTT) resumption. Explanation: TLS 1.3 introduces the Zero Round-Trip Time (0-RTT) resumption feature, which allows clients to send encrypted data to the server in the first message of the handshake process when resuming a previous session, thereby reducing the handshake to a single round-trip.

What is the primary difference between signature-based and anomaly-based IDS/IPS systems?
a) Signature-based systems encrypt data, whereas anomaly-based systems do not.
b) Signature-based systems rely on known patterns, while anomaly-based systems detect deviations from normal behavior.

c) Anomaly-based systems are used only in military applications, while signature-based systems are used in commercial applications.
d) Signature-based systems are used for physical security, whereas anomaly-based systems are used for cyber security.

Answer: b) Signature-based systems rely on known patterns, while anomaly-based systems detect deviations from normal behavior. Explanation: Signature-based IDS/IPS systems detect attacks based on specific patterns or signatures associated with known threats, similar to antivirus software. Anomaly-based IDS/IPS systems, on the other hand, monitor network traffic and compare it against a baseline of normal activity, alerting on deviations that might indicate a security incident.

Where is an IDS most effectively placed within a network infrastructure?
a) Directly behind the firewall to inspect incoming traffic
b) On the external side of the firewall to monitor potential attacks
c) Inside the core network to monitor internal traffic
d) Parallel to the firewall to reduce latency

Answer: a) Directly behind the firewall to inspect incoming traffic. Explanation: Placing an IDS directly behind the firewall allows it to inspect incoming traffic that has been allowed through the firewall, making it effective in detecting potential threats before they reach internal resources. This placement helps in identifying malicious activities that might have passed through the firewall's initial filtering.

For an IPS to prevent attacks in real-time, where should it be deployed?
a) In offline mode for detailed forensic analysis
b) In line with the network traffic to actively block or mitigate threats
c) Outside the network perimeter to scan incoming traffic
d) After the data has reached its destination for post-event analysis

Answer: b) In line with the network traffic to actively block or mitigate threats. Explanation: An IPS (Intrusion Prevention System) needs to be deployed in line with the network traffic so it can actively monitor, block, or mitigate threats in real-time. By being in the path of the traffic, the IPS can take immediate action to prevent malicious activities from affecting the network.

How does an anomaly-based IDS/IPS system determine what is considered normal network behavior?
a) By using a predefined list of known good IP addresses
b) Through an initial learning phase to establish a baseline of normal activity
c) By manual configuration of acceptable protocols and ports
d) Using a static database of common network applications

Answer: b) Through an initial learning phase to establish a baseline of normal activity. Explanation: Anomaly-based IDS/IPS systems use an initial learning phase to analyze the typical operation of the network and establish a baseline of normal behavior. This baseline is then used to detect deviations or anomalies that may indicate malicious activity, allowing the system to alert on or block such deviations.

What is a key consideration when placing an IPS in a high-availability network environment?
a) Ensuring the IPS does not introduce a single point of failure
b) Placing the IPS outside the network perimeter for maximum visibility
c) Configuring the IPS to allow all traffic by default
d) Using multiple IPS units in series for increased security

Answer: a) Ensuring the IPS does not introduce a single point of failure. Explanation: In a high-availability network environment, it's crucial to ensure that the IPS does not introduce a single point of failure. This can be achieved by implementing failover strategies, redundant IPS units, or bypass mechanisms that ensure network traffic continuity in the event the IPS fails or needs to be taken offline for maintenance.

Which feature of an IPS can be used to dynamically adapt to evolving threats?

a) Hard-coded signatures that rarely need updates
b) Manual intervention for each detected threat
c) Heuristic analysis for detecting unknown threats
d) A fixed threshold for anomaly detection

Answer: c) Heuristic analysis for detecting unknown threats. Explanation: Heuristic analysis is a feature of some IPS systems that allows them to dynamically adapt to evolving threats by using advanced algorithms to detect suspicious behavior patterns. This method can identify previously unknown threats based on characteristics or behaviors that are typical of malicious activity, rather than relying solely on known signatures.

In deploying an IDS in a distributed network, what strategy helps in managing the detection capabilities effectively?
a) Deploying a single, centralized IDS to monitor all network traffic
b) Using a consistent set of signatures across all network segments
c) Implementing different IDS policies tailored to each network segment's specific needs
d) Placing IDS sensors randomly to increase unpredictability

Answer: c) Implementing different IDS policies tailored to each network segment's specific needs. Explanation: In a distributed network, implementing different IDS policies tailored to the specific needs and characteristics of each network segment helps in managing detection capabilities effectively. This approach ensures that the IDS is sensitive to the unique traffic patterns and threat landscape of each segment, improving the accuracy and effectiveness of threat detection.

What advantage does deploying an IPS provide over an IDS in terms of incident response?
a) An IPS requires less frequent updates than an IDS.
b) An IPS can only detect known threats, making it more reliable.
c) An IPS can take proactive actions to block or mitigate threats in real-time.
d) An IPS is easier to bypass than an IDS, reducing false positives.

Answer: c) An IPS can take proactive actions to block or mitigate threats in real-time. Explanation: The key advantage of an IPS over an IDS is its ability to take proactive actions to block or mitigate threats in real-time, rather than just detecting and alerting on them. This capability allows for immediate incident response, reducing the potential impact of attacks on the network.

When considering the deployment of an IDS/IPS solution, why is it important to regularly update its signatures or detection algorithms?
a) To ensure compatibility with legacy systems
b) To decrease the system's processing requirements
c) To maintain the system's ability to detect new and evolving threats
d) To reduce the frequency of alerts generated by the system

Answer: c) To maintain the system's ability to detect new and evolving threats. Explanation: Regularly updating the signatures or detection algorithms of an IDS/IPS solution is crucial to maintain its ability to detect new and evolving threats. As attackers continually develop new techniques and malware, keeping the detection mechanisms up-to-date ensures that the system can recognize and respond to the latest threats effectively.

What distinguishes a stateful firewall from a packet filtering firewall?
a. Stateful firewalls encrypt traffic, whereas packet filtering firewalls do not.
b. Stateful firewalls analyze the full content of packets, while packet filtering firewalls only look at headers.
c. Stateful firewalls track the state of active connections and make decisions based on the context of the traffic.
d. Packet filtering firewalls require manual configuration for each packet type, whereas stateful firewalls automatically adapt.

Answer: c. Stateful firewalls track the state of active connections and make decisions based on the context of the traffic. Explanation: Stateful firewalls monitor the state of active connections and consider the state of the traffic within the context of the session, which allows them to make more informed decisions about allowing or blocking traffic compared to packet filtering firewalls that make decisions based solely on predefined rules and packet headers.

Which feature is commonly found in Next-Generation Firewalls (NGFW)?
a. Outbound email filtering
b. Integrated antivirus software
c. Application-level inspection and control
d. Automatic security patch deployment

Answer: c. Application-level inspection and control. Explanation: NGFWs offer application-level inspection and control, which allows them to identify and manage traffic based on application types, such as social media or instant messaging, providing a more granular approach to traffic management and security compared to traditional firewalls.

How does a packet filtering firewall determine whether to allow or block traffic?
a. By the geographical location of the source IP address
b. Through behavioral analysis of the packet contents
c. Based on predetermined rules that examine packet headers
d. By establishing a secure VPN tunnel for packet inspection

Answer: c. Based on predetermined rules that examine packet headers. Explanation: Packet filtering firewalls make decisions to allow or block traffic based on a set of predetermined rules that examine the packet headers, including source and destination IP addresses, port numbers, and protocol types, without maintaining state information about connections.

In what scenario might a stateful firewall be preferred over a packet filtering firewall?
a. When minimal resource usage is a primary concern
b. In environments requiring detailed audit logs of individual packet contents
c. For networks needing dynamic protection based on the context of traffic flows
d. When the primary goal is to block traffic from specific countries

Answer: c. For networks needing dynamic protection based on the context of traffic flows. Explanation: Stateful firewalls are preferred in environments where dynamic protection is needed, as they can make decisions based on the context of traffic flows by tracking the state of connections, providing more nuanced and effective security compared to the static rule set of packet filtering firewalls.

What is a characteristic feature of Next-Generation Firewalls (NGFW) compared to traditional firewalls?
a. The ability to function without any predefined security rules
b. Support for older, legacy encryption protocols for compatibility
c. Integrated intrusion prevention systems (IPS)
d. Lower processing and throughput capabilities to enhance security

Answer: c. Integrated intrusion prevention systems (IPS). Explanation: NGFWs are characterized by their integrated intrusion prevention systems (IPS), which provide additional security measures such as monitoring and blocking potentially harmful traffic in real-time, a feature that goes beyond traditional firewall capabilities.

Why might an organization choose to deploy a packet filtering firewall for a specific segment of its network?
a. To provide detailed web content filtering and malware scanning
b. For high-throughput environments where basic filtering is sufficient
c. To enforce data loss prevention (DLP) policies
d. When application-level inspection is deemed unnecessary

Answer: b. For high-throughput environments where basic filtering is sufficient. Explanation: Organizations might choose to deploy packet filtering firewalls in segments of their network where high throughput is essential, and basic filtering based on IP addresses, ports, and protocols is deemed sufficient, due to their relatively lower processing overhead compared to stateful or NGFW solutions.

What limitation is commonly associated with traditional stateful firewalls when dealing with modern web applications?
a. Inability to encrypt web traffic
b. Lack of support for IPv6 traffic
c. Difficulty in inspecting encrypted HTTPS traffic
d. The requirement for frequent manual updates to security rules

Answer: c. Difficulty in inspecting encrypted HTTPS traffic. Explanation: Traditional stateful firewalls often struggle with inspecting encrypted HTTPS traffic, as they cannot easily decrypt and analyze the contents of encrypted packets, limiting their effectiveness in identifying threats within secure web applications.

How do NGFWs handle encrypted traffic differently from traditional firewalls?
a. By blocking all encrypted traffic by default
b. Through the use of SSL/TLS decryption and inspection
c. By relying solely on endpoint protection solutions for encrypted traffic
d. Using external decryption services before analysis

Answer: b. Through the use of SSL/TLS decryption and inspection. Explanation: NGFWs handle encrypted traffic through SSL/TLS decryption and inspection capabilities, allowing them to decrypt, analyze, and re-encrypt traffic, thereby maintaining security measures even for encrypted communications, a feature not typically found in traditional firewalls.

In what way does a stateful firewall enhance security for dynamic applications compared to a packet filtering firewall?
a. By allowing all outbound traffic without inspection
b. Utilizing user identity information for traffic filtering
c. Maintaining a session table to track active connections and their states
d. Requiring two-factor authentication for all network access requests

Answer: c. Maintaining a session table to track active connections and their states. Explanation: Stateful firewalls enhance security for dynamic applications by maintaining a session table that tracks active connections and their states. This allows the firewall to understand the context of each session and apply security policies dynamically, which is particularly useful for applications with complex, stateful interactions.

What next-generation feature enables NGFWs to block malware from entering a network?
a. Traditional packet filtering techniques
b. Deep packet inspection combined with real-time threat intelligence
c. Fixed bandwidth allocation for each network device
d. Manual submission of suspicious files for analysis

Answer: b. Deep packet inspection combined with real-time threat intelligence. Explanation: NGFWs utilize deep packet inspection (DPI) combined with real-time threat intelligence to identify and block malware from entering a network. This allows NGFWs to inspect the contents of packets at a granular level and compare them against databases of known threats for more effective and proactive malware prevention.

What is the primary purpose of IPsec in a VPN configuration?
a) To provide a user-friendly interface for VPN clients
b) To ensure data integrity and confidentiality between VPN endpoints
c) To increase the speed of data transmission over the VPN
d) To assign dynamic IP addresses to VPN clients

Answer: b) To ensure data integrity and confidentiality between VPN endpoints. Explanation: IPsec (Internet Protocol Security) is used in VPN configurations to ensure data integrity and confidentiality between VPN endpoints. It provides secure encrypted tunnels for data packets, protecting the data as it travels over an insecure network like the internet.

Which VPN tunneling protocol is known for providing a secure and encrypted connection but has been deprecated due to known vulnerabilities?
a) L2TP
b) PPTP
c) SSTP
d) IPsec

Answer: b) PPTP (Point-to-Point Tunneling Protocol). Explanation: PPTP has been widely used for creating VPN tunnels but is known to have several security vulnerabilities that make it susceptible to attacks. Due to these vulnerabilities, it has been deprecated in favor of more secure protocols.

In a site-to-site VPN, what is typically used to establish the connection between two networks?
a) Individual VPN software clients on each device
b) VPN gateways at each site
c) A single centralized VPN server
d) Direct peer-to-peer connections between users

Answer: b) VPN gateways at each site. Explanation: In a site-to-site VPN, VPN gateways are used at each site to establish and manage the VPN connection between the two networks. These gateways handle the encryption, decryption, and routing of traffic between the networks, allowing for seamless connectivity as if they were on the same local network.

How does L2TP differ from PPTP in terms of security when used in VPNs?
a) L2TP inherently provides encryption, while PPTP does not.
b) L2TP is less secure due to its older encryption algorithms.
c) L2TP typically relies on IPsec for encryption, while PPTP uses its own encryption mechanism.
d) PPTP provides stronger authentication methods than L2TP.

Answer: c) L2TP typically relies on IPsec for encryption, while PPTP uses its own encryption mechanism. Explanation: L2TP (Layer 2 Tunneling Protocol) itself does not provide encryption. It is often combined with IPsec to provide encryption and secure the VPN tunnel. PPTP (Point-to-Point Tunneling Protocol), on the other hand, includes its own encryption mechanism, which is considered less secure than the encryption provided by IPsec.

What is a key advantage of using a remote access VPN?
a) It allows entire networks to connect to each other over the internet.
b) It enables individual users to securely access the corporate network from any location.
c) It eliminates the need for any form of encryption.
d) It provides a permanent, always-on connection between two sites.

Answer: b) It enables individual users to securely access the corporate network from any location. Explanation: A remote access VPN allows individual users to securely connect to and access resources on the corporate network from remote locations, such as from home or while traveling, providing flexibility and security for mobile users or teleworkers.

Which component is essential in a VPN to authenticate and encrypt data packets as they travel through the tunnel?
a) VPN concentrator
b) Network access server (NAS)
c) DHCP server
d) Proxy server

Answer: a) VPN concentrator. Explanation: A VPN concentrator is a device specifically designed to create and manage VPN communication infrastructures. It is essential for authenticating users, establishing tunnels, and encrypting data packets to ensure secure communication between VPN clients and the network.

When comparing IPsec tunnel mode to transport mode, what is a primary difference in their use cases?

a) Tunnel mode encrypts only the payload, while transport mode encrypts the entire packet.
b) Tunnel mode is used for site-to-site VPNs, while transport mode is used for remote access VPNs.
c) Transport mode is typically used in applications requiring high throughput.
d) Tunnel mode is only used for non-IP traffic, while transport mode supports all traffic types.

Answer: b) Tunnel mode is used for site-to-site VPNs, while transport mode is used for remote access VPNs. Explanation: IPsec tunnel mode encapsulates the entire IP packet and is commonly used in site-to-site VPNs to connect two different networks. Transport mode, on the other hand, only encrypts the payload of the IP packet and is often used in remote access VPN scenarios where end-to-end communication needs to be secured.

What role does IKE (Internet Key Exchange) play in an IPsec VPN?
a) It is responsible for user authentication.
b) It dynamically manages keys and negotiates the IPsec security associations.
c) It compresses data to improve VPN throughput.
d) It routes packets through the VPN tunnel.

Answer: b) It dynamically manages keys and negotiates the IPsec security associations. Explanation: IKE (Internet Key Exchange) is a component of IPsec used for dynamically managing keys and negotiating the security associations used in the VPN. IKE automates the process of key exchange and ensures that the VPN connection remains secure over time.

Why might an organization choose to implement a site-to-site VPN over a remote access VPN?
a) To connect individual remote users to the corporate network
b) To avoid using encryption for sensitive data
c) To provide a secure connection between two or more fixed locations
d) To reduce the overall cost of the network infrastructure

Answer: c) To provide a secure connection between two or more fixed locations. Explanation: An organization might choose to implement a site-to-site VPN to securely connect two or more fixed locations, such as office buildings or campuses, over the internet. This type of VPN is used to extend the network's reach, allowing different sites to share resources as if they were on the same local network.

What is a primary security concern when implementing PPTP for a VPN, leading many to consider alternative protocols?
a) It requires additional hardware to be installed.
b) Its encryption and authentication methods are considered weak and vulnerable.
c) It does not support modern operating systems.
d) It is incompatible with NAT (Network Address Translation).

Answer: b) Its encryption and authentication methods are considered weak and vulnerable. Explanation: PPTP has been criticized for its weak encryption and authentication methods, making it vulnerable to various security attacks. Due to these vulnerabilities, many organizations and security professionals recommend using alternative VPN protocols that offer stronger security measures.

Which biometric authentication method measures unique patterns in the iris of the eye?
a. Fingerprint scanning
b. Voice recognition
c. Iris recognition
d. Hand geometry

Answer: c. Iris recognition. Explanation: Iris recognition involves capturing and analyzing the unique patterns found in the iris of an individual's eye. This method is known for its high level of accuracy and security, as the patterns in the iris are complex and unique to each person.

In the context of smart card technology, what is a contactless smart card?

a. A card that requires physical contact with the reader to transfer data
b. A card that uses RFID technology to communicate with the reader wirelessly
c. A card that can only be read by specific, proprietary card readers
d. A card that uses magnetic stripe technology for data storage and transfer

Answer: b. A card that uses RFID technology to communicate with the reader wirelessly. Explanation: Contactless smart cards utilize Radio-Frequency Identification (RFID) technology to enable wireless communication with a card reader. This allows for faster and more convenient transactions, as the card does not need to make physical contact with the reader.

What is the primary security concern associated with biometric authentication systems?
a. They are easily duplicated
b. They require regular password changes
c. False acceptance or false rejection rates
d. They only work in conjunction with a PIN

Answer: c. False acceptance or false rejection rates. Explanation: Biometric authentication systems are evaluated based on their false acceptance rate (FAR), the likelihood of incorrectly accepting an unauthorized user, and their false rejection rate (FRR), the likelihood of incorrectly rejecting an authorized user. Balancing these rates is a primary security concern to ensure both security and usability.

How do smart cards contribute to a multi-factor authentication system?
a. By serving as a physical token that must be combined with something the user knows (like a PIN)
b. By providing a backup password storage system
c. By encrypting data on the user's device
d. By scanning the user's biometric data

Answer: a. By serving as a physical token that must be combined with something the user knows (like a PIN). Explanation: Smart cards contribute to multi-factor authentication systems by acting as a physical token that the user must possess. When combined with something the user knows, such as a Personal Identification Number (PIN) or a password, it provides a more secure authentication method by requiring two different factors.

What advantage does facial recognition offer over other biometric systems?
a. It is the most inexpensive biometric system to implement
b. It requires the least amount of user interaction
c. It has the lowest false acceptance rate
d. It is the least intrusive and fastest for user verification

Answer: d. It is the least intrusive and fastest for user verification. Explanation: Facial recognition is considered less intrusive compared to other biometric systems like fingerprint or iris scans, as it can often be performed at a distance and without physical contact, making the verification process faster and more seamless for users.

In what scenario might a logical access control system use a smart card for authentication?
a. To grant access to a secure physical facility
b. To verify a user's identity before allowing access to computer systems and networks
c. To track employee time and attendance
d. To encrypt email communications

Answer: b. To verify a user's identity before allowing access to computer systems and networks. Explanation: Smart cards are often used in logical access control systems to verify a user's identity before granting access to sensitive computer systems and networks. The smart card, typically requiring a PIN for use, provides a secure method of authenticating the user's identity in the digital realm.

What is a potential drawback of using fingerprint scanners for biometric authentication?
a. They are significantly more expensive than other biometric systems

b. They can be less reliable if a user's fingerprints are worn or damaged
c. They are slower than password-based systems
d. They cannot be used in conjunction with other authentication methods

Answer: b. They can be less reliable if a user's fingerprints are worn or damaged. Explanation: Fingerprint scanners can sometimes be less reliable if a user's fingerprints are worn, damaged, or temporarily altered (due to cuts, dirt, or other factors), which can lead to higher false rejection rates compared to other biometric systems.

How might a next-generation firewall (NGFW) incorporate biometric authentication data?
a. By directly scanning and storing biometric data within the firewall
b. By integrating with external biometric authentication systems for identity verification
c. By using biometric data as the sole method for network access control
d. By encrypting biometric data and using it as a key for network encryption

Answer: b. By integrating with external biometric authentication systems for identity verification. Explanation: NGFWs can enhance security by integrating with external biometric authentication systems to verify identities before allowing access to network resources. This integration allows the firewall to make more informed access control decisions based on biometric verification, without directly handling biometric data itself.

What role do smart cards play in physical access control systems?
a. They are primarily used for decorative purposes on employee badges
b. They store biometric templates for comparison at access points
c. They act as electronic keys to unlock doors and gates when authenticated
d. They are used to visually verify an individual's authorization to access a facility

Answer: c. They act as electronic keys to unlock doors and gates when authenticated. Explanation: In physical access control systems, smart cards function as electronic keys. When a user presents a smart card to a card reader at a secure entrance, the card is authenticated, and if authorized, the system unlocks the door or gate, granting the user access.

Which factor is crucial for the effectiveness of biometric systems in high-security environments?
a. The system's ability to operate in low-light conditions
b. The use of a single biometric modality for simplicity
c. The integration of biometric data with employee productivity software
d. The combination of biometric verification with another form of authentication

Answer: d. The combination of biometric verification with another form of authentication. Explanation: In high-security environments, the effectiveness of biometric systems is greatly enhanced by combining biometric verification with another form of authentication, such as a PIN, password, or smart card. This multi-factor authentication approach significantly increases security by requiring multiple forms of verification before granting access.

In which access control model do users have the flexibility to set their own policies on objects they own?
a) Discretionary Access Control (DAC)
b) Mandatory Access Control (MAC)
c) Role-Based Access Control (RBAC)
d) Attribute-Based Access Control (ABAC)

Answer: a) Discretionary Access Control (DAC). Explanation: In the Discretionary Access Control (DAC) model, the owner of the resource (object) has the flexibility to set policies on who can access it. This model allows users to grant or restrict access to their resources based on their discretion.

Which access control model is characterized by the use of security labels and clearance levels?
a) Discretionary Access Control (DAC)
b) Mandatory Access Control (MAC)
c) Role-Based Access Control (RBAC)
d) Attribute-Based Access Control (ABAC)

Answer: b) Mandatory Access Control (MAC). Explanation: The Mandatory Access Control (MAC) model is characterized by the use of security labels (which contain classification levels and categories) and clearance levels for users. Access decisions are based on these labels and the clearance level of the user, not on the discretion of the object's owner.

In an organization where permissions are assigned based on job functions, which access control model is being used?
a) Discretionary Access Control (DAC)
b) Mandatory Access Control (MAC)
c) Role-Based Access Control (RBAC)
d) Attribute-Based Access Control (ABAC)

Answer: c) Role-Based Access Control (RBAC). Explanation: Role-Based Access Control (RBAC) assigns permissions to users based on their role within an organization. Access rights are grouped by role name, and users are assigned roles based on their responsibilities and qualifications.

Which access control model uses policies that evaluate attributes of the user, environment, and resource to make access decisions?
a) Discretionary Access Control (DAC)
b) Mandatory Access Control (MAC)
c) Role-Based Access Control (RBAC)
d) Attribute-Based Access Control (ABAC)

Answer: d) Attribute-Based Access Control (ABAC). Explanation: Attribute-Based Access Control (ABAC) uses policies that evaluate attributes (characteristics) of the user, the environment, and the resource to make access decisions. This model provides a high level of flexibility and granularity in access control.

How does RBAC differ from ABAC in terms of policy complexity and administration?
a) RBAC is more flexible and scalable than ABAC.
b) RBAC relies on a fixed number of roles, while ABAC can evaluate a potentially unlimited number of attributes.
c) ABAC is easier to manage and requires less administrative effort than RBAC.
d) ABAC policies are simpler and more straightforward than RBAC policies.

Answer: b) RBAC relies on a fixed number of roles, while ABAC can evaluate a potentially unlimited number of attributes. Explanation: RBAC is based on predefined roles within an organization, which simplifies administration but can lack the flexibility of ABAC. ABAC evaluates multiple attributes, providing more granular control but potentially increasing complexity and administrative overhead.

In what scenario is DAC considered less secure than MAC?
a) When users need to share resources frequently
b) In environments requiring high levels of data confidentiality and integrity
c) In small, tightly-knit teams with high levels of trust
d) When rapid changes in access permissions are required

Answer: b) In environments requiring high levels of data confidentiality and integrity. Explanation: DAC is considered less secure than MAC in environments requiring high levels of data confidentiality and integrity because DAC allows users to control access to their own resources, potentially leading to insecure permissions or accidental sharing.

What advantage does RBAC have over DAC in a large organization with many users and complex access requirements?
a) RBAC allows users to directly manage access to their own files.
b) RBAC simplifies management by grouping permissions into roles based on job functions.
c) RBAC uses security labels to enforce access controls.
d) RBAC is less resource-intensive and easier to implement.

Answer: b) RBAC simplifies management by grouping permissions into roles based on job functions. Explanation: In large organizations with complex access requirements, RBAC offers the advantage of simplifying management by grouping permissions into roles. This approach is more scalable and manageable than DAC, where permissions are set individually by resource owners.

Which access control model can dynamically adjust permissions based on changing attributes, such as location or time?
a) Discretionary Access Control (DAC)
b) Mandatory Access Control (MAC)
c) Role-Based Access Control (RBAC)
d) Attribute-Based Access Control (ABAC)

Answer: d) Attribute-Based Access Control (ABAC). Explanation: ABAC can dynamically adjust permissions based on changing attributes, such as the user's location, the time of access, or the state of the resource. This allows for more flexible and context-aware access control policies.

How does the implementation of MAC contribute to the security of government and military information systems?
a) By allowing users to share data freely within the organization
b) By enforcing access controls based on user roles and responsibilities
c) By mandating predefined access controls that cannot be altered by end-users
d) By simplifying access control lists for easier management

Answer: c) By mandating predefined access controls that cannot be altered by end-users. Explanation: In government and military information systems, MAC contributes to security by mandating predefined access controls based on security labels that cannot be altered by end-users. This ensures that access decisions are enforced uniformly, according to official classification levels and need-to-know principles.

For a healthcare application that needs to enforce access based on the patient-doctor relationship, department, and time of day, which access control model is most suitable?
a) Discretionary Access Control (DAC)
b) Mandatory Access Control (MAC)
c) Role-Based Access Control (RBAC)
d) Attribute-Based Access Control (ABAC)

Answer: d) Attribute-Based Access Control (ABAC). Explanation: ABAC is most suitable for a healthcare application needing to enforce complex access controls based on multiple factors such as the patient-doctor relationship, department, and time of day. ABAC's use of diverse attributes allows for the creation of nuanced access control policies that can address the multifaceted requirements of healthcare applications.

What is the primary benefit of Single Sign-On (SSO) in a multi-application environment?
a. It encrypts all user data stored within the applications.
b. It allows users to access multiple applications with one set of credentials, reducing password fatigue.
c. It increases the complexity of the authentication process to enhance security.
d. It requires users to remember multiple sets of credentials, increasing security.

Answer: b. It allows users to access multiple applications with one set of credentials, reducing password fatigue. Explanation: SSO simplifies the user experience by allowing individuals to use a single set of credentials to access multiple applications, reducing the need to remember numerous usernames and passwords, thereby minimizing password fatigue and the likelihood of insecure password practices.

How does Multi-Factor Authentication (MFA) enhance security compared to traditional username and password authentication?
a. By using a single, complex password for all systems
b. By requiring additional verification methods beyond just a password, such as a token or biometric data
c. By frequently changing passwords
d. By encrypting the password with a stronger algorithm

Answer: b. By requiring additional verification methods beyond just a password, such as a token or biometric data. Explanation: MFA enhances security by requiring two or more independent credentials for authentication, which typically includes something the user knows (password), something the user has (security token), and something the user is (biometric verification). This layered defense makes it more difficult for unauthorized users to gain access.

What is Federated Identity Management primarily used for?
a. To consolidate user identities across different systems within a single organization
b. To enable users to access systems across different organizations using their home organization's credentials
c. To encrypt user data on federated servers
d. To create a single sign-on experience across all devices a user owns

Answer: b. To enable users to access systems across different organizations using their home organization's credentials. Explanation: Federated Identity Management allows users to use a single set of credentials to access resources across multiple domains or organizations, simplifying access management and improving user experience while maintaining security.

In what scenario is SSO particularly beneficial?
a. When an organization wants to increase the complexity of its password policy
b. When users need to access a large number of applications, reducing the need to remember multiple passwords
c. When an organization has a small number of applications with similar access requirements
d. When there is a need to use different passwords for each application for auditing purposes

Answer: b. When users need to access a large number of applications, reducing the need to remember multiple passwords. Explanation: SSO is particularly beneficial in environments where users need to access multiple applications or systems. It simplifies the login process by reducing the number of passwords users must remember and manage, thereby improving user convenience and reducing the risk of weak password practices.

What distinguishes MFA from Two-Factor Authentication (2FA)?
a. MFA requires more than two authentication factors, whereas 2FA is limited to exactly two.
b. MFA is used exclusively in banking, while 2FA is used in all other industries.
c. MFA is less secure than 2FA because it introduces more potential points of failure.
d. MFA encrypts authentication data, whereas 2FA does not.

Answer: a. MFA requires more than two authentication factors, whereas 2FA is limited to exactly two. Explanation: The key difference between MFA and 2FA is that MFA may involve two or more independent credentials for authenticating a user, while 2FA specifically involves exactly two. This makes MFA more versatile in strengthening security by potentially incorporating additional layers of authentication.

How does SSO improve user experience in cloud-based services?
a. By requiring users to log in each time they switch between services
b. By allowing users to access multiple services with a single login, streamlining the authentication process
c. By increasing the time it takes to switch between services
d. By requiring users to use more complex passwords for each service

Answer: b. By allowing users to access multiple services with a single login, streamlining the authentication process. Explanation: SSO improves user experience in cloud-based services by enabling a single authentication process to access multiple services or applications. This streamlines the user's experience by reducing the number of login prompts and the need to remember multiple sets of credentials.

What feature of federated identity management systems can help reduce administrative overhead for managing user accounts?
a. The ability to automatically synchronize password changes across all federated systems
b. The ability for each system to maintain its own set of user credentials independently

c. The centralization of user identity management, allowing users to authenticate across different systems using one set of credentials
d. The requirement for users to use a different password for each system for increased security

Answer: c. The centralization of user identity management, allowing users to authenticate across different systems using one set of credentials. Explanation: Federated identity management reduces administrative overhead by centralizing user identity management. This allows users to authenticate across different systems using a single set of credentials, thereby simplifying the administration of user accounts and access rights.

In the deployment of MFA, what role does a one-time password (OTP) play?
a. It serves as the primary password for all user accounts.
b. It provides a dynamic authentication factor that changes with each login attempt.
c. It replaces the need for any other form of authentication.
d. It is used as a backup password in case the user forgets their main password.

Answer: b. It provides a dynamic authentication factor that changes with each login attempt. Explanation: In MFA, an OTP serves as a dynamic authentication factor that is valid for only one login session or transaction. This adds an additional layer of security, as the password changes with each login attempt, making it more difficult for unauthorized users to gain access.

How does federated identity management enhance collaboration between different organizations?
a. By allowing each organization to enforce its own access controls independently
b. By requiring users to maintain separate credentials for each organization
c. By enabling seamless and secure access to shared resources using their existing credentials
d. By creating a centralized database of user credentials shared among organizations

Answer: c. By enabling seamless and secure access to shared resources using their existing credentials. Explanation: Federated identity management enhances collaboration by allowing users from different organizations to access shared resources securely using their existing credentials from their home organization. This facilitates seamless collaboration without the need for multiple sets of credentials or complex cross-organization identity management solutions.

What is the primary function of OAuth 2.0 in session management?
a) Encrypting session data to prevent interception
b) Providing a framework for third-party services to exchange web resource access on behalf of a user
c) Authenticating users directly to the service they wish to access
d) Storing session tokens in a centralized database for validation

Answer: b) Providing a framework for third-party services to exchange web resource access on behalf of a user. Explanation: OAuth 2.0 is primarily used to provide a secure delegated access framework, allowing third-party services to exchange web resource access tokens on behalf of the user. This allows users to grant limited access to their resources from one service to another without exposing their credentials.

How does OpenID Connect enhance OAuth 2.0 for authentication purposes?
a) By providing an additional encryption layer for OAuth tokens
b) By extending OAuth 2.0 with an identity layer for user authentication
c) By replacing the token-based mechanism with a certificate-based approach
d) By introducing a decentralized authentication mechanism

Answer: b) By extending OAuth 2.0 with an identity layer for user authentication. Explanation: OpenID Connect builds on OAuth 2.0 by adding an identity layer on top of the authorization framework. This provides a standardized way for clients to verify the identity of the end-user based on the authentication performed by an authorization server, as well as to obtain basic profile information about the end-user.

In OAuth 2.0, what is the role of an access token?
a) To encrypt user credentials for secure transmission

b) To serve as a digital signature for data integrity
c) To grant the holder a specific scope of access to resources for a limited time
d) To authenticate the client to the authorization server

Answer: c) To grant the holder a specific scope of access to resources for a limited time. Explanation: In OAuth 2.0, an access token is used to grant the holder (usually a third-party application) a specific scope of access to the user's resources hosted by a resource server for a limited time, without revealing the user's credentials.

What is the significance of the "state" parameter in an OAuth 2.0 authorization request?
a) It indicates the current state of the token's validity.
b) It is used as a nonce to prevent CSRF attacks.
c) It specifies the state of the resource server.
d) It determines the stateful or stateless nature of the session.

Answer: b) It is used as a nonce to prevent CSRF attacks. Explanation: The "state" parameter in an OAuth 2.0 authorization request is used as a nonce (a value that is used only once) to prevent Cross-Site Request Forgery (CSRF) attacks. It helps in maintaining the state between the client's request and the callback, ensuring that the response is directed to the same user who initiated the request.

Which OAuth 2.0 grant type is most suitable for clients executing on the device of the resource owner?
a) Client Credentials Grant
b) Authorization Code Grant
c) Resource Owner Password Credentials Grant
d) Implicit Grant

Answer: d) Implicit Grant. Explanation: The Implicit Grant type is most suitable for clients executing on the device of the resource owner where the client application is incapable of keeping a client secret due to the application environment. This grant type is used by mobile apps or web applications (running in a browser), where the token is returned directly to the client without an intermediate authorization code exchange step.

How does the Authorization Code Grant type in OAuth 2.0 enhance security for web applications?
a) By allowing the direct transmission of access tokens to user agents
b) By requiring client authentication for token exchange
c) By using short-lived authorization codes to exchange for access tokens
d) By storing user credentials in the authorization request

Answer: c) By using short-lived authorization codes to exchange for access tokens. Explanation: The Authorization Code Grant type enhances security by using short-lived authorization codes that the client can exchange for an access token. This adds an extra layer of security since the authorization code is transmitted through the user's browser, reducing the risk of interception, and the client must authenticate itself to exchange the code for an access token.

In what scenario is the Resource Owner Password Credentials Grant type considered appropriate in OAuth 2.0?
a) When the resource owner has a high level of trust in the client
b) For third-party access without user intervention
c) When client applications are running on an untrusted device
d) For stateless client applications

Answer: a) When the resource owner has a high level of trust in the client. Explanation: The Resource Owner Password Credentials Grant type is considered appropriate when the resource owner has a high level of trust in the client application, such as the device's operating system or a highly privileged application. This grant type allows the client application to request and receive access tokens directly by presenting the resource owner's credentials.

What mechanism does OpenID Connect use to provide the client with the authenticated identity of the user?
a) ID Token
b) Access Token

c) Refresh Token
d) Authorization Code

Answer: a) ID Token. Explanation: OpenID Connect uses an ID Token to provide the client with the authenticated identity of the user. The ID Token is a JSON Web Token (JWT) that contains claims about the authentication of the end-user by the Authorization Server, along with any other requested claims.

What is a primary security concern when using the Implicit Grant in OAuth 2.0, and how is it mitigated?
a) The access token is exposed in the redirection URI, which can be mitigated by using short-lived tokens and HTTPS.
b) The client secret is transmitted in the clear, which can be mitigated by using client-side encryption.
c) The refresh tokens are not supported, which can be mitigated by issuing long-lived access tokens.
d) The authorization server cannot authenticate the client, which can be mitigated by using dynamic registration.

Answer: a) The access token is exposed in the redirection URI, which can be mitigated by using short-lived tokens and HTTPS. Explanation: A primary security concern with the Implicit Grant is that the access token is exposed in the redirection URI, which can be intercepted by unauthorized parties. This is mitigated by using short-lived access tokens and ensuring that the communication is encrypted using HTTPS to protect the token in transit.

What is a key feature of Identity as a Service (IDaaS) solutions like Azure AD and Okta?
a. They primarily provide hardware-based authentication mechanisms.
b. They offer centralized identity management and authentication services over the cloud.
c. They are limited to on-premises deployment models only.
d. They require manual synchronization with each service provider.

Answer: b. They offer centralized identity management and authentication services over the cloud. Explanation: IDaaS solutions such as Azure AD and Okta provide centralized cloud-based services for identity management and authentication, enabling organizations to manage user identities and access controls across multiple applications and services efficiently.

How does Azure AD support multi-factor authentication (MFA) for enhancing security?
a. By requiring biometric data for every sign-in attempt
b. By allowing users to choose their preferred method of MFA from a list of options
c. By mandating hardware tokens for all users
d. By integrating with third-party antivirus software

Answer: b. By allowing users to choose their preferred method of MFA from a list of options. Explanation: Azure AD supports MFA by providing various options for secondary authentication methods, such as phone calls, text messages, or app notifications, enhancing security by requiring additional verification beyond just the username and password.

In what way does Okta's Single Sign-On (SSO) feature improve user experience?
a. By increasing the number of passwords users must remember
b. By enabling users to access multiple applications with a single set of credentials
c. By requiring users to log in each time they switch applications
d. By limiting access to only one application per user

Answer: b. By enabling users to access multiple applications with a single set of credentials. Explanation: Okta's SSO feature significantly improves user experience by allowing individuals to access multiple applications using just one set of credentials, eliminating the need to remember and manage multiple usernames and passwords.

What advantage does federated identity provided by IDaaS solutions offer to organizations?
a. It restricts user access to a single application.
b. It enables seamless and secure access across different organizational boundaries using a common set of credentials.
c. It eliminates the need for passwords.
d. It decentralizes identity management, requiring each application to manage its own user identities.

Answer: b. It enables seamless and secure access across different organizational boundaries using a common set of credentials. Explanation: Federated identity management in IDaaS solutions allows users to access resources across different organizations securely using their home organization's credentials, facilitating collaboration and streamlining access without compromising security.

How do IDaaS solutions like Azure AD and Okta integrate with enterprise applications?
a. By using proprietary protocols that require extensive application modification
b. Through standard protocols like SAML and OAuth for seamless integration
c. By requiring dedicated VPN connections for each application
d. Through manual entry of user credentials for each application integration

Answer: b. Through standard protocols like SAML and OAuth for seamless integration. Explanation: IDaaS solutions typically integrate with enterprise applications using standard protocols such as Security Assertion Markup Language (SAML) and Open Authorization (OAuth), enabling seamless and secure integration without the need for extensive modifications to the applications.

What role does Azure AD play in managing access to Microsoft 365 services?
a. It acts solely as a backup authentication system.
b. It serves as the primary identity and access management service, controlling user access to Microsoft 365 services.
c. It provides physical security for Microsoft 365 data centers.
d. It is used only for external user access to Microsoft 365.

Answer: b. It serves as the primary identity and access management service, controlling user access to Microsoft 365 services. Explanation: Azure AD is the primary service for identity and access management in Microsoft 365, managing user identities, credentials, and access policies to ensure that only authorized users can access Microsoft 365 services.

In the context of IDaaS, what is the significance of conditional access policies?
a. They mandate the use of a single device for access to all services.
b. They allow unrestricted access to all users within an organization.
c. They enforce security requirements based on the user's context, such as location or device security status.
d. They require all users to undergo periodic security training.

Answer: c. They enforce security requirements based on the user's context, such as location or device security status. Explanation: Conditional access policies in IDaaS platforms enable organizations to enforce access controls based on specific conditions or contexts, such as the user's location, device compliance status, or risk level, providing a more dynamic and context-aware approach to security.

What feature of IDaaS solutions aids in regulatory compliance and audit requirements?
a. Unlimited data storage
b. Automated password resets
c. Detailed logging and reporting of authentication and access events
d. Mandatory use of public Wi-Fi for access

Answer: c. Detailed logging and reporting of authentication and access events. Explanation: IDaaS solutions offer comprehensive logging and reporting capabilities that record detailed information about authentication attempts and access events. This feature aids organizations in meeting regulatory compliance requirements and facilitates audits by providing clear, accessible records of access control activities.

How does the integration of IDaaS solutions with API security gateways benefit an organization?
a. By reducing internet bandwidth usage
b. By providing a centralized platform for managing both identities and API access controls
c. By automatically generating API documentation
d. By converting all APIs to use SOAP protocols for enhanced security

Answer: b. By providing a centralized platform for managing both identities and API access controls.

What primary function does a Password Manager serve in credential management?
a) Generates and stores complex passwords for various accounts securely
b) Directly manages user permissions and access rights within a system
c) Encrypts data transmission between clients and servers
d) Monitors and audits user activities within a network

Answer: a) Generates and stores complex passwords for various accounts securely. Explanation: A Password Manager primarily generates, stores, and manages complex passwords for various user accounts securely. It helps users maintain strong, unique passwords for different services without the need to remember each one, enhancing security by reducing the risk of password reuse or simple passwords that are easily compromised.

How does Privileged Access Management (PAM) enhance security within an organization?
a) By encrypting all data stored within the organization's databases
b) By providing a framework for managing and monitoring access to privileged accounts
c) By implementing firewall rules to control network traffic
d) By scanning emails for phishing attempts

Answer: b) By providing a framework for managing and monitoring access to privileged accounts. Explanation: PAM enhances security by providing a comprehensive framework for managing and monitoring access to privileged accounts within an organization. It controls, audits, and monitors privileged access, ensuring that only authorized users can access critical systems and data, thus reducing the risk of security breaches.

In credential management, what is the purpose of using a Master Password with a Password Manager?
a) To serve as a unique identifier for the user across different platforms
b) To encrypt user data on the organization's servers
c) To act as the single password a user needs to remember to unlock access to all stored credentials
d) To reset forgotten passwords for various accounts

Answer: c) To act as the single password a user needs to remember to unlock access to all stored credentials. Explanation: The Master Password in a Password Manager is the single, strong password that a user must remember to gain access to all their stored credentials within the manager. This approach secures multiple complex passwords behind one master password, simplifying password management while maintaining security.

What feature of PAM solutions helps mitigate the risk of pass-the-hash attacks?
a) Password rotation
b) Multi-factor authentication
c) Session monitoring
d) Credential vaulting

Answer: d) Credential vaulting. Explanation: Credential vaulting in PAM solutions securely stores privileged account credentials and mitigates the risk of pass-the-hash attacks by preventing direct access to hashed passwords. It typically involves isolating, managing, and monitoring the use of privileged account credentials, ensuring they are not exposed or left vulnerable on user endpoints or within scripts.

How do Password Managers typically secure stored passwords and sensitive information?
a) Using a shared key system accessible by all users within the organization
b) Through end-to-end encryption with the keys held by the service provider
c) By employing strong, client-side encryption where only the user has the key
d) Storing passwords in plain text for ease of recovery

Answer: c) By employing strong, client-side encryption where only the user has the key. Explanation: Password Managers secure stored passwords and sensitive information through strong client-side encryption, where the encryption key is derived from the user's master password. This means only the user has access to the key to decrypt their data, ensuring that even the service provider cannot access the stored passwords.

What is a primary benefit of implementing Just-In-Time (JIT) access within a PAM solution?
a) It permanently elevates a user's access level to reduce administrative overhead
b) It allows unlimited access to resources for improved productivity
c) It provides privileged access only when needed, reducing the attack surface
d) It encrypts all communications within the network

Answer: c) It provides privileged access only when needed, reducing the attack surface. Explanation: Just-In-Time (JIT) access within a PAM solution grants privileged access only when necessary and for a limited duration. This approach minimizes the attack surface by reducing the number of accounts with always-on privileges and the time those privileges are available, thereby enhancing security.

In what way does Single Sign-On (SSO) integration with a Password Manager benefit users?
a) By requiring users to memorize more complex passwords for each account
b) By allowing users to access multiple applications with one set of credentials
c) By disabling password storage for enhanced security
d) By broadcasting user passwords across the network for easier access

Answer: b) By allowing users to access multiple applications with one set of credentials. Explanation: SSO integration with a Password Manager simplifies the user experience by allowing users to access multiple applications and services with a single set of credentials. This reduces password fatigue and the security risks associated with managing multiple passwords, while still maintaining secure access to resources.

Which practice enhances the security of a Master Password used with a Password Manager?
a) Using easily guessable personal information
b) Regularly changing the Master Password based on a fixed schedule
c) Combining it with multi-factor authentication (MFA)
d) Storing a written copy of the Master Password in a secure location

Answer: c) Combining it with multi-factor authentication (MFA). Explanation: Combining the Master Password with multi-factor authentication (MFA) significantly enhances its security. MFA requires a second form of verification beyond just the Master Password, such as a fingerprint or a one-time code, adding an additional layer of security to protect against unauthorized access.

What role does session management play in a PAM solution?
a) It tracks and records user actions during privileged sessions for auditing and review
b) It increases the session timeout duration for convenience
c) It limits the encryption strength to speed up system performance
d) It broadcasts session activity to all users for transparency

Answer: a) It tracks and records user actions during privileged sessions for auditing and review. Explanation: Session management in a PAM solution involves tracking, recording, and potentially controlling user actions during privileged sessions. This provides an audit trail that can be reviewed for suspicious activity, compliance purposes, and forensic analysis, enhancing accountability and security.

What is the primary goal of a code review in the context of security control testing?
a. To enhance the readability and maintainability of the code
b. To identify security vulnerabilities and weaknesses in the source code
c. To assess the performance efficiency of the application
d. To evaluate the user interface design for usability concerns

Answer: b. To identify security vulnerabilities and weaknesses in the source code. Explanation: The primary goal of a code review in security control testing is to meticulously examine the source code to identify any security vulnerabilities, weaknesses, or violations of programming best practices that could potentially be exploited by attackers, thereby enhancing the security posture of the application.

What does misuse case testing involve in software security?
a. Testing the application with typical use cases to ensure it meets user requirements
b. Analyzing the software's ability to handle large volumes of data
c. Evaluating the system's response to unexpected or malicious input
d. Checking compatibility with various operating systems and platforms

Answer: c. Evaluating the system's response to unexpected or malicious input. Explanation: Misuse case testing involves creating scenarios that simulate malicious or unexpected input and actions to evaluate how the system responds, identifying potential vulnerabilities or behaviors that could be exploited by attackers, thereby ensuring the application's resilience against misuse.

In automated code review tools, what is typically used to identify security issues?
a. A database of known good code examples for comparison
b. A checklist of user interface best practices
c. Heuristics and patterns that signify common vulnerabilities
d. Real-time user feedback during application use

Answer: c. Heuristics and patterns that signify common vulnerabilities. Explanation: Automated code review tools often rely on heuristics and patterns that are indicative of common security vulnerabilities and coding flaws. These tools scan the codebase for these patterns to identify potential security issues that need to be addressed.

How does dynamic application security testing (DAST) differ from static code review?
a. DAST requires the application to be running and tests it from the outside, simulating external attacks.
b. DAST focuses solely on the user experience aspects of security.
c. DAST is performed manually by security experts, while static code review is automated.
d. DAST evaluates the application's documentation rather than its code.

Answer: a. DAST requires the application to be running and tests it from the outside, simulating external attacks. Explanation: Dynamic Application Security Testing (DAST) differs from static code review by requiring the application to be in a running state and testing it from an external perspective, akin to how an attacker might, to identify vulnerabilities that manifest during its operation.

What role does peer review play in the code review process?
a. Solely to assess the algorithmic efficiency of the code
b. To provide a platform for developers to share coding tips and tricks
c. To have another developer examine the code for potential security issues and improvements
d. To compare the application's features with competitor products

Answer: c. To have another developer examine the code for potential security issues and improvements. Explanation: Peer review in the code review process involves one or more fellow developers examining the code to identify potential security issues, bugs, and areas for improvement, leveraging their expertise to enhance the code's security and quality.

Which aspect of security is specifically targeted by misuse case testing?
a. Usability under high traffic conditions
b. Compliance with data protection regulations
c. Vulnerability to specific types of cyberattacks
d. Efficiency of data processing and storage mechanisms

Answer: c. Vulnerability to specific types of cyberattacks. Explanation: Misuse case testing specifically targets the application's vulnerability to particular types of cyberattacks by simulating malicious actions or inputs, aiming to uncover how the system behaves under such conditions and identify potential security weaknesses.

In what way can automated code review tools be limited?
a. They can only review code written in older programming languages.
b. They may not always understand the context of the application, potentially missing business logic vulnerabilities.
c. They are less effective than manual reviews at identifying syntax errors.
d. They can only be used during the initial development phase of the application.

Answer: b. They may not always understand the context of the application, potentially missing business logic vulnerabilities. Explanation: While automated code review tools are effective at scanning for known patterns and vulnerabilities, they can be limited by their lack of understanding of the application's specific context and business logic, potentially missing vulnerabilities that require human insight to identify.

What is a key benefit of incorporating misuse case testing in the software development lifecycle?
a. It ensures that the application can handle peak user loads.
b. It helps in identifying and mitigating security risks early in the development process.
c. It guarantees compliance with all relevant software development standards.
d. It simplifies the code deployment process on various platforms.

Answer: b. It helps in identifying and mitigating security risks early in the development process. Explanation: Incorporating misuse case testing early in the software development lifecycle allows teams to identify and address security risks and vulnerabilities proactively, reducing the potential for security issues in the released version of the application.

How do smart IDE (Integrated Development Environment) plugins aid in the code review process for security?
a. By automatically writing secure code to replace vulnerable code
b. By providing real-time security guidance and identifying potential vulnerabilities as the code is written
c. By generating comprehensive user manuals that include security guidelines
d. By creating virtual environments for penetration testing within the IDE

Answer: b. By providing real-time security guidance and identifying potential vulnerabilities as the code is written. Explanation: Smart IDE plugins enhance the code review process by offering real-time security guidance to developers, identifying potential vulnerabilities and security anti-patterns as the code is being written, thereby helping to prevent the introduction of security flaws from the outset.

Why is it important for misuse case testing to include scenarios based on current threat intelligence?
a. To ensure the testing environment is as realistic as possible
b. To focus testing efforts on vulnerabilities that are currently being exploited in the wild
c. To comply with international software testing standards
d. To increase the processing speed of the application under test

Answer: b. To focus testing efforts on vulnerabilities that are currently being exploited in the wild. Explanation: Incorporating current threat intelligence into misuse case testing ensures that the testing scenarios are relevant and focused on vulnerabilities and attack vectors that are actively being exploited, thereby enhancing the application's resilience against real-world threats.

What is the primary purpose of conducting security audits within an organization?
a) To test the physical security controls like locks and access cards
b) To assess and ensure compliance with security policies and regulations
c) To deploy new security technologies and tools
d) To train employees on security best practices

Answer: b) To assess and ensure compliance with security policies and regulations. Explanation: The primary purpose of conducting security audits is to assess and ensure that the organization's security controls are effective and comply with internal security policies, as well as external regulatory requirements. Audits help identify security weaknesses and ensure that security measures are properly implemented and followed.

In log management, what does the term "log rotation" refer to?
a) The process of changing log file encryption keys periodically
b) The practice of regularly updating log file formats for better readability
c) The method of archiving old log files and starting new ones at regular intervals
d) The technique of rotating log files between different servers for load balancing

Answer: c) The method of archiving old log files and starting new ones at regular intervals. Explanation: Log rotation is a log management practice where old log files are archived, and new log files are started at regular intervals. This prevents log files from becoming too large, which can make them difficult to manage and analyze, and ensures that disk space is used efficiently.

Which type of security audit focuses on evaluating the technical controls and security measures of an organization's information systems?
a) Financial audit
b) Operational audit
c) Compliance audit
d) Technical audit

Answer: d) Technical audit. Explanation: A technical audit specifically focuses on evaluating the technical controls and security measures implemented within an organization's information systems. It assesses the effectiveness of these controls in protecting against unauthorized access, data breaches, and other cyber threats.

What is a Security Information and Event Management (SIEM) system primarily used for in log management?
a) To encrypt log files for secure storage
b) To generate log files from various sources
c) To collect, analyze, and report on log data from different systems and devices
d) To rotate and archive log files at specified intervals

Answer: c) To collect, analyze, and report on log data from different systems and devices. Explanation: A Security Information and Event Management (SIEM) system is primarily used to collect, analyze, and report on log data from various systems and devices across an organization's network. SIEM systems help in detecting anomalies, monitoring security events, and providing insights for incident response and forensic analysis.

During a security audit, what is the significance of reviewing change management logs?
a) To verify that hardware components are up-to-date
b) To ensure that all software installations are authorized
c) To confirm that changes to systems and applications are properly authorized, documented, and tested
d) To check the efficiency of the organization's data backup procedures

Answer: c) To confirm that changes to systems and applications are properly authorized, documented, and tested. Explanation: Reviewing change management logs during a security audit is significant because it helps confirm that all changes made to systems and applications have been properly authorized, documented, and tested before implementation. This ensures that changes do not introduce new vulnerabilities and that the integrity of the system is maintained.

How does centralized log management enhance security monitoring?
a) By allowing logs to be stored indefinitely without the need for rotation
b) By distributing logs across multiple servers to prevent data loss
c) By consolidating log data from various sources for more efficient analysis and correlation
d) By encrypting log data at rest and in transit between devices

Answer: c) By consolidating log data from various sources for more efficient analysis and correlation. Explanation: Centralized log management enhances security monitoring by consolidating log data from various sources into a single repository. This allows for more efficient analysis, correlation of events across different systems, and quicker detection of potential security incidents.

What role do audit trails play in incident response?
a) They provide a real-time feed of security events for immediate action
b) They offer a historical record of events to assist in investigating and understanding security incidents
c) They act as a preventive measure by blocking malicious activities
d) They serve as the primary method for restoring lost data

Answer: b) They offer a historical record of events to assist in investigating and understanding security incidents. Explanation: Audit trails play a crucial role in incident response by providing a historical record of events, which assists in investigating and understanding the sequence of actions that led to a security incident. This information is vital for determining the scope of the incident, identifying the affected systems, and taking appropriate remedial actions.

In the context of log management, what is the purpose of log normalization?
a) To ensure all log files conform to the same size and format for storage efficiency
b) To convert log data from different sources into a common format for easier analysis
c) To rotate and archive log files based on a predefined schedule
d) To increase the speed of log file encryption and decryption processes

Answer: b) To convert log data from different sources into a common format for easier analysis. Explanation: Log normalization involves converting log data from different sources and formats into a common, standardized format. This process facilitates easier and more effective analysis, especially when dealing with logs from diverse systems and applications.

What is the importance of setting appropriate log retention policies?
a) To ensure compliance with regulatory requirements and manage storage costs
b) To increase the processing speed of log analysis tools
c) To prevent unauthorized users from accessing old log files
d) To automatically escalate security events based on log data age

Answer: a) To ensure compliance with regulatory requirements and manage storage costs. Explanation: Setting appropriate log retention policies is important to ensure that an organization complies with regulatory requirements regarding data retention and to manage storage costs effectively. Retention policies determine how long log data should be kept before being archived or deleted, balancing the need for historical data with the cost and feasibility of storage.

What is a Key Performance Indicator (KPI) in the context of information security?
a. A confidential metric only accessible to senior management
b. A financial benchmark related to the cost of security investments
c. A measurable value that demonstrates how effectively a company is achieving key security objectives
d. A static measure that remains unchanged over time

Answer: c. A measurable value that demonstrates how effectively a company is achieving key security objectives. Explanation: In information security, a KPI is a quantifiable measure used to evaluate the success of an organization in meeting pre-defined security objectives. KPIs help in understanding the effectiveness of security measures and controls implemented within the organization.

Which of the following would be considered a Key Risk Indicator (KRI) in cybersecurity?
a. The percentage increase in sales due to a new online platform
b. The number of failed login attempts detected on a system per week
c. The average time employees spend on cybersecurity training
d. The total expenditure on cybersecurity technologies

Answer: b. The number of failed login attempts detected on a system per week. Explanation: A KRI is a metric used to indicate how risky an activity is and in cybersecurity, the number of failed login attempts can serve as a KRI, as it might indicate attempted unauthorized access, thereby signaling a potential security threat.

What security metric would be most useful for measuring the effectiveness of an incident response team?
a. The total number of security awareness training sessions conducted
b. The average time from detection to containment of a security incident
c. The percentage of employees with access to sensitive data
d. The annual budget allocated to the cybersecurity department

Answer: b. The average time from detection to containment of a security incident. Explanation: The average time from the detection of a security incident to its containment is a crucial metric for assessing the effectiveness of an incident response team, as it reflects the team's ability to quickly and effectively mitigate threats.

In a scenario where a new firewall is deployed, which metric would best indicate its effectiveness?
a. The number of new connections established through the firewall
b. The decrease in the volume of malicious traffic detected
c. The increase in internet bandwidth usage
d. The number of user complaints regarding network access

Answer: b. The decrease in the volume of malicious traffic detected. Explanation: The effectiveness of a new firewall can be best gauged by the decrease in the volume of detected malicious traffic. This indicates that the firewall is successfully blocking potential threats from entering the network.

How can Mean Time to Detect (MTTD) be categorized in security metrics?
a. As a KPI focusing on the proactive aspects of security monitoring
b. As a financial metric related to the cost of security breaches
c. As a KRI indicating the potential for future security breaches
d. As a metric unrelated to security, focusing on system performance

Answer: a. As a KPI focusing on the proactive aspects of security monitoring. Explanation: MTTD is a KPI that measures the effectiveness of an organization's security monitoring capabilities by quantifying the average time it takes to detect a security threat or incident, highlighting the proactive measures in place to identify potential security issues.

What role do KRIs play in cybersecurity risk management?
a. They provide real-time data on the financial impact of a cyberattack
b. They indicate the potential likelihood or impact of a cybersecurity risk materializing
c. They measure the effectiveness of cybersecurity training programs
d. They track the performance of individual security team members

Answer: b. They indicate the potential likelihood or impact of a cybersecurity risk materializing. Explanation: KRIs are critical in cybersecurity risk management as they help in identifying potential indicators of risk, allowing organizations to take preemptive measures to mitigate or prevent the materialization of these risks.

In the context of security controls, what would be an example of a KPI related to access control mechanisms?
a. The number of systems patched within the standard patching cycle
b. The percentage of users who have completed access control training
c. The ratio of successful to unsuccessful authentication attempts
d. The total number of access control policies documented

Answer: c. The ratio of successful to unsuccessful authentication attempts. Explanation: A KPI related to access control mechanisms could be the ratio of successful to unsuccessful authentication attempts, as it provides insight into the effectiveness of the access control system in allowing legitimate access while preventing unauthorized attempts.

What metric would be most relevant for assessing the performance of a security awareness program?
a. The number of third-party applications integrated with the company's systems
b. The decrease in phishing simulation click rates over time
c. The total number of firewall rules implemented

d. The increase in encrypted traffic on the company's network

Answer: b. The decrease in phishing simulation click rates over time. Explanation: The decrease in click rates during phishing simulations over time is a relevant metric for assessing the effectiveness of a security awareness program. It indicates an improvement in employees' ability to recognize and avoid phishing attempts, demonstrating the program's impact on their behavior.

How would a security team use the metric 'Mean Time to Repair (MTTR)'?
a. To calculate the average cost of repairing systems after a security breach
b. To determine the average time taken to restore a system to full functionality after a security incident
c. To assess the efficiency of the network infrastructure
d. To gauge the effectiveness of the firewall in blocking threats

Answer: b. To determine the average time taken to restore a system to full functionality after a security incident. Explanation: MTTR is used by security teams to quantify the average time required to repair and restore a system or service to full functionality following a security incident, providing insight into the efficiency and effectiveness of the incident response and recovery processes.

When evaluating the security posture of an organization, why is it important to distinguish between KPIs and KRIs?
a. KPIs measure performance towards achieving specific security objectives, while KRIs indicate potential risk exposure levels, providing a more comprehensive view of security health.
b. KPIs are only used by management, whereas KRIs are used by technical staff.
c. KRIs are financial metrics, whereas KPIs are technical metrics.
d. KPIs are related to external threats, while KRIs focus on internal processes.

Answer: a. KPIs measure performance towards achieving specific security objectives, while KRIs indicate potential risk exposure levels, providing a more comprehensive view of security health. Explanation: Distinguishing between KPIs and KRIs is crucial because KPIs focus on measuring the effectiveness and performance of security initiatives, while KRIs provide insights into the potential and actual exposures to risks, together offering a holistic view of an organization's security posture.

What is the primary goal of conducting phishing simulation exercises within an organization?
a) To identify and remove malicious software from corporate systems
b) To test the efficiency of the email filtering system
c) To enhance employees' ability to recognize and respond to phishing attempts
d) To penalize employees who fail to identify phishing emails

Answer: c) To enhance employees' ability to recognize and respond to phishing attempts. Explanation: The primary goal of conducting phishing simulation exercises is to train and sensitize employees to recognize and appropriately respond to phishing attempts. By simulating real-life phishing scenarios, organizations can effectively gauge and improve their workforce's awareness and preparedness against such cyber threats.

How should the effectiveness of a phishing simulation campaign be measured?
a) By the number of phishing emails sent to employees
b) By the percentage of employees who report the simulated phishing email
c) By the complexity of the phishing email content
d) By the time it takes for IT staff to respond to reported phishing emails

Answer: b) By the percentage of employees who report the simulated phishing email. Explanation: The effectiveness of a phishing simulation campaign is best measured by the percentage of employees who correctly identify and report the simulated phishing email. This metric indicates the level of security awareness among employees and the effectiveness of the training they have received.

What key factor should be considered when designing phishing simulation emails?
a) The emails should be easy to identify as phishing to boost employee confidence

b) The emails should closely mimic real phishing attempts in tone, style, and content
c) The emails should contain obvious grammatical and spelling errors
d) The emails should always target high-profile employees to assess organizational risk

Answer: b) The emails should closely mimic real phishing attempts in tone, style, and content. Explanation: When designing phishing simulation emails, it is crucial that they closely mimic real phishing attempts in terms of tone, style, and content. This realism ensures that the simulation is an effective training tool, preparing employees to recognize and respond to actual phishing threats they may encounter.

In the context of phishing simulations, what is a "click rate"?
a) The speed at which IT personnel respond to phishing incident reports
b) The percentage of emails that reach the recipient's inbox
c) The ratio of employees who clicked on a link or attachment in the simulation email
d) The frequency of phishing simulation emails sent per year

Answer: c) The ratio of employees who clicked on a link or attachment in the simulation email. Explanation: In the context of phishing simulations, the "click rate" refers to the ratio or percentage of employees who clicked on a link or attachment in the simulated phishing email. This metric is used to assess the vulnerability of the workforce to phishing attacks and the effectiveness of the current security awareness training.

What follow-up action is recommended for employees who fall for a phishing simulation?
a) Immediate termination of employment
b) Temporary suspension of email privileges
c) Targeted security awareness training
d) Public disclosure of their names for accountability

Answer: c) Targeted security awareness training. Explanation: For employees who fall for a phishing simulation, the recommended follow-up action is to provide targeted security awareness training. This approach focuses on educating the employee about the specific mistake made and how to avoid such pitfalls in the future, reinforcing the organization's commitment to continuous improvement in security awareness.

How can phishing simulations contribute to a culture of security within an organization?
a) By fostering an environment of distrust and skepticism among employees
b) By encouraging a proactive mindset and vigilance towards cybersecurity threats
c) By shifting the responsibility of security from IT departments to individual employees
d) By implementing strict penalties for those who fail phishing tests

Answer: b) By encouraging a proactive mindset and vigilance towards cybersecurity threats. Explanation: Phishing simulations contribute to a culture of security by encouraging a proactive mindset and vigilance towards cybersecurity threats among employees. Regular simulations and training foster an environment where employees are continuously alert to the possibility of phishing attacks, reinforcing the importance of cybersecurity in their daily activities.

What is an important consideration when scheduling phishing simulation campaigns?
a) Conducting simulations outside of working hours to avoid disrupting productivity
b) Ensuring simulations are frequent enough to keep security awareness high without causing fatigue
c) Limiting simulations to once a year to maintain the element of surprise
d) Focusing simulations exclusively on new employees who are more likely to be vulnerable

Answer: b) Ensuring simulations are frequent enough to keep security awareness high without causing fatigue. Explanation: An important consideration when scheduling phishing simulation campaigns is to ensure that they are frequent enough to maintain high levels of security awareness but not so frequent that they cause fatigue or desensitization among employees. Finding the right balance is crucial for the effectiveness of the training program.

What role does feedback play in phishing simulation exercises?
a) It is used to penalize departments with the highest failure rates

b) It provides actionable insights and learning opportunities for participants
c) It is only given to employees who successfully identify the phishing email
d) It serves as a metric for reducing the frequency of future simulations

Answer: b) It provides actionable insights and learning opportunities for participants. Explanation: Feedback plays a critical role in phishing simulation exercises by providing actionable insights and learning opportunities for participants. Whether an employee falls for the simulation or successfully identifies it, constructive feedback helps them understand their actions, reinforces correct behaviors, and educates them on how to improve their response to future threats.

What is the significance of including a variety of phishing tactics in simulation campaigns?
a) To ensure that the simulations are legally compliant
b) To reduce the overall cost of the security awareness program
c) To expose employees to a wide range of phishing techniques and increase their ability to detect threats
d) To identify which employees are most likely to be targeted in real attacks

Answer: c) To expose employees to a wide range of phishing techniques and increase their ability to detect threats. Explanation: Including a variety of phishing tactics in simulation campaigns is significant because it exposes employees to a broad spectrum of phishing techniques. This diversity in training helps employees become more adept at recognizing a wide range of threats, enhancing their overall ability to detect and respond to different types of phishing attacks.

What is the primary purpose of digital forensics in cybersecurity investigations?
a. To repair systems damaged by cyberattacks
b. To identify, collect, examine, and preserve digital evidence to support or refute a hypothesis or theory
c. To monitor network traffic in real-time for threat detection
d. To implement security controls post-incident

Answer: b. To identify, collect, examine, and preserve digital evidence to support or refute a hypothesis or theory. Explanation: Digital forensics involves a detailed process aimed at identifying, collecting, examining, and preserving digital evidence from various digital devices and sources. This evidence is crucial for understanding how a cybersecurity incident occurred, who was responsible, and how similar incidents can be prevented in the future, as well as for legal proceedings if necessary.

In the context of incident response, what does the chain of custody refer to?
a. The chronological documentation showing the seizure, custody, control, transfer, analysis, and disposition of digital evidence
b. The hierarchical command structure within the incident response team
c. The sequence of events that led to the incident
d. The timeline of system recovery and restoration post-incident

Answer: a. The chronological documentation showing the seizure, custody, control, transfer, analysis, and disposition of digital evidence. Explanation: The chain of custody in digital forensics and incident response is a critical process that involves documenting each step of handling digital evidence, from collection to analysis to presentation in court, ensuring the integrity and reliability of the evidence by tracking who had possession of it and when.

What is a critical first step in preserving digital evidence during an incident response?
a. Immediately isolating affected systems from the network
b. Making a forensic image of affected systems
c. Notifying law enforcement agencies
d. Conducting a root cause analysis

Answer: b. Making a forensic image of affected systems. Explanation: Making a forensic image or copy of affected systems is a critical first step in preserving digital evidence. This ensures that all data is captured in its current state, including potentially volatile memory, which is crucial for a thorough investigation while allowing affected systems to be restored and brought back online with minimal disruption.

Which of the following scenarios best illustrates the use of misuse case testing in digital investigations?
a. Simulating a DDoS attack to test network resilience
b. Conducting a phishing campaign to assess employee awareness
c. Analyzing system logs for indicators of compromise post-incident
d. Creating scenarios based on potential attacker behaviors to identify system vulnerabilities

Answer: d. Creating scenarios based on potential attacker behaviors to identify system vulnerabilities. Explanation: Misuse case testing involves creating scenarios that simulate potential attacker actions or behaviors to identify and address system vulnerabilities before an incident occurs. This proactive approach helps in strengthening the security posture of systems and can provide valuable insights during investigations by understanding how an attacker might exploit the system.

How does maintaining a detailed chain of custody benefit the incident response process?
a. By ensuring rapid system recovery
b. By providing a clear audit trail of evidence handling, which is crucial for legal admissibility and integrity
c. By outlining the steps needed for system patching
d. By assigning specific roles and responsibilities within the incident response team

Answer: b. By providing a clear audit trail of evidence handling, which is crucial for legal admissibility and integrity. Explanation: Maintaining a detailed chain of custody provides a clear and auditable trail of how evidence was handled, preserved, and analyzed, which is essential for ensuring the integrity and admissibility of evidence in legal proceedings. It helps to establish trust in the evidence collected and the investigation's findings.

In incident response, what role does an intrusion detection system (IDS) play in digital forensics?
a. It serves as the primary tool for evidence collection and analysis
b. It acts as a preventive measure by blocking all intrusions
c. It provides alerts and logs that can serve as initial indicators of compromise for further investigation
d. It is used exclusively for restoring systems to their pre-incident state

Answer: c. It provides alerts and logs that can serve as initial indicators of compromise for further investigation. Explanation: An intrusion detection system (IDS) is instrumental in the early stages of an incident response, as it provides alerts and logs that can indicate suspicious activities or potential security breaches. These initial indicators of compromise are crucial for guiding further forensic analysis and investigation efforts.

What distinguishes a live forensic analysis from traditional static forensic analysis?
a. Live forensics is performed only on systems that are offline
b. Live forensics involves analyzing volatile data in memory on a running system
c. Static forensic analysis is used for immediate incident response
d. Live forensics does not require documentation or a chain of custody

Answer: b. Live forensics involves analyzing volatile data in memory on a running system. Explanation: Live forensic analysis is distinguished by its focus on collecting and analyzing data from a system that is still operational, particularly volatile data stored in memory, which can be lost if the system is powered down. This approach is crucial for capturing real-time data that provides insights into the state of the system and ongoing activities at the time of the incident.

Why is it important to have predefined security metrics and KPIs for incident response?
a. To ensure that all incidents are reported to law enforcement
b. To measure the effectiveness and efficiency of the incident response process
c. To calculate the financial impact of every incident
d. To determine the technical skills required for the incident response team

Answer: b. To measure the effectiveness and efficiency of the incident response process. Explanation: Predefined security metrics and KPIs are essential for evaluating how effectively and efficiently the incident response process is being conducted. These metrics can include time to detect and respond to incidents, impact mitigation effectiveness, and overall improvements in security posture, providing valuable feedback for continuous improvement.

In the aftermath of a cybersecurity incident, how can KRIs be utilized?
a. To predict the exact time of the next incident
b. To assess the potential future impact of similar incidents and guide risk management strategies
c. To replace the need for an incident response plan
d. To provide detailed technical analysis of the malware used

Answer: b. To assess the potential future impact of similar incidents and guide risk management strategies. Explanation: Key Risk Indicators (KRIs) can be utilized in the aftermath of a cybersecurity incident to evaluate the potential impact of similar future incidents. By analyzing these indicators, organizations can better understand their risk exposure and refine their risk management and incident response strategies to mitigate potential threats more effectively.

What is a primary security concern when provisioning resources in a public cloud environment?
a) Reduced physical access to the data center
b) Incompatibility with on-premises infrastructure
c) Data residency and sovereignty issues
d) Increased network latency

Answer: c) Data residency and sovereignty issues. Explanation: When provisioning resources in a public cloud environment, a primary security concern is data residency and sovereignty. This pertains to the laws and regulations that govern the physical location of data, which can affect privacy, compliance, and legal jurisdiction. Organizations must ensure that their use of cloud resources complies with relevant laws related to where their data is stored and processed.

In virtualization, what is the role of the hypervisor?
a) To act as an intermediary between virtual machines and physical hardware
b) To encrypt data at rest and in transit
c) To provide a graphical interface for managing network configurations
d) To monitor and filter network traffic for virtual machines

Answer: a) To act as an intermediary between virtual machines and physical hardware. Explanation: The hypervisor, in a virtualized environment, acts as an intermediary between virtual machines (VMs) and the physical hardware. It allows multiple VMs to share physical resources such as CPU, memory, and storage, while managing resource allocation and ensuring isolation among VMs.

Which cloud deployment model allows for shared infrastructure among various organizations while maintaining separation between user environments?
a) Public cloud
b) Private cloud
c) Community cloud
d) Hybrid cloud

Answer: c) Community cloud. Explanation: The community cloud deployment model is designed to allow multiple organizations to share infrastructure and resources while maintaining separation between their user environments. This model is often used by organizations with similar requirements and concerns, such as regulatory compliance, to achieve economies of scale without compromising security or privacy.

What is the main advantage of using containerization in cloud environments?
a) Increased data encryption capabilities
b) Enhanced physical security of the data centers

c) Improved application portability and efficiency
d) Simplified network configuration and management

Answer: c) Improved application portability and efficiency. Explanation: The main advantage of using containerization in cloud environments is improved application portability and efficiency. Containers package an application and its dependencies into a single executable unit, which can run consistently across different computing environments, making applications easier to deploy, scale, and manage.

How does multi-tenancy in cloud services impact security considerations?
a) It eliminates the need for data encryption.
b) It reduces the complexity of compliance auditing.
c) It necessitates robust isolation mechanisms to prevent data leaks between tenants.
d) It guarantees absolute data privacy by default.

Answer: c) It necessitates robust isolation mechanisms to prevent data leaks between tenants. Explanation: Multi-tenancy, where multiple customers (tenants) share the same infrastructure or application instances, necessitates robust isolation mechanisms to ensure that actions and data of one tenant do not impact or leak into another tenant's environment. This is crucial for maintaining privacy and security in a shared environment.

What is the significance of "elasticity" in cloud computing?
a) The ability to physically expand data center facilities
b) The capability to automatically scale resources to match demand
c) The feature that allows for manual network reconfigurations
d) The process of transitioning all resources to renewable energy sources

Answer: b) The capability to automatically scale resources to match demand. Explanation: Elasticity in cloud computing refers to the capability of cloud resources to be automatically scaled up or down to match demand. This allows for efficient resource utilization, ensuring that users have access to the resources they need when they need them, and avoid paying for idle capacity.

In the context of virtualized networking, what does "network function virtualization" (NFV) entail?
a) Replacing physical network appliances with software-based virtual appliances
b) Encrypting all data packets transmitted across the network
c) Creating physical network connections between virtual machines
d) Increasing the bandwidth of virtual network connections

Answer: a) Replacing physical network appliances with software-based virtual appliances. Explanation: Network Function Virtualization (NFV) involves replacing traditional physical network appliances, such as routers and firewalls, with software-based virtual appliances that can run on standard hardware. This increases flexibility and scalability while reducing dependency on specialized hardware.

When provisioning cloud resources, what practice helps in managing costs effectively?
a) Always choosing the highest available capacity for resources
b) Utilizing auto-scaling features to adjust resources based on load
c) Consolidating all workloads onto a single large instance
d) Disabling monitoring and logging to save on storage costs

Answer: b) Utilizing auto-scaling features to adjust resources based on load. Explanation: Utilizing auto-scaling features in cloud environments helps manage costs effectively by automatically adjusting the amount of computational resources based on the current load. This ensures that you pay only for the resources you need at any given time, avoiding overprovisioning and underutilization.

What is a key security feature to look for in a cloud service provider's offering?
a) The physical location of their data centers
b) The availability of detailed usage logs for auditing and compliance

c) The color scheme of their web management console
d) The speed of their customer service response

Answer: b) The availability of detailed usage logs for auditing and compliance. Explanation: A key security feature to look for in a cloud service provider's offering is the availability of detailed usage logs, which are crucial for security auditing, compliance, and forensics. These logs provide visibility into the actions performed within the cloud environment, helping to track access, changes, and potential security incidents.

In a virtualized environment, what is the purpose of using VLANs (Virtual Local Area Networks)?
a) To physically separate network hardware
b) To encrypt data stored on virtual machines
c) To segment network traffic for improved performance and security
d) To increase the processing power available to virtual machines

Answer: c) To segment network traffic for improved performance and security. Explanation: In a virtualized environment, VLANs are used to segment network traffic, which can improve both performance and security. By creating distinct broadcast domains within a shared network infrastructure, VLANs can isolate sensitive traffic, enforce security policies, and reduce network congestion.

What is the primary function of Data Loss Prevention (DLP) systems in corporate environments?
a. To monitor, detect, and prevent unauthorized access to or use of sensitive data
b. To encrypt data transmissions across the network
c. To facilitate secure data backup and recovery processes
d. To manage user access rights and permissions

Answer: a. To monitor, detect, and prevent unauthorized access to or use of sensitive data. Explanation: DLP systems are designed to protect sensitive data from unauthorized access, use, or transfer. They monitor data at rest, in use, and in motion across the network and endpoints, detecting and preventing potential data breaches or leaks by enforcing security policies.

How does endpoint protection differ from traditional antivirus software?
a. Endpoint protection only secures mobile devices, while antivirus is for desktops
b. Endpoint protection offers comprehensive security features, including antivirus, firewall, and intrusion detection
c. Traditional antivirus focuses on data encryption, whereas endpoint protection does not
d. Endpoint protection is used exclusively for data at rest; traditional antivirus is for data in transit

Answer: b. Endpoint protection offers comprehensive security features, including antivirus, firewall, and intrusion detection. Explanation: Endpoint protection platforms provide a broader range of security capabilities compared to traditional antivirus software. These include not only antivirus but also firewall services, intrusion detection and prevention, and more, offering a holistic approach to securing endpoints against various types of cyber threats.

In the context of DLP, what is content-aware protection primarily concerned with?
a. The physical security of data storage devices
b. Monitoring and protecting data based on its content and context
c. The speed of data transmission across networks
d. The version control of documents and files

Answer: b. Monitoring and protecting data based on its content and context. Explanation: Content-aware DLP focuses on understanding the content and context of the data, such as the presence of sensitive information within files or communications. This allows for more precise control and protection measures to be applied based on the actual content of the data, rather than just file types or transmission methods.

What role does encryption play in Data Loss Prevention strategies?
a. It serves as the only necessary component for complete data protection
b. It obfuscates data to prevent unauthorized users from understanding it, even if accessed

c. It detects anomalies in data usage patterns
d. It restricts access to data based on user roles and permissions

Answer: b. It obfuscates data to prevent unauthorized users from understanding it, even if accessed. Explanation: Encryption is a critical component of DLP strategies because it transforms readable data into an unreadable format for unauthorized users. Even if sensitive data is accessed or transferred without authorization, encryption ensures that the data remains unintelligible and secure from unauthorized use.

How do endpoint protection platforms (EPP) typically identify malware?
a. By creating a whitelist of approved software
b. Through user behavior analytics exclusively
c. Using signatures, heuristics, and behavioral analysis
d. Solely based on the reputation of IP addresses

Answer: c. Using signatures, heuristics, and behavioral analysis. Explanation: EPPs use a combination of signature-based detection, heuristic analysis, and behavioral analysis to identify and block malware. Signature-based detection compares files against a database of known threats, heuristic analysis looks for characteristic features of malware, and behavioral analysis monitors the behavior of software for suspicious activities, providing comprehensive malware protection.

What is the benefit of integrating DLP with cloud access security brokers (CASBs)?
a. It limits DLP to only function with on-premises data storage solutions
b. It extends DLP policies and protections to cloud applications and services
c. It replaces the need for encryption in cloud environments
d. It focuses DLP efforts solely on endpoint devices

Answer: b. It extends DLP policies and protections to cloud applications and services. Explanation: Integrating DLP with CASBs allows organizations to extend their data protection policies and controls to cloud environments, ensuring that sensitive data is monitored and protected not only on-premises but also in cloud applications and services, providing a cohesive data security posture across all platforms.

In what scenario might an organization prioritize deploying an Endpoint Detection and Response (EDR) solution?
a. When their primary concern is enhancing the physical security of their data centers
b. To improve visibility and response capabilities to advanced threats on endpoints
c. When they need to ensure compliance with data residency regulations
d. For the sole purpose of monitoring employee productivity

Answer: b. To improve visibility and response capabilities to advanced threats on endpoints. Explanation: EDR solutions are prioritized when an organization needs to enhance its ability to detect, investigate, and respond to advanced threats on endpoints. EDR provides comprehensive visibility into endpoint activities, offering tools for threat hunting, investigation, and remediation, which are crucial for defending against sophisticated cyber threats.

What distinguishes cloud-based endpoint protection from on-premises solutions?
a. Cloud-based solutions are less secure due to their internet dependency
b. On-premises solutions offer real-time global threat intelligence updates
c. Cloud-based solutions provide scalability and real-time updates from global threat intelligence
d. On-premises solutions do not require internet access to function

Answer: c. Cloud-based solutions provide scalability and real-time updates from global threat intelligence. Explanation: Cloud-based endpoint protection platforms offer significant advantages in terms of scalability and the ability to receive real-time threat intelligence updates from across the globe. This ensures that the protection is always up-to-date and can scale with the organization's needs without the hardware and maintenance requirements of on-premises solutions.

How does the implementation of a DLP program impact an organization's compliance posture?

a. It ensures automatic compliance with all industry standards without further action required
b. It may complicate compliance efforts by introducing too many controls
c. It supports compliance by preventing unauthorized disclosure of regulated data
d. It has no impact on compliance but improves the organization's security

Answer: c. It supports compliance by preventing unauthorized disclosure of regulated data. Explanation: Implementing a DLP program significantly bolsters an organization's compliance posture by establishing controls and mechanisms to prevent the unauthorized disclosure of sensitive or regulated data. This is crucial for adhering to various privacy and data protection regulations, thereby reducing the risk of compliance violations and associated penalties.

What is the primary goal of incident response as outlined by NIST SP 800-61?
a) To prosecute the perpetrators of cyber incidents
b) To minimize damage and recover from incidents rapidly
c) To monitor employee activities continuously
d) To encrypt all data stored on affected systems

Answer: b) To minimize damage and recover from incidents rapidly. Explanation: According to NIST SP 800-61, the primary goal of incident response is to minimize damage caused by incidents and to recover from them as rapidly and efficiently as possible. This includes containing the incident, eradicating the cause, recovering systems to normal operation, and preventing recurrence.

Which phase of the NIST SP 800-61 incident response lifecycle involves identifying whether an incident has occurred?
a) Preparation
b) Detection and Analysis
c) Containment, Eradication, and Recovery
d) Post-Incident Activity

Answer: b) Detection and Analysis. Explanation: The Detection and Analysis phase of the NIST SP 800-61 incident response lifecycle involves identifying and investigating potential security incidents to determine whether an incident has indeed occurred, understanding its nature, and devising an appropriate response strategy.

In the context of NIST SP 800-61, what is a crucial first step in preparing for incident response?
a) Purchasing the latest cybersecurity tools
b) Developing and documenting incident response policies and procedures
c) Waiting for an incident to occur to understand the threat landscape
d) Hiring a third-party incident response team after an incident

Answer: b) Developing and documenting incident response policies and procedures. Explanation: A crucial first step in preparing for incident response, as per NIST SP 800-61, is to develop and document comprehensive incident response policies, plans, and procedures. This foundational work ensures that the organization has a clear, structured approach to dealing with incidents, including roles, responsibilities, and actions to be taken.

How does "Containment" in the incident response process contribute to minimizing damage?
a) By immediately shutting down all systems
b) By isolating affected systems to prevent the spread of the incident
c) By encrypting all data to protect confidentiality
d) By publicly announcing the incident to warn other organizations

Answer: b) By isolating affected systems to prevent the spread of the incident. Explanation: Containment strategies in the incident response process, such as isolating affected systems or segments of the network, are critical for preventing the spread of an incident. This step ensures that the impact is limited, protecting unaffected systems and data from compromise.

What role does "Eradication" play in the incident response process according to NIST SP 800-61?
a) To enhance the encryption of data across the network

b) To identify all affected parties and inform them of the incident
c) To remove the cause of the incident and eliminate vulnerabilities
d) To conduct training sessions for employees on incident prevention

Answer: c) To remove the cause of the incident and eliminate vulnerabilities. Explanation: Eradication involves eliminating the root cause of the incident, such as removing malware from infected systems and addressing the vulnerabilities that were exploited. This step is crucial to ensure that the incident is fully resolved and similar incidents are prevented in the future.

During the "Recovery" phase of incident response, what is a key activity?
a) Assigning blame to the individuals responsible for the incident
b) Restoring systems and data from clean backups
c) Purchasing new hardware to replace all affected systems
d) Implementing stronger network firewalls across the board

Answer: b) Restoring systems and data from clean backups. Explanation: A key activity during the Recovery phase of incident response is restoring affected systems and data from clean, verified backups. This ensures that business operations can be resumed safely and efficiently, with minimal data loss.

Why is "Lessons Learned" an essential part of the Post-Incident Activity phase?
a) To fulfill insurance reporting requirements
b) To provide a record for legal proceedings
c) To improve the incident response process for future incidents
d) To allocate the budget for security investments

Answer: c) To improve the incident response process for future incidents. Explanation: The "Lessons Learned" session is an essential part of the Post-Incident Activity phase because it provides an opportunity to review and analyze the incident response process, identifying what was successful and what could be improved. This feedback loop is crucial for refining the organization's incident response capabilities.

In NIST SP 800-61, how is "incident categorization" beneficial in the incident response process?
a) It determines the financial compensation for damages.
b) It helps in prioritizing incidents based on their impact and severity.
c) It is used to choose which incidents to ignore.
d) It serves as the basis for legal action against perpetrators.

Answer: b) It helps in prioritizing incidents based on their impact and severity. Explanation: Incident categorization is beneficial because it helps responders prioritize their efforts based on the impact and severity of incidents. By categorizing incidents, organizations can allocate resources more effectively, ensuring that the most critical incidents are addressed promptly.

What is the importance of having an "Incident Response Team" as outlined in NIST SP 800-61?
a) To handle all IT support tickets within the organization
b) To manage the organization's social media presence during an incident
c) To coordinate and manage the response to security incidents
d) To conduct daily data backup operations

Answer: c) To coordinate and manage the response to security incidents. Explanation: An Incident Response Team is crucial as it is responsible for coordinating and managing the response to security incidents. This team is composed of members with various skills and responsibilities, ensuring a comprehensive and effective approach to incident management.

How does "communication" play a role in the effectiveness of incident response according to NIST SP 800-61?
a) It is only necessary for communicating with external media outlets.
b) It involves sharing detailed incident information with the public immediately.

c) It ensures that all stakeholders are informed and that response efforts are coordinated.
d) It is limited to communication within the Incident Response Team.

Answer: c) It ensures that all stakeholders are informed and that response efforts are coordinated. Explanation: Effective communication is vital in incident response to ensure that all stakeholders, including management, IT staff, and possibly affected users, are kept informed about the incident and the response efforts. Coordinated communication helps in managing the incident efficiently and maintaining trust.

What is the primary purpose of conducting Disaster Recovery Plan (DRP) testing?
a. To evaluate the IT team's response time
b. To verify the effectiveness and efficiency of the DRP in a simulated disaster scenario
c. To train new employees on the company's software systems
d. To assess the financial impact of potential disasters

Answer: b. To verify the effectiveness and efficiency of the DRP in a simulated disaster scenario. Explanation: DRP testing is crucial to ensure that the disaster recovery plan is effective and efficient in restoring business operations after a disaster. Through simulation or actual enactment, the organization can identify any weaknesses or gaps in the plan and make necessary adjustments.

Which backup strategy involves creating a copy of only the data that has changed since the last full backup?
a. Differential backup
b. Incremental backup
c. Mirror backup
d. Full backup

Answer: b. Incremental backup. Explanation: Incremental backup strategy involves backing up only the data that has changed since the last backup, whether it was a full or another incremental backup. This method minimizes storage requirements and speeds up the backup process but may require a longer recovery time as all incremental backups since the last full backup must be restored.

In disaster recovery, what is the significance of the Recovery Point Objective (RPO)?
a. It defines the maximum tolerable length of time that data might be lost due to a disaster
b. It specifies the minimum internet bandwidth required during the recovery process
c. It identifies the key personnel involved in the disaster recovery process
d. It outlines the specific steps to be taken immediately following a disaster

Answer: a. It defines the maximum tolerable length of time that data might be lost due to a disaster. Explanation: The Recovery Point Objective (RPO) is a critical metric in disaster recovery planning that defines the maximum age of files or data in backup storage that an organization can tolerate losing in the event of a disaster. It essentially determines how frequently data backups should be performed.

What distinguishes a hot site in disaster recovery environments?
a. It is a location where backups are stored at room temperature
b. It is a fully equipped data center ready to take over operations at a moment's notice
c. It is a backup site that requires significant setup time before becoming operational
d. It is a location used exclusively for DRP testing, not actual disaster recovery

Answer: b. It is a fully equipped data center ready to take over operations at a moment's notice. Explanation: A hot site is a fully equipped and operational data center that an organization can immediately switch to in the event of a disaster at the primary site. It includes hardware, software, telecommunications, and personnel required to resume business operations with minimal downtime.

How does a cold site differ from a hot site in disaster recovery planning?
a. A cold site is only used for data storage, while a hot site includes operational servers and networks
b. A cold site is a fully operational duplicate of the primary site, whereas a hot site requires setup and configuration

c. A cold site is an alternative facility that is not immediately operational and requires setup, whereas a hot site is fully equipped and ready
d. A cold site is used for short-term recovery, while a hot site is for long-term recovery

Answer: c. A cold site is an alternative facility that is not immediately operational and requires setup, whereas a hot site is fully equipped and ready. Explanation: A cold site provides the necessary infrastructure but lacks the equipment and resources found in a hot site. It offers a cost-effective solution for disaster recovery but requires more time to become operational after a disaster.

What role does a Business Impact Analysis (BIA) play in DRP testing?
a. It determines the budget for the DRP testing process
b. It identifies critical systems and processes and their tolerance for downtime and data loss
c. It assesses the performance of the IT department
d. It focuses solely on the recovery of physical assets

Answer: b. It identifies critical systems and processes and their tolerance for downtime and data loss. Explanation: A Business Impact Analysis (BIA) is a foundational element of DRP testing that identifies critical business functions and the impact of their disruption. It helps in prioritizing the recovery of systems and processes based on their criticality and the acceptable downtime and data loss (RTO and RPO).

What is the purpose of a tabletop exercise in disaster recovery testing?
a. To physically relocate staff to an alternate recovery site
b. To perform a technical, system-level recovery test of specific components
c. To discuss and walk through the disaster recovery plan in a simulated scenario without actual system recovery
d. To test the electrical systems in a backup data center

Answer: c. To discuss and walk through the disaster recovery plan in a simulated scenario without actual system recovery. Explanation: Tabletop exercises are discussion-based sessions where team members walk through the disaster recovery plan in a simulated disaster scenario. This type of exercise is crucial for identifying gaps in the plan and ensuring that all team members understand their roles and responsibilities during an actual disaster.

Which backup strategy is most effective for ensuring quick recovery times?
a. Full backup performed weekly
b. Incremental backup performed daily
c. Differential backup performed nightly
d. Continuous data protection

Answer: d. Continuous data protection. Explanation: Continuous data protection (CDP) involves continuously capturing or tracking data modifications, allowing for the recovery of data to any point in time. This strategy offers the most efficient recovery process since it minimizes data loss and enables rapid restoration.

How do Recovery Time Objective (RTO) and Recovery Point Objective (RPO) influence backup strategies?
a. Lower RTO and RPO values typically require more frequent backups and potentially higher costs
b. Higher RTO values suggest a need for less secure backup storage facilities
c. Lower RPO values indicate that backups can be stored offsite only
d. Higher RTO values allow for longer data retention periods in backups

Answer: a. Lower RTO and RPO values typically require more frequent backups and potentially higher costs. Explanation: Lower RTO and RPO values indicate a need for quicker recovery times and less data loss in the event of a disaster, which usually necessitates more frequent backups and advanced technologies, potentially increasing the cost of the disaster recovery solution.

In the context of DRP testing, what is a full interruption test?
a. A test where only non-critical systems are shut down to simulate a disaster
b. A test that involves a complete shutdown of the primary site to validate the disaster recovery plan

c. A test conducted exclusively on the backup systems without affecting the primary site
d. A theoretical test where the disaster scenario is only discussed, not enacted

Answer: b. A test that involves a complete shutdown of the primary site to validate the disaster recovery plan. Explanation: A full interruption test is one of the most comprehensive forms of DRP testing, where the primary site's operations are fully halted to simulate a disaster. This test validates the effectiveness of the disaster recovery plan in a real-world scenario but carries significant risk and potential disruption to business operations.

Which principle of Crime Prevention Through Environmental Design (CPTED) focuses on ensuring that public and private spaces are clearly delineated?
a) Territorial Reinforcement
b) Natural Surveillance
c) Natural Access Control
d) Maintenance

Answer: a) Territorial Reinforcement. Explanation: Territorial Reinforcement is a CPTED principle that emphasizes the clear demarcation between public and private spaces through the use of physical elements like landscaping, fencing, and signage. This distinction enhances a sense of ownership and encourages individuals to protect the area, thereby deterring unauthorized access and potential criminal activities.

How does "Natural Surveillance" in CPTED principles contribute to physical security?
a) By using artificial intelligence to monitor CCTV footage
b) By designing the environment to maximize visibility and observation
c) By deploying security personnel in all areas
d) By implementing biometric access controls

Answer: b) By designing the environment to maximize visibility and observation. Explanation: Natural Surveillance, a key principle of CPTED, involves designing the physical environment in a way that maximizes visibility and natural observation. Features such as strategic placement of windows, lighting, and landscape design can enable individuals in the area to see and be seen, which acts as a deterrent to criminal behavior by increasing the perceived risk of detection.

In the context of physical security, what role does security lighting play?
a) To solely provide aesthetic enhancements to a building
b) To illuminate areas for the purpose of deterring criminal activity and increasing safety
c) To highlight escape routes exclusively during emergencies
d) To signal alarms during security breaches

Answer: b) To illuminate areas for the purpose of deterring criminal activity and increasing safety. Explanation: Security lighting plays a critical role in physical security by illuminating areas around buildings and landscapes, thereby deterring criminal activity and increasing safety. Proper lighting can enhance natural surveillance, make individuals feel safer, and reduce dark spots where criminals might hide, thus contributing to a more secure environment.

Which CPTED principle emphasizes the importance of keeping landscapes and buildings well-maintained to deter criminal activity?
a) Maintenance
b) Target Hardening
c) Activity Support
d) Natural Access Control

Answer: a) Maintenance. Explanation: The Maintenance principle of CPTED underscores the significance of keeping landscapes, buildings, and public spaces well-maintained to deter criminal activity. A well-maintained area signals that the space is actively used and monitored, which can discourage potential offenders by suggesting a lower likelihood of successful criminal behavior.

What is the purpose of "Natural Access Control" in CPTED?
a) To use physical barriers and security personnel to restrict movement
b) To guide people entering and leaving a space through strategically designed routes
c) To control access through electronic means like key cards and biometrics
d) To monitor all access points with surveillance cameras

Answer: b) To guide people entering and leaving a space through strategically designed routes. Explanation: Natural Access Control, a principle of CPTED, aims to guide people entering and leaving a space through the use of strategically designed and placed physical elements, such as paths, landscaping, and signage. This approach helps to deter access to unauthorized areas by subtly directing traffic flow and reducing opportunities for criminal activity.

How can "Activity Support" in CPTED principles enhance physical security in a community space?
a) By installing more CCTV cameras in the area
b) By promoting activities that encourage legitimate use of the space
c) By increasing the height of perimeter fencing
d) By implementing stricter access control measures

Answer: b) By promoting activities that encourage legitimate use of the space. Explanation: Activity Support, one of the CPTED principles, enhances physical security by promoting legitimate activities in community spaces that encourage the presence of more people. This increased occupancy and ownership of the space can deter criminal activities, as the area is more likely to be under natural surveillance by the community members themselves.

What is the impact of "Target Hardening" in physical security measures?
a) Making targets less appealing to potential offenders by increasing the effort required to commit a crime
b) Reducing the need for security personnel by automating all security processes
c) Focusing exclusively on electronic security measures
d) Encouraging the public to participate in neighborhood watch programs

Answer: a) Making targets less appealing to potential offenders by increasing the effort required to commit a crime. Explanation: Target Hardening involves implementing physical security measures that make a target less appealing to potential offenders by increasing the effort required to commit a crime. This can include measures such as locks, security doors, window bars, and alarm systems, which act as barriers to entry or access, thereby deterring criminal actions.

How does integrating security lighting with natural surveillance contribute to CPTED objectives?
a) By providing a backup power source for surveillance cameras
b) By ensuring areas are well-lit, thereby enhancing the ability of people to observe and be observed
c) By using lighting to distract potential intruders
d) By replacing the need for physical security measures

Answer: b) By ensuring areas are well-lit, thereby enhancing the ability of people to observe and be observed. Explanation: Integrating security lighting with natural surveillance contributes to CPTED objectives by ensuring that areas are well-lit, which enhances the ability of people within the space to observe and be observed. Good lighting reduces hiding spots for potential offenders and increases the likelihood of detection, thereby acting as a deterrent to criminal behavior.

In what way does environmental design influence the effectiveness of security lighting?
a) By determining the color scheme of the lighting fixtures
b) By affecting where and how lighting can be most effectively deployed to enhance visibility and deterrence
c) By dictating the wattage of light bulbs used in outdoor fixtures
d) By limiting the use of lighting to daylight hours to conserve energy

Answer: b) By affecting where and how lighting can be most effectively deployed to enhance visibility and deterrence. Explanation: Environmental design significantly influences the effectiveness of security lighting by determining the most strategic locations and methods for deploying lighting to enhance visibility and deterrence. Proper design takes into account factors such as landscaping, architectural features, and the natural flow of people to ensure that lighting contributes effectively to the overall security strategy, eliminating shadows and dark areas where potential offenders could hide.

What is the primary goal of integrating security practices into the Software Development Life Cycle (SDLC) through DevSecOps?
a. To ensure that security testing is only conducted in the deployment phase
b. To make security a shared responsibility across development, security, and operations teams
c. To reduce the cost of software development
d. To eliminate the need for a dedicated cybersecurity team

Answer: b. To make security a shared responsibility across development, security, and operations teams. Explanation: DevSecOps integrates security practices within the SDLC, making security a shared responsibility among development, security, and operations teams. This approach ensures that security considerations are integrated from the inception of the project through development, deployment, and operations, facilitating early detection and mitigation of security vulnerabilities.

Which of the following is a fundamental principle of secure coding practices?
a. Writing code that is only tested for functionality, not security
b. Assuming that external libraries and dependencies are secure without verification
c. Validating and sanitizing all inputs to prevent injection attacks
d. Relying on security through obscurity as a primary defense mechanism

Answer: c. Validating and sanitizing all inputs to prevent injection attacks. Explanation: A fundamental principle of secure coding is to validate and sanitize all inputs to prevent injection attacks such as SQL injection, cross-site scripting (XSS), and command injection. This involves checking all input data against expected formats and values and sanitizing data to remove or neutralize any potential malicious content.

In DevSecOps, what role does Continuous Integration/Continuous Deployment (CI/CD) play in enhancing security?
a. It eliminates the need for manual code reviews and security assessments
b. It allows for the automated integration of security tests and checks within the development pipeline
c. It guarantees that software will be free from vulnerabilities
d. It focuses solely on the speed of deployment, not on security

Answer: b. It allows for the automated integration of security tests and checks within the development pipeline. Explanation: In DevSecOps, CI/CD pipelines are enhanced by integrating automated security tests and checks, allowing for continuous assessment of the codebase. This integration helps in identifying and mitigating security vulnerabilities early in the development process, thus enhancing the overall security posture of the software.

What is a security benefit of using code obfuscation techniques?
a. It makes the source code easier to understand and audit for security vulnerabilities
b. It can help protect intellectual property and hinder reverse engineering efforts
c. It ensures that the code is completely free from security vulnerabilities
d. It replaces the need for encryption in data protection

Answer: b. It can help protect intellectual property and hinder reverse engineering efforts. Explanation: Code obfuscation is a technique used to make source code more difficult to understand and read, which can help protect intellectual property and deter reverse engineering efforts. While it does not eliminate vulnerabilities, it adds a layer of complexity for anyone trying to analyze the code for malicious purposes.

How does threat modeling contribute to security in the SDLC?
a. By providing a framework for bypassing security controls in the testing phase
b. By identifying potential threats and vulnerabilities early in the development process
c. By focusing exclusively on physical security threats to the development environment
d. By eliminating the need for security testing and code review

Answer: b. By identifying potential threats and vulnerabilities early in the development process. Explanation: Threat modeling is a proactive approach used in the SDLC to identify potential security threats, vulnerabilities, and attack vectors that could impact the software. By understanding these potential threats early in the development process, teams can design and implement appropriate security controls and mitigation strategies to enhance the software's security posture.

What is the significance of using static application security testing (SAST) tools in secure coding practices?
a. They replace the need for manual security testing by developers
b. They dynamically analyze running code to identify runtime vulnerabilities
c. They analyze source code at rest to identify vulnerabilities early in the development phase
d. They only need to be used once the software is in the production environment

Answer: c. They analyze source code at rest to identify vulnerabilities early in the development phase. Explanation: SAST tools analyze the source code at rest, without the need for the code to be running. These tools help in identifying vulnerabilities and coding errors early in the development phase, allowing developers to address security issues before the software progresses further in the SDLC.

In the context of SDLC, why is it important to conduct security code reviews?
a. To ensure that the code meets performance benchmarks only
b. To identify and remediate security vulnerabilities and weaknesses in the codebase
c. To assess the software's user interface design
d. To validate the software's compliance with financial regulations

Answer: b. To identify and remediate security vulnerabilities and weaknesses in the codebase. Explanation: Security code reviews are a critical part of the SDLC as they involve a systematic examination of the source code to identify security vulnerabilities, weaknesses, and violations of secure coding practices. This process helps in remediating issues before the software is deployed, thereby enhancing its security.

What role does patch management play in secure software development?
a. It is unrelated to software development and focuses only on hardware vulnerabilities
b. It ensures that all software dependencies and libraries are kept up to date with the latest security patches
c. It solely focuses on updating the software's documentation
d. It is only necessary for open-source software

Answer: b. It ensures that all software dependencies and libraries are kept up to date with the latest security patches. Explanation: Patch management is a critical security practice within software development that involves regularly updating and patching software dependencies and libraries to address known vulnerabilities. Keeping these components up to date helps in mitigating potential security risks that could be exploited by attackers.

How does implementing a secure software development framework like OWASP SAMM benefit organizations?
a. By providing a checklist for bypassing standard security protocols
b. By offering a structured approach to integrating security throughout the SDLC
c. By focusing solely on post-deployment security incident management
d. By eliminating the need for security training for developers

Answer: b. By offering a structured approach to integrating security throughout the SDLC. Explanation: Implementing a secure software development framework like OWASP SAMM (Software Assurance Maturity Model) provides organizations with a structured and comprehensive approach to integrating security practices and considerations throughout all phases of the SDLC. This holistic approach helps in systematically addressing security from the requirements phase to design, implementation, testing, and maintenance, thereby enhancing the overall security of the software developed.

What security practice ensures that a container image is free from vulnerabilities before deployment?
a) Running the container as a root user

b) Using the latest version of all software within the container
c) Conducting a thorough vulnerability scan of the container image
d) Storing container images in a public repository

Answer: c) Conducting a thorough vulnerability scan of the container image. Explanation: Conducting a thorough vulnerability scan of the container image before deployment is a critical security practice. This process identifies known vulnerabilities within the image, allowing developers to address potential security issues before the container is deployed in a production environment, thereby reducing the risk of exploitation.

In the context of development environment security, what is the primary purpose of code signing?
a) To encrypt the source code
b) To increase code execution speed
c) To verify the integrity and authenticity of the code
d) To compress the code for faster transmission

Answer: c) To verify the integrity and authenticity of the code. Explanation: The primary purpose of code signing is to verify the integrity and authenticity of the code. By signing code with a digital signature, developers can assure that the code has not been altered or tampered with after it was signed, and that it was indeed created by the stated developer, thereby enhancing trust and security in the software distribution process.

How can "least privilege" principles be applied to enhance container security?
a) By granting all containers root access to simplify management
b) By assigning minimal necessary permissions to containers for their specific tasks
c) By allowing containers unrestricted network access for greater connectivity
d) By using a single container for all applications to reduce complexity

Answer: b) By assigning minimal necessary permissions to containers for their specific tasks. Explanation: Applying "least privilege" principles to container security involves assigning only the minimal necessary permissions and resources that a container needs to perform its specific tasks. This limits the potential impact of a security breach, as compromised containers have restricted access and capabilities, thereby reducing the attack surface.

What is a security benefit of immutability in container environments?
a) Containers can be easily modified after deployment.
b) Containers are prevented from being deleted.
c) Once deployed, containers cannot be changed, reducing the risk of runtime attacks.
d) Containers can auto-update their software without approval.

Answer: c) Once deployed, containers cannot be changed, reducing the risk of runtime attacks. Explanation: The principle of immutability in container environments means that once a container is deployed, it cannot be changed or modified. This reduces the risk of runtime attacks and unauthorized changes, as any alterations require the deployment of a new container version. This approach enhances security by ensuring consistency and predictability in the deployment environment.

What is the role of a "secure container registry" in development environments?
a) To serve as a public backup location for all container images
b) To provide an execution environment for containers
c) To securely store and manage access to container images
d) To monitor and report container performance metrics

Answer: c) To securely store and manage access to container images. Explanation: A secure container registry plays a crucial role in development environments by securely storing and managing access to container images. It ensures that images are scanned for vulnerabilities, access is controlled through authentication and authorization mechanisms, and that the integrity of images is maintained, thereby protecting the container ecosystem from compromised or malicious images.

How does "namespace isolation" contribute to container security?
a) By encrypting the data exchanged between containers
b) By limiting the visibility and communication between containers
c) By increasing the storage capacity available to each container
d) By synchronizing the execution of multiple containers

Answer: b) By limiting the visibility and communication between containers. Explanation: Namespace isolation is a security feature that limits the visibility and communication between containers by providing each container with its own isolated view of the underlying system, including file systems, network interfaces, and process tables. This prevents processes within one container from interacting with processes in other containers or the host system, thereby enhancing security through isolation.

What security challenge is addressed by implementing container orchestration tools?
a) The complexity of managing multiple containers and their configurations
b) The need for manual code reviews for each container
c) The requirement for containers to run on dedicated physical servers
d) The inability of containers to communicate over encrypted channels

Answer: a) The complexity of managing multiple containers and their configurations. Explanation: Container orchestration tools address the security challenge of managing multiple containers and their configurations across different environments. These tools automate the deployment, scaling, networking, and management of containerized applications, ensuring that security policies and configurations are consistently applied, thereby reducing the risk of misconfigurations and vulnerabilities.

In what way does "container scanning" enhance security within a CI/CD pipeline?
a) By performing syntax checking of the application code
b) By detecting vulnerabilities in container images before they are deployed
c) By monitoring the runtime behavior of containers for performance optimization
d) By ensuring compliance with coding standards

Answer: b) By detecting vulnerabilities in container images before they are deployed. Explanation: Container scanning is a security practice that involves analyzing container images for known vulnerabilities and misconfigurations before they are deployed. Integrating this process within a CI/CD pipeline ensures that any security issues are identified and addressed early in the development cycle, enhancing the overall security posture of the containerized applications.

How does "secrets management" in containerized environments enhance security?
a) By publicly sharing encryption keys for transparency
b) By hardcoding sensitive information within container images
c) By securely storing, accessing, and managing sensitive information like passwords and keys
d) By generating easily guessable passwords for convenience

Answer: c) By securely storing, accessing, and managing sensitive information like passwords and keys. Explanation: Secrets management in containerized environments involves securely storing, accessing, and managing sensitive information such as passwords, tokens, and encryption keys. This prevents hardcoding secrets in container images or source code, reducing the risk of exposure or compromise. Secure secrets management systems ensure that only authorized containers and services can access the necessary secrets, enhancing the security of the environment.

What is the significance of "network segmentation" in the context of container security?
a) To combine all container traffic onto a single network for simplicity
b) To allocate the same IP address to all containers
c) To divide the network into smaller, controlled segments to limit lateral movement in case of compromise
d) To increase bandwidth by allowing all containers to communicate without restrictions

Answer: c) To divide the network into smaller, controlled segments to limit lateral movement in case of compromise. Explanation: Network segmentation in the context of container security involves dividing the network into smaller, controlled segments or subnetworks. This practice limits the ability of an attacker to move laterally within the network if a container or segment is compromised, thereby containing the potential impact of a breach and enhancing the overall security posture.

What distinguishes Static Application Security Testing (SAST) from Dynamic Application Security Testing (DAST)?
a. SAST analyzes running applications in their operational environment, whereas DAST analyzes the code at rest.
b. SAST requires the application to be fully developed, whereas DAST can be performed at any stage of development.
c. SAST analyzes the source code for vulnerabilities before the application is run, whereas DAST tests the running application from an external perspective.
d. SAST is exclusively used for web applications, whereas DAST is used for all types of software.

Answer: c. SAST analyzes the source code for vulnerabilities before the application is run, whereas DAST tests the running application from an external perspective. Explanation: SAST involves analyzing the application's source, byte, or binary code for vulnerabilities without executing the program, making it effective early in the development lifecycle. DAST, on the other hand, involves testing the application while it is running, typically from an external viewpoint, to identify vulnerabilities that manifest during its execution.

In the context of secure software development, how does SAST contribute to the early detection of security issues?
a. By simulating user interactions with the application to uncover runtime errors
b. By examining the source code for patterns that indicate security vulnerabilities
c. By monitoring network traffic to and from the application for suspicious activity
d. By assessing the physical security measures of the hardware running the application

Answer: b. By examining the source code for patterns that indicate security vulnerabilities. Explanation: SAST tools scan the application's source code to identify patterns and coding practices that could lead to security vulnerabilities. This allows developers to address potential security issues early in the development process, before the application is compiled and run, thereby improving the overall security posture of the software.

What is a primary advantage of DAST in testing web applications?
a. It does not require access to the source code to identify vulnerabilities.
b. It can be used to evaluate the application's performance under load.
c. It simplifies the code review process by automating it completely.
d. It ensures compliance with coding standards and best practices.

Answer: a. It does not require access to the source code to identify vulnerabilities. Explanation: DAST tools test the application from the outside in, as an attacker would, without needing access to the source code. This makes DAST particularly useful for identifying vulnerabilities in web applications that are already running, providing a real-world perspective on the application's security posture.

How do SAST tools enhance the effectiveness of a secure development lifecycle?
a. By providing real-time protection against attacks on the application
b. By facilitating automated code reviews and identifying security issues during the development phase
c. By conducting penetration testing on the live application
d. By encrypting sensitive data within the application's source code

Answer: b. By facilitating automated code reviews and identifying security issues during the development phase. Explanation: SAST tools are integrated into the development process to automatically review the code for potential security issues as it is being written. This enables developers to address vulnerabilities early in the development lifecycle, reducing the risk of security flaws in the released version of the software.

What challenge is commonly associated with DAST?
a. It can only be performed by developers who have an in-depth understanding of the application's source code.
b. It may produce false positives and false negatives due to its external testing approach.
c. It requires extensive manual effort to set up and configure the testing environment.
d. It is less effective than manual testing methods in identifying complex security vulnerabilities.

Answer: b. It may produce false positives and false negatives due to its external testing approach. Explanation: DAST involves testing the application from an external perspective without knowledge of the internal workings, which can lead to false positives (incorrectly identifying non-issues as vulnerabilities) and false negatives (failing to detect actual vulnerabilities). This necessitates additional manual review to validate the findings.

Why is it important to integrate both SAST and DAST in the software development lifecycle?
a. To ensure that all security vulnerabilities, both static code issues and runtime vulnerabilities, are identified and mitigated.
b. Because SAST alone is sufficient to guarantee the security of the application.
c. To reduce the reliance on manual security testing and code review processes.
d. Because DAST can replace the need for secure coding practices and standards.

Answer: a. To ensure that all security vulnerabilities, both static code issues and runtime vulnerabilities, are identified and mitigated. Explanation: Integrating both SAST and DAST provides a comprehensive approach to security testing, covering both vulnerabilities that can be detected in the source code without executing the program (SAST) and those that become apparent only when the application is running in its operational environment (DAST).

In what scenario might DAST be particularly advantageous over SAST?
a. When the source code is not available, such as with third-party applications or services.
b. During the initial stages of development when the application's features are not yet fully defined.
c. For assessing adherence to coding standards and best practices.
d. When evaluating the application's compatibility with various development frameworks.

Answer: a. When the source code is not available, such as with third-party applications or services. Explanation: DAST does not require access to the application's source code to conduct testing, making it particularly useful for assessing the security of third-party applications or services where the source code may not be accessible. This allows for the identification of vulnerabilities from an outsider's perspective.

What is a limitation of SAST that organizations need to consider?
a. It can only be conducted after the application has been deployed.
b. It may not identify vulnerabilities that only manifest during runtime or through specific user interactions.
c. It eliminates the need for any other form of security testing.
d. It is only applicable to applications developed in certain programming languages.

Answer: b. It may not identify vulnerabilities that only manifest during runtime or through specific user interactions. Explanation: While SAST is effective in identifying vulnerabilities within the source code, it may not detect issues that only become apparent when the application is running or through specific user interactions. This underscores the need for complementary testing approaches, such as DAST.

How do SAST tools typically identify potential security vulnerabilities in the code?
a. By executing the application in a controlled environment and monitoring for abnormal behavior
b. By analyzing the source code for patterns and signatures associated with common vulnerabilities
c. By manually reviewing the source code line by line
d. By monitoring the network traffic generated by the application

Answer: b. By analyzing the source code for patterns and signatures associated with common vulnerabilities. Explanation: SAST tools analyze the source code statically, looking for patterns, signatures, and coding practices that are known to be associated with common security vulnerabilities. This allows for the identification and remediation of potential security issues early in the development process.

What secure coding practice is essential to protect against Injection flaws, as identified in the OWASP Top 10?
a) Using encryption for all sensitive data
b) Employing regular expression matching for input validation
c) Implementing strong access control measures

d) Validating, sanitizing, and parameterizing all user-supplied input

Answer: d) Validating, sanitizing, and parameterizing all user-supplied input. Explanation: To protect against Injection flaws, such as SQL, NoSQL, OS, and LDAP injection, it is essential to validate, sanitize, and parameterize all user-supplied input. This practice prevents attackers from injecting malicious code into the system that can be executed by the interpreter, leading to data loss, corruption, or unauthorized access.

Which mitigation strategy is recommended for preventing Cross-Site Scripting (XSS) attacks, as highlighted in the OWASP Top 10?
a) Enabling full disk encryption
b) Applying the principle of least privilege
c) Encoding data on output
d) Segmenting the network

Answer: c) Encoding data on output. Explanation: To prevent Cross-Site Scripting (XSS) attacks, it is recommended to encode data on output. This involves escaping user-supplied input so that browsers interpret it only as data, not as executable code. Encoding prevents malicious scripts from being injected into web pages, where they can be executed in the context of the user's session.

In addressing security concerns related to authentication, what does OWASP recommend?
a) Limiting password length to eight characters to streamline user access
b) Using default credentials for all administrative accounts
c) Implementing multi-factor authentication (MFA) where possible
d) Storing passwords in clear text to simplify password recovery processes

Answer: c) Implementing multi-factor authentication (MFA) where possible. Explanation: OWASP recommends implementing multi-factor authentication (MFA) to enhance security around user authentication. MFA adds additional verification layers beyond just the password, making unauthorized access significantly more difficult. This can include something the user knows (password), something the user has (security token), or something the user is (biometric verification).

What is a key secure coding guideline to mitigate the risk associated with sensitive data exposure, according to the OWASP Top 10?
a) Disabling server-side logging of all data
b) Avoiding encryption to simplify data access
c) Applying appropriate access controls to all resources
d) Encrypting sensitive data at rest and in transit

Answer: d) Encrypting sensitive data at rest and in transit. Explanation: To mitigate the risk associated with sensitive data exposure, a key secure coding guideline is to encrypt sensitive data both at rest and in transit. This ensures that even if data is intercepted or accessed without authorization, it remains unreadable and protected against unauthorized disclosure or modification.

How can software developers prevent security misconfigurations, one of the risks listed in the OWASP Top 10?
a) By leaving software in its default configuration to maintain vendor support
b) By regularly updating and patching systems, software, and libraries
c) By using only one configuration setting for all environments
d) By disabling all logging and monitoring capabilities

Answer: b) By regularly updating and patching systems, software, and libraries. Explanation: Preventing security misconfigurations involves regularly updating and patching systems, software, and libraries to address known vulnerabilities. It also includes configuring security settings appropriately for each environment and conducting regular security reviews to ensure that configurations remain secure over time.

What approach is recommended to prevent Broken Access Control issues, as per the OWASP Top 10?

a) Enforcing default permissions for all users
b) Denying all access by default and only allowing permissions per user basis
c) Removing all access controls to simplify user experience
d) Granting administrative privileges to all users to ensure accessibility

Answer: b) Denying all access by default and only allowing permissions per user basis. Explanation: To prevent Broken Access Control issues, the recommended approach is to deny all access by default and only allow permissions on a per-user basis, following the principle of least privilege. This ensures that users have access only to the resources and data necessary for their roles, reducing the risk of unauthorized access and data breaches.

In the context of the CWE/SANS TOP 25, what is a fundamental practice to defend against software vulnerabilities?
a) Compiling code with all warnings disabled to streamline development
b) Conducting comprehensive code reviews and static analysis to identify vulnerabilities
c) Relying solely on automated vulnerability scanners without manual review
d) Using older software versions for their proven stability

Answer: b) Conducting comprehensive code reviews and static analysis to identify vulnerabilities. Explanation: A fundamental practice to defend against software vulnerabilities, as highlighted in the CWE/SANS TOP 25, is conducting comprehensive code reviews and static analysis. This proactive approach allows developers to identify and remediate potential security issues in the codebase before the software is deployed, significantly reducing the risk of exploitation.

What strategy should be employed to address the use of components with known vulnerabilities, a risk identified in the OWASP Top 10?
a) Maintaining an inventory of all components and their versions
b) Using components that are no longer supported or updated
c) Ignoring security warnings related to third-party components
d) Prioritizing feature development over security updates

Answer: a) Maintaining an inventory of all components and their versions. Explanation: To address the use of components with known vulnerabilities, it is essential to maintain an inventory of all components used within the application, including their versions. This enables the organization to track vulnerabilities associated with these components and apply necessary patches or updates promptly to mitigate potential risks.

How does the implementation of Content Security Policy (CSP) headers contribute to secure coding practices?
a) By increasing the speed of website loading times
b) By defining approved sources of content that browsers are allowed to load
c) By automatically updating web content to the latest versions
d) By compressing content to reduce bandwidth usage

Answer: b) By defining approved sources of content that browsers are allowed to load. Explanation: The implementation of Content Security Policy (CSP) headers is a secure coding practice that contributes to mitigating risks such as Cross-Site Scripting (XSS). CSP headers allow web developers to define a whitelist of approved sources of content that browsers are allowed to load on a webpage, thereby preventing the execution of unauthorized or malicious scripts.

Which component of the NIST Cybersecurity Framework focuses on developing and implementing safeguards to ensure delivery of critical services?
a) Identify
b) Protect
c) Detect
d) Respond

Answer: b) Protect. Explanation: The "Protect" function of the NIST Cybersecurity Framework focuses on developing and implementing appropriate safeguards to ensure the delivery of critical infrastructure services. This includes access control, data security, maintenance of security policies, and protective technology to mitigate the risk of cybersecurity events.

In the context of ISO 27001, what is the purpose of a Statement of Applicability (SoA)?
a) To document the financial impact of a security breach
b) To list all the ISO 27001 controls that have been implemented
c) To declare which controls are not applicable and justify why
d) Both b and c

Answer: d) Both b and c. Explanation: The Statement of Applicability (SoA) in ISO 27001 serves to document all the controls from Annex A that are relevant to the organization, explaining which controls have been implemented and justifying why certain controls may not be applicable. This is a core component of an ISO 27001-compliant Information Security Management System (ISMS).

What is the primary focus of PCI DSS compliance?
a) Protecting healthcare information
b) Ensuring the security of cardholder data in payment systems
c) Governing the management of IT services
d) Setting standards for governmental cybersecurity practices

Answer: b) Ensuring the security of cardholder data in payment systems. Explanation: The primary focus of the Payment Card Industry Data Security Standard (PCI DSS) is to ensure the security of cardholder data within payment systems. PCI DSS provides a set of security standards designed to protect credit card information during and after financial transactions to prevent credit card fraud.

Which framework primarily aims at providing a structured approach to aligning IT processes with business objectives?
a) NIST Cybersecurity Framework
b) ISO 27001
c) COBIT
d) HIPAA

Answer: c) COBIT. Explanation: COBIT (Control Objectives for Information and Related Technologies) is a framework designed to provide a structured approach to aligning IT processes and goals with the broader business objectives. COBIT helps organizations govern and manage their IT environments effectively, ensuring that IT supports business strategies and objectives.

How does the "Identify" function in the NIST Cybersecurity Framework assist organizations?
a) By recognizing the immediate impact of a detected cybersecurity event
b) By outlining the specific steps to contain a cybersecurity incident
c) By establishing an understanding of managing cybersecurity risk to systems, assets, and data
d) By detailing recovery plans after a cybersecurity incident

Answer: c) By establishing an understanding of managing cybersecurity risk to systems, assets, and data. Explanation: The "Identify" function of the NIST Cybersecurity Framework assists organizations by establishing an understanding of managing cybersecurity risk to systems, assets, data, and capabilities. This function helps in identifying the business context, resources that support critical functions, and related cybersecurity risks, enabling an organization to focus and prioritize its efforts.

Under HIPAA, what is required to protect the privacy and security of health information?
a) Implementation of end-to-end encryption for all data
b) Regular auditing of financial transactions
c) Adoption of physical, technical, and administrative safeguards
d) Mandatory two-factor authentication for all users

Answer: c) Adoption of physical, technical, and administrative safeguards. Explanation: HIPAA (Health Insurance Portability and Accountability Act) requires the adoption of physical, technical, and administrative safeguards to protect the privacy and security of protected health information (PHI). These safeguards are designed to protect PHI from unauthorized access, disclosure, alteration, and destruction, ensuring the confidentiality, integrity, and availability of health information.

In aligning frameworks to business goals, what is a critical first step?
a) Selecting the most expensive security solutions
b) Understanding the organization's business context and risk appetite
c) Implementing all controls from selected frameworks indiscriminately
d) Focusing solely on technological controls

Answer: b) Understanding the organization's business context and risk appetite. Explanation: A critical first step in aligning frameworks to business goals is understanding the organization's business context, operations, and risk appetite. This understanding allows for the strategic selection and implementation of framework components that directly support business objectives and manage risk according to the organization's tolerance levels.

What is a key advantage of integrating the "Respond" function of the NIST Cybersecurity Framework into an organization's incident response plan?
a) It eliminates the need for a dedicated incident response team.
b) It provides guidelines for proactive prevention of all cybersecurity incidents.
c) It ensures a timely and effective response to cybersecurity incidents, minimizing impact.
d) It guarantees legal immunity in the event of a data breach.

Answer: c) It ensures a timely and effective response to cybersecurity incidents, minimizing impact. Explanation: Integrating the "Respond" function of the NIST Cybersecurity Framework into an organization's incident response plan ensures a timely and effective response to cybersecurity incidents. This function focuses on response planning, communication, analysis, mitigation, and improvements to minimize the impact of incidents and support rapid recovery.

For an organization subject to both ISO 27001 and HIPAA regulations, what approach ensures compliance with both frameworks?
a) Focusing exclusively on the stricter requirements of HIPAA
b) Implementing ISO 27001 controls without considering HIPAA requirements
c) Developing a compliance strategy that addresses the overlapping and unique requirements of both frameworks
d) Adopting a generic set of security controls not specific to either framework

Answer: c) Developing a compliance strategy that addresses the overlapping and unique requirements of both frameworks. Explanation: For an organization subject to both ISO 27001 and HIPAA regulations, developing a compliance strategy that addresses the overlapping and unique requirements of both frameworks ensures comprehensive compliance. This approach involves identifying where the frameworks align and where specific actions are needed to meet the distinct requirements of each, thereby efficiently managing compliance efforts.

Which of the following best defines a security policy within an organization?
a) A detailed step-by-step instruction on how to configure a firewall
b) A high-level overall plan that embodies the beliefs of an organization
c) A technical specification of the allowed and disallowed software
d) A daily checklist for system administrators to follow

Answer: b) A high-level overall plan that embodies the beliefs of an organization. Explanation: A security policy is a high-level document that sets out the organization's overall plan and beliefs regarding security. It outlines the importance of security, the organization's stance on various security issues, and sets the direction for the procedures and standards that follow.

How does a security standard differ from a security policy?

a) A standard is less formal and often undocumented.
b) A standard provides specific technical guidelines derived from the policy.
c) A standard is a legal document, while a policy is not.
d) A standard is a daily operational task list.

Answer: b) A standard provides specific technical guidelines derived from the policy. Explanation: Security standards provide specific, mandatory technical guidelines, configurations, and rules that must be followed, derived from the broader directives outlined in a security policy. While the policy provides the overarching principles, standards offer the concrete specifications needed to implement those principles.

What is the primary purpose of security procedures in the context of information security governance?
a) To define the organization's stance on security
b) To list recommended software and tools
c) To provide detailed, step-by-step actions to implement standards
d) To outline the consequences of non-compliance with policies

Answer: c) To provide detailed, step-by-step actions to implement standards. Explanation: Security procedures are detailed, step-by-step instructions that describe exactly how to implement the organization's security standards and policies. Procedures are practical guides for employees and IT staff to follow, ensuring consistency and compliance in the day-to-day operations of security measures.

In what way do security guidelines differ from security policies, standards, and procedures?
a) Guidelines are mandatory rules that must be followed.
b) Guidelines offer recommendations that can be tailored to specific circumstances.
c) Guidelines provide step-by-step implementation instructions.
d) Guidelines are legally binding documents.

Answer: b) Guidelines offer recommendations that can be tailored to specific circumstances. Explanation: Security guidelines are advisory in nature and offer recommendations that can be tailored to specific situations or environments. Unlike policies, standards, and procedures, which are more prescriptive and often mandatory, guidelines allow for some discretion based on professional judgment.

What is a critical first step in the development of security policies and standards?
a) Conducting a detailed software audit
b) Establishing a punitive framework for policy violations
c) Assessing the organization's risk and security posture
d) Implementing a new security software suite

Answer: c) Assessing the organization's risk and security posture. Explanation: A critical first step in the development of security policies and standards is to conduct a thorough assessment of the organization's risk and security posture. This involves identifying assets, threats, vulnerabilities, and existing controls to understand the organization's specific security needs and requirements, which will inform the development of relevant and effective policies and standards.

How should an organization ensure the ongoing effectiveness of its security policies and standards?
a) By keeping them unchanged to maintain consistency
b) Through regular review and updates in response to evolving threats and business needs
c) By reducing the frequency of audits to minimize disruptions
d) By limiting access to policies and standards to senior management only

Answer: b) Through regular review and updates in response to evolving threats and business needs. Explanation: To ensure the ongoing effectiveness of security policies and standards, an organization should regularly review and update them to reflect evolving security threats, technological advancements, and changing business objectives. This dynamic approach helps maintain the relevance and effectiveness of the security framework.

What role do security policies play in integrating security into business processes?

a) They automate security controls within business software.
b) They ensure that security considerations are embedded in business decisions and operations.
c) They eliminate the need for security training among staff.
d) They delegate all security responsibilities to the IT department.

Answer: b) They ensure that security considerations are embedded in business decisions and operations. Explanation: Security policies play a crucial role in integrating security into business processes by ensuring that security considerations are an integral part of business decisions and operations. Well-defined policies communicate the importance of security across the organization and guide how security is to be considered in daily activities, project planning, and strategic decision-making.

In the context of compliance, how do security standards support adherence to regulations such as HIPAA or PCI DSS?
a) By providing general security advice applicable to any organization
b) By offering specific technical requirements that align with regulatory mandates
c) By encouraging voluntary compliance without specific directives
d) By focusing solely on the physical security controls

Answer: b) By offering specific technical requirements that align with regulatory mandates. Explanation: Security standards support compliance with regulations like HIPAA or PCI DSS by offering specific technical requirements and controls that align with the mandates of these regulations. By adhering to relevant security standards, organizations can ensure they meet the regulatory requirements, thereby achieving compliance.

When reviewing and maintaining security policies, what factor is crucial to consider for ensuring alignment with business goals?
a) The personal preferences of the security team
b) The latest trends in security technology
c) The evolving nature of business strategies and objectives
d) The popularity of policies among employees

Answer: c) The evolving nature of business strategies and objectives. Explanation: When reviewing and maintaining security policies, it is crucial to consider the evolving nature of business strategies and objectives to ensure that the policies remain aligned with the business goals. Security policies should support and enable the organization's mission, adapting to changes in business direction, growth, and the external environment to remain relevant and effective.

Which principle from the (ISC)² Code of Ethics requires professionals to provide diligent and competent service to principals?
a) Protect society, the common good, and necessary public trust and confidence.
b) Act honorably, honestly, justly, responsibly, and legally.
c) Provide diligent and competent service to principals.
d) Advance and protect the profession.

Answer: c) Provide diligent and competent service to principals. Explanation: The (ISC)² Code of Ethics mandates that professionals should provide diligent and competent service to their principals, meaning they should carry out their duties with care and proficiency, ensuring that the services they provide meet the professional standards required and that they act in the best interests of those they are serving.

In the context of ethical decision-making in cybersecurity, what does it mean to "advance and protect the profession" according to the (ISC)² Code of Ethics?
a) To engage in continuous learning and contribute to the knowledge base of the cybersecurity field.
b) To ensure that all cybersecurity measures are legally compliant.
c) To prioritize the security of the organization over everything else.
d) To report any unethical behavior within the industry to the authorities.

Answer: a) To engage in continuous learning and contribute to the knowledge base of the cybersecurity field. Explanation: "Advance and protect the profession," as per the (ISC)² Code of Ethics, involves committing to continuous professional development, sharing knowledge with peers, and upholding the integrity of the profession. This includes contributing to the body of knowledge, mentoring emerging professionals, and acting in a manner that reflects well on the profession.

When faced with an ethical dilemma, a cybersecurity professional should FIRST consider:
a) The potential for personal gain.
b) The impact of their actions on public trust and the common good.
c) The preferences of their colleagues.
d) The most technologically advanced solution.

Answer: b) The impact of their actions on public trust and the common good. Explanation: When faced with an ethical dilemma, the (ISC)² Code of Ethics emphasizes considering the impact of actions on public trust and the common good as a priority. This principle underscores the responsibility of cybersecurity professionals to protect society and the integrity of information and information systems, placing societal welfare above personal or organizational interests.

What is an appropriate action for a cybersecurity professional who discovers a colleague exploiting system vulnerabilities for personal gain?
a) Ignore the issue if the exploitation does not directly impact their own work.
b) Confront the colleague privately and offer to participate in exchange for a share of the gains.
c) Report the activity to appropriate management or authorities in line with organizational policies and legal requirements.
d) Use the knowledge of the exploitation to bargain for a promotion or raise.

Answer: c) Report the activity to appropriate management or authorities in line with organizational policies and legal requirements. Explanation: Ethical conduct, as guided by the (ISC)² Code of Ethics, requires professionals to act honorably and legally, which includes reporting unethical or illegal activities such as the exploitation of system vulnerabilities. This ensures the maintenance of trust, the common good, and the protection of the profession's integrity.

How does "acting honorably, honestly, justly, responsibly, and legally" guide cybersecurity professionals in handling confidential information?
a) By encouraging the sharing of confidential information to promote transparency.
b) By mandating the disclosure of all confidential information to the public to prevent misinformation.
c) By upholding the confidentiality and privacy of information entrusted to them, except when legally obligated to disclose.
d) By using confidential information as leverage in organizational negotiations.

Answer: c) By upholding the confidentiality and privacy of information entrusted to them, except when legally obligated to disclose. Explanation: This principle from the (ISC)² Code of Ethics emphasizes the importance of handling confidential information with care and integrity, ensuring it is protected from unauthorized access or disclosure. Professionals are expected to respect privacy and confidentiality, revealing such information only when required by law or ethical duties.

In applying the (ISC)² Code of Ethics, what should a professional do when they observe a legal but unethical practice within their organization?
a) Adopt the practice as a new standard for personal conduct.
b) Report the practice to the media to gain public sympathy.
c) Seek to change the practice through appropriate internal channels.
d) Resign immediately without addressing the practice.

Answer: c) Seek to change the practice through appropriate internal channels. Explanation: Observing a legal but unethical practice requires a professional to act in a manner that seeks to rectify the situation while adhering to the principles of acting honorably and responsibly. This involves raising concerns through the appropriate channels within the organization, such as speaking to a supervisor, the ethics committee, or the human resources department, to address and seek change in the practice.

What ethical principle is at stake when a cybersecurity professional considers implementing a security measure that significantly invades personal privacy?
a) The necessity to advance technological capabilities at all costs.
b) The balance between protecting society and respecting individual rights and freedoms.
c) The priority of organizational goals over societal norms.
d) The obligation to use the most cost-effective security solutions.

Answer: b) The balance between protecting society and respecting individual rights and freedoms. Explanation: Implementing security measures that significantly invade personal privacy raises ethical concerns about balancing the need to protect society, information, and systems with the obligation to respect individual rights and freedoms. Ethical decision-making in such scenarios requires careful consideration of the impact on privacy and the common good, guided by principles of justice and responsibility.

When integrating security into business processes, how should a cybersecurity professional align their actions with the (ISC)² Code of Ethics?
a) By ensuring security measures enhance the profitability of the business at any cost.
b) By implementing security controls that covertly monitor all employee activities.
c) By making security decisions that contribute to the safety and welfare of society.
d) By prioritizing the development of proprietary security solutions over open standards.

Answer: c) By making security decisions that contribute to the safety and welfare of society. Explanation: Integrating security into business processes in alignment with the (ISC)² Code of Ethics involves making decisions that not only protect the organization's assets but also contribute to the broader welfare of society. This includes considering the societal impact of security measures and ensuring they uphold ethical standards, public trust, and respect for privacy and individual rights.

In dealing with ethical dilemmas related to cybersecurity, what framework can professionals use to ensure their decisions are ethically sound?
a) A framework that maximizes personal career advancement.
b) A decision-making framework that considers legal compliance, organizational policy, professional ethics, and societal impact.
c) A framework based solely on the potential financial outcomes of decisions.
d) A decision-making process that prioritizes the interests of the cybersecurity community over those of the general public.

Answer: b) A decision-making framework that considers legal compliance, organizational policy, professional ethics, and societal impact. Explanation: An ethically sound decision-making framework in cybersecurity encompasses considerations of legal compliance, adherence to organizational policies, alignment with professional ethical standards such as the (ISC)² Code of Ethics, and the potential impact on society. This comprehensive approach ensures that decisions are made responsibly, considering multiple facets of ethical conduct.

What is the most effective method for preventing SQL injection attacks in web applications?
a. Using complex and obscure table names
b. Implementing strict network firewall rules
c. Employing prepared statements and parameterized queries
d. Conducting frequent user training sessions on cybersecurity

Answer: c. Employing prepared statements and parameterized queries. Explanation: Prepared statements and parameterized queries are the most effective method for preventing SQL injection attacks, as they enforce a clear separation between the code and the data. This approach ensures that user input is treated as data only and not as part of the SQL command, effectively neutralizing the injection attack vector.

How does Database Activity Monitoring (DAM) contribute to database security?

a. By encrypting database files on disk
b. By monitoring and analyzing all database activities in real-time for suspicious behavior
c. By restricting physical access to the database servers
d. By automatically updating database software to the latest version

Answer: b. By monitoring and analyzing all database activities in real-time for suspicious behavior. Explanation: Database Activity Monitoring (DAM) enhances database security by continuously monitoring, recording, and analyzing all database activities in real-time. This allows for the detection of suspicious or anomalous behavior, unauthorized access attempts, and potential security breaches, enabling timely intervention.

What role does data masking play in protecting sensitive information in databases?
a. It physically relocates sensitive data to secure offsite storage.
b. It replaces sensitive data with realistic but fictional data, suitable for development and testing environments.
c. It strengthens database firewalls to prevent unauthorized access.
d. It increases the complexity of SQL queries to confuse potential attackers.

Answer: b. It replaces sensitive data with realistic but fictional data, suitable for development and testing environments. Explanation: Data masking is a technique used to protect sensitive information in databases by obscuring it with realistic but fictional data. This method is particularly useful in development and testing environments, where using real data can pose a risk, but where maintaining operational realism is important.

In preventing SQL injection, why is input validation considered a critical step?
a. It ensures all user inputs match the expected format, reducing the risk of malicious injections.
b. It encrypts user inputs to make them unreadable to attackers.
c. It increases the speed of database queries by optimizing input data.
d. It automatically corrects any SQL syntax errors in user inputs.

Answer: a. It ensures all user inputs match the expected format, reducing the risk of malicious injections. Explanation: Input validation is crucial for preventing SQL injection attacks as it involves verifying that all user-supplied data adheres to the expected format and content. By ensuring that inputs cannot contain SQL control characters or commands, this step significantly reduces the risk of injection attacks.

How does implementing least privilege access control benefit database security?
a. By ensuring that all users have administrator access to simplify management
b. By granting users only the access rights necessary for their role, minimizing potential damage from breaches
c. By allowing unrestricted data access to facilitate easier data analysis
d. By encrypting all data transactions, making them unreadable

Answer: b. By granting users only the access rights necessary for their role, minimizing potential damage from breaches. Explanation: The principle of least privilege enhances database security by ensuring that users, applications, and services are granted only the minimum levels of access—or permissions—needed to perform their functions. This minimizes the potential damage from accidents, errors, or unauthorized use, as it limits the access scope for potential attackers.

What is the significance of regular database vulnerability assessments?
a. To ensure the database management system (DBMS) remains up-to-date with the latest features
b. To identify and remediate vulnerabilities in the database system before they can be exploited
c. To evaluate the database's performance under high load conditions
d. To assess the database's compliance with international data handling standards

Answer: b. To identify and remediate vulnerabilities in the database system before they can be exploited. Explanation: Regular database vulnerability assessments are critical for identifying security weaknesses, misconfigurations, and other vulnerabilities within the database system. By detecting these issues early, organizations can remediate them before attackers can exploit them, thereby strengthening the overall security posture of the database environment.

Why is database encryption a crucial aspect of data protection strategies?
a. It speeds up database query response times
b. It ensures data integrity by preventing unauthorized data modifications
c. It protects sensitive data at rest and in transit from unauthorized access
d. It simplifies database administration tasks

Answer: c. It protects sensitive data at rest and in transit from unauthorized access. Explanation: Database encryption is a vital data protection strategy as it secures sensitive data by converting it into an unreadable format for unauthorized users, both at rest (stored data) and in transit (data being transferred). This ensures that even if the data is intercepted or accessed without authorization, it remains confidential and undecipherable without the proper encryption keys.

How do regular database backups contribute to an organization's disaster recovery plan?
a. By ensuring that all deleted data is permanently unrecoverable for security
b. By providing a recent copy of data that can be restored in the event of data loss or corruption
c. By increasing the storage requirements to improve database performance
d. By encrypting all backup data to prevent unauthorized access

Answer: b. By providing a recent copy of data that can be restored in the event of data loss or corruption. Explanation: Regular database backups are a cornerstone of effective disaster recovery planning. They ensure that, in the event of incidents such as data corruption, system failures, or malicious attacks leading to data loss, a recent and intact copy of the database can be restored, minimizing downtime and data loss.

What is the purpose of a database audit trail in the context of security?
a. To provide a real-time backup of all transactions
b. To track and log all database activities for analysis and forensic purposes
c. To enhance the database's performance by optimizing query execution
d. To serve as the primary data source for business intelligence applications

Answer: b. To track and log all database activities for analysis and forensic purposes. Explanation: A database audit trail is essential for security as it involves tracking, logging, and storing all database activities. This enables security analysts to monitor for suspicious activities, conduct forensic analyses in the event of a security incident, ensure compliance with regulatory requirements, and establish accountability by tracing actions back to individual users.

What defines the 'Confidential' data classification level?
a) Information that can be freely shared with the public without any repercussions.
b) Information whose unauthorized disclosure could cause moderate harm to an individual, group, or organization.
c) Information whose unauthorized disclosure could cause significant harm to an individual, group, or organization.
d) Information intended only for use within the organization and whose unauthorized disclosure could cause serious harm.

Answer: c) Information whose unauthorized disclosure could cause significant harm to an individual, group, or organization. Explanation: The 'Confidential' data classification level is assigned to information whose unauthorized disclosure, alteration, or destruction could cause a significant level of harm to an individual, group, or organization. This classification typically covers sensitive business information, personal data subject to privacy laws, and other information that requires strict access controls and protection measures.

In the context of data handling, what is the primary responsibility of a data owner?
a) Performing regular backups of the data
b) Implementing encryption on all classified data
c) Deciding on the classification level of the data and ensuring appropriate handling
d) Physically securing servers and storage devices containing the data

Answer: c) Deciding on the classification level of the data and ensuring appropriate handling. Explanation: The primary responsibility of a data owner is to determine the appropriate classification level for the data under their control and to ensure that the data is handled in accordance with its classification. This includes implementing and enforcing policies and procedures related to access control, data sharing, and data protection to maintain the confidentiality, integrity, and availability of the data.

What is the role of data custodians in the context of data classification and handling?
a) To define the classification levels for data within the organization
b) To decide who has access to different types of data
c) To manage and protect the data as per the guidelines set by the data owner
d) To create data and generate content for organizational needs

Answer: c) To manage and protect the data as per the guidelines set by the data owner. Explanation: Data custodians are responsible for the technical management and protection of data, implementing the controls, and handling procedures as defined by the data owner. Their role involves ensuring the security and proper maintenance of the data, including storage, backup, and access controls, in alignment with its classification level and the organization's policies.

How should 'Restricted' classified data be handled compared to 'Confidential' data?
a) With less stringent controls since it is intended for public release.
b) With the same level of controls as 'Confidential' data.
c) With more stringent controls due to its higher sensitivity.
d) Without any special controls, as it is equivalent to 'Public' data.

Answer: c) With more stringent controls due to its higher sensitivity. Explanation: 'Restricted' classified data is considered more sensitive than 'Confidential' data and therefore requires more stringent handling controls and protection measures. This level of classification is typically reserved for the most sensitive information, whose unauthorized disclosure could cause severe harm or damage. As such, access is highly restricted and controlled, often limited to a small group of authorized individuals.

What is the significance of data labeling in the context of data classification?
a) It serves as a decorative aspect to make data more visually appealing.
b) It identifies the data's sensitivity level and informs users of the handling requirements.
c) It encrypts the data based on its classification level.
d) It tracks changes and edits made to the data over time.

Answer: b) It identifies the data's sensitivity level and informs users of the handling requirements. Explanation: Data labeling is a critical aspect of data classification, as it involves marking or tagging the data with its classification level (e.g., Public, Confidential, Restricted). This practice informs users and systems of the sensitivity of the data and the associated handling requirements, ensuring that the data is treated and protected appropriately according to its classification.

In what way does aligning data classification standards with business goals benefit an organization?
a) It ensures that all data is treated as 'Confidential' for maximum security.
b) It allows for unrestricted data sharing to promote transparency.
c) It enables efficient data management and protection aligned with organizational priorities.
d) It mandates the encryption of all data, regardless of its sensitivity.

Answer: c) It enables efficient data management and protection aligned with organizational priorities. Explanation: Aligning data classification standards with business goals ensures that data management and protection efforts are in sync with the organization's priorities and risk management strategies. This alignment helps in efficiently allocating resources to protect critical data and supports compliance, operational efficiency, and strategic decision-making.

What approach should be taken when a piece of data does not clearly fit into established classification levels?
a) Automatically classify it at the highest level for maximum security.
b) Leave the data unclassified until a new classification level is created.
c) Classify the data at the lowest level to facilitate easier access and sharing.
d) Consult with relevant stakeholders to determine the most appropriate classification.

Answer: d) Consult with relevant stakeholders to determine the most appropriate classification. Explanation: When a piece of data does not clearly fit into established classification levels, the best approach is to consult with relevant stakeholders, including data owners, security professionals, and possibly legal or compliance teams, to determine the most appropriate classification. This collaborative approach ensures that the data is classified in a manner that balances the need for protection with operational requirements.

How does periodic review of data classification standards support organizational security?
a) By ensuring that all data remains classified at the highest level for maximum protection.
b) By adapting to changes in the business environment, regulatory requirements, and threat landscape.
c) By allowing for the gradual declassification of data over time.
d) By reducing the overall amount of classified data to minimize storage costs.

Answer: b) By adapting to changes in the business environment, regulatory requirements, and threat landscape. Explanation: Periodic review of data classification standards is crucial for adapting to changes in the business environment, regulatory requirements, and evolving threat landscape. This ensures that classification standards remain relevant and effective in protecting sensitive information while supporting business needs and compliance obligations.

What is a key consideration when integrating security into business processes concerning data handling?
a) Ensuring that all business processes are automated to reduce human error.
b) Bypassing security controls to expedite business operations.
c) Balancing the need for data protection with operational efficiency and usability.
d) Implementing security controls that are only visible to the IT department.

Answer: c) Balancing the need for data protection with operational efficiency and usability. Explanation: When integrating security into business processes, especially concerning data handling, a key consideration is to balance the need for data protection with operational efficiency and usability. This involves implementing security measures that effectively protect data while ensuring that business processes remain agile and user-friendly, supporting productivity and compliance without unnecessary hindrance.

In the event of a data breach, how does a well-defined data classification system benefit the organization's response?
a) By ensuring that all data is considered compromised, simplifying the response process.
b) By providing clear guidelines on the prioritization and protection of affected data based on its sensitivity.
c) By eliminating the need for an incident response plan.
d) By automatically restoring all data to its original state.

Answer: b) By providing clear guidelines on the prioritization and protection of affected data based on its sensitivity. Explanation: A well-defined data classification system benefits the organization's response to a data breach by providing clear guidelines on how to prioritize and protect the affected data.

What is the primary difference between qualitative and quantitative risk assessment methodologies?
a. Qualitative focuses on numerical data, while quantitative relies on subjective judgments.
b. Qualitative assesses risk based on subjective analysis, while quantitative uses numerical data and models.
c. Qualitative is used exclusively in financial risk analysis, whereas quantitative is used in cybersecurity.
d. Qualitative methodologies are legally required, while quantitative are optional best practices.

Answer: b. Qualitative assesses risk based on subjective analysis, while quantitative uses numerical data and models. Explanation: Qualitative risk assessment methodologies rely on subjective judgments to categorize risks based on their severity and impact, often using categories like low, medium, and high. Quantitative methodologies, on the other hand, employ numerical data and statistical models to quantify risk in terms of potential financial impact, making it possible to prioritize risks based on calculated potential losses.

In risk management, what does it mean to 'accept' a risk?
a. Implementing controls to reduce the risk to an acceptable level
b. Transferring the risk to another party through insurance or other means
c. Deciding not to take any action on the risk due to its low priority or impact

d. Completely eliminating the risk from the organization

Answer: c. Deciding not to take any action on the risk due to its low priority or impact. Explanation: Accepting a risk means acknowledging the risk's existence but deciding not to take any specific action to mitigate it, either because its probability and impact are deemed acceptable or because the cost of mitigation exceeds the benefit. This decision usually follows a thorough risk assessment process.

Which risk treatment option involves implementing controls to reduce the likelihood or impact of a risk?
a. Avoidance
b. Acceptance
c. Mitigation
d. Transfer

Answer: c. Mitigation. Explanation: Risk mitigation involves taking steps to reduce either the likelihood of a risk event occurring or its impact on the organization, or both. This can include implementing security controls, developing policies and procedures, or making changes to business processes.

What is a key component of developing a risk management plan?
a. Identifying potential tax benefits
b. Establishing a communication plan for stakeholders
c. Choosing an insurance provider
d. Deciding on office locations

Answer: b. Establishing a communication plan for stakeholders. Explanation: A key component of developing a risk management plan is establishing a communication plan that outlines how risk information will be communicated to stakeholders. This ensures that all relevant parties are informed about the risks, the decisions made regarding risk treatment, and any actions required to manage those risks.

In a quantitative risk assessment, what does the Annual Loss Expectancy (ALE) represent?
a. The total cost of insurance premiums for the year
b. The expected annual cost of all security incidents combined
c. The maximum potential loss from a single risk event in a year
d. The projected annual revenue of the company

Answer: b. The expected annual cost of all security incidents combined. Explanation: The Annual Loss Expectancy (ALE) is a key concept in quantitative risk assessment that represents the expected monetary loss from a specific risk over the course of a year. It is calculated by multiplying the Single Loss Expectancy (SLE), the cost of a single occurrence of a risk event, by the Annual Rate of Occurrence (ARO), the expected number of times that event will occur in a year.

When a company chooses to 'transfer' a risk, what are they typically doing?
a. Moving the risk to a different department within the company
b. Outsourcing the risk to a third-party vendor
c. Purchasing insurance to cover potential losses from the risk
d. Ignoring the risk in hopes that it will not materialize

Answer: c. Purchasing insurance to cover potential losses from the risk. Explanation: Transferring a risk usually involves shifting the financial impact of the risk to another party, most commonly through the purchase of insurance. This doesn't eliminate the risk but manages its financial impact by having the insurance company cover potential losses.

How does 'avoidance' as a risk treatment strategy differ from 'mitigation'?
a. Avoidance eliminates the risk entirely, while mitigation only reduces its impact or likelihood.
b. Avoidance increases the risk's impact, while mitigation decreases its likelihood.
c. Avoidance involves accepting the risk, while mitigation involves transferring it.

d. Avoidance is a short-term strategy, while mitigation is used for long-term risk management.

Answer: a. Avoidance eliminates the risk entirely, while mitigation only reduces its impact or likelihood. Explanation: Risk avoidance involves changing plans or strategies to completely eliminate a risk or its impact, essentially avoiding any scenario where the risk could materialize. Mitigation, on the other hand, accepts the presence of the risk but takes steps to reduce its likelihood of occurring or minimize its impact if it does occur.

What role does a risk register play in risk management plans?
a. It serves as a formal contract for insurance policies.
b. It lists all identified risks, their assessments, and plans for their treatment.
c. It records the daily activities of the risk management team.
d. It is used to register complaints and issues from employees about workplace safety.

Answer: b. It lists all identified risks, their assessments, and plans for their treatment. Explanation: A risk register is a crucial tool in risk management that documents all identified risks, along with information about their assessment (likelihood and impact), the strategies chosen for their treatment (accept, mitigate, avoid, transfer), and any actions taken or planned to address them. It helps in tracking and monitoring the status of each risk over time.

When implementing risk treatment measures, why is it important to consider residual risk?
a. Residual risk is the risk that remains after all insurance policies are applied.
b. Residual risk refers to the new risks introduced by implementing treatment measures.
c. Residual risk is the risk that remains after treatment measures have been applied.
d. Residual risk is always negligible and does not require consideration.

Answer: c. Residual risk is the risk that remains after treatment measures have been applied. Explanation: Residual risk is the level of risk that remains after risk treatment measures have been implemented. It is important to consider and assess residual risk to ensure that it is within acceptable levels, as not all risks can be fully eliminated. Understanding residual risk helps organizations make informed decisions about further action, if necessary.

How can risk management plans be integrated into organizational processes and decision-making?
a. By limiting the risk management process to the IT department only
b. By making risk management a part of strategic business planning and operational activities
c. By focusing risk management efforts solely on financial risks
d. By addressing risks only after they have materialized and caused damage

Answer: b. By making risk management a part of strategic business planning and operational activities. Explanation: Integrating risk management into organizational processes and decision-making involves embedding risk considerations into strategic business planning, project management, and daily operational activities. This holistic approach ensures that risk management is not siloed but is a continuous and integral part of how the organization functions, enabling proactive and informed decision-making across all levels.

What secure data handling procedure should be implemented during the creation phase of the data lifecycle?
a) Data should be immediately classified according to its sensitivity level to determine appropriate handling requirements.
b) Data should be encrypted only if it is intended for external sharing.
c) Data creation should be unrestricted to encourage innovation and ease of use.
d) All data should be made publicly accessible to ensure transparency.

Answer: a) Data should be immediately classified according to its sensitivity level to determine appropriate handling requirements. Explanation: At the creation phase of the data lifecycle, it is crucial to classify data according to its sensitivity level. This initial classification dictates the handling requirements, access controls, and security measures needed to protect the data throughout its lifecycle, ensuring that sensitive information is adequately safeguarded from the outset.

During the storage phase of the data lifecycle, what is a key security measure?

a) Storing all data in a single, centralized location for ease of access.
b) Employing physical storage methods only, as they are less vulnerable to cyber attacks.
c) Implementing encryption for data at rest, particularly for sensitive or critical information.
d) Allowing open access to storage locations to facilitate collaboration.

Answer: c) Implementing encryption for data at rest, particularly for sensitive or critical information. Explanation: During the storage phase, implementing encryption for data at rest is a key security measure. This protects sensitive or critical information from unauthorized access and potential breaches, ensuring that even if physical or logical access controls are bypassed, the data remains unreadable and secure.

What secure practice is essential when accessing data?
a) Implementing a single, static password for all users to simplify access management.
b) Ensuring open access to all data to promote a culture of transparency and trust.
c) Utilizing strong authentication and access controls based on the principle of least privilege.
d) Logging access only for high-sensitivity data to reduce the volume of log files.

Answer: c) Utilizing strong authentication and access controls based on the principle of least privilege. Explanation: When accessing data, it is essential to utilize strong authentication methods and implement access controls based on the principle of least privilege. This ensures that individuals can access only the data necessary for their specific roles and responsibilities, minimizing the risk of unauthorized access or data breaches.

In the context of data use, what guideline ensures secure handling?
a) Allowing unrestricted data manipulation to support a wide range of business activities.
b) Implementing strict controls that prevent data from being used outside of its intended purpose.
c) Encouraging the use of personal devices for work-related tasks to enhance flexibility.
d) Prohibiting the use of data analytics tools to prevent data misinterpretation.

Answer: b) Implementing strict controls that prevent data from being used outside of its intended purpose. Explanation: Secure data handling during the use phase involves implementing strict controls to ensure that data is not used outside of its intended purpose. This includes applying data usage policies, monitoring data access and manipulation, and employing data loss prevention techniques to safeguard against unauthorized use and ensure compliance with data protection regulations.

What is a best practice for secure data sharing?
a) Sharing data through public cloud services without access controls to maximize availability.
b) Using unencrypted email attachments for convenience and speed of delivery.
c) Utilizing secure, encrypted channels and ensuring recipients are authorized to access the data.
d) Distributing physical copies of sensitive data to avoid digital vulnerabilities.

Answer: c) Utilizing secure, encrypted channels and ensuring recipients are authorized to access the data. Explanation: A best practice for secure data sharing is to utilize secure, encrypted channels for transmission and to verify that the recipients are authorized to access the shared data. This approach protects data in transit and ensures that sensitive information is only accessible to intended, authorized parties, thereby reducing the risk of data leaks or breaches.

During the archiving phase, what strategy contributes to data security?
a) Archiving all data indefinitely to preserve organizational knowledge.
b) Selectively archiving data based on its business value and legal retention requirements, and securely deleting unnecessary data.
c) Storing archived data on easily accessible, unencrypted media for quick retrieval.
d) Outsourcing data archiving to third-party vendors without conducting due diligence.

Answer: b) Selectively archiving data based on its business value and legal retention requirements, and securely deleting unnecessary data. Explanation: During the archiving phase, selectively archiving data based on its business value and compliance with legal retention requirements is a key strategy. Additionally, securely deleting data that is no longer necessary helps manage storage costs, reduces clutter, and minimizes the risk associated with retaining unnecessary sensitive information.

What is a fundamental requirement for secure data destruction?
a) Documenting the destruction process for all data, regardless of its classification level.
b) Utilizing basic delete functions provided by the operating system for efficiency.
c) Conducting data destruction in a public setting to ensure transparency.
d) Employing methods that render data unrecoverable, such as physical destruction or cryptographic erasure for digital data.

Answer: d) Employing methods that render data unrecoverable, such as physical destruction or cryptographic erasure for digital data. Explanation: A fundamental requirement for secure data destruction is to employ methods that render the data unrecoverable. This can include physical destruction of storage media for non-digital data or cryptographic erasure and secure wiping techniques for digital data, ensuring that sensitive information cannot be reconstructed or retrieved.

How can organizations ensure secure data handling during the data lifecycle?
a) By limiting security training to IT staff only.
b) By integrating security practices into every phase of the data lifecycle and ensuring all employees are aware of their responsibilities.
c) By focusing solely on external threats and neglecting internal security risks.
d) By adopting a reactive approach to security, addressing issues only after a breach occurs.

Answer: b) By integrating security practices into every phase of the data lifecycle and ensuring all employees are aware of their responsibilities. Explanation: Organizations can ensure secure data handling throughout the data lifecycle by integrating security practices into every phase—from creation to destruction—and making sure that all employees, not just IT staff,

Which encryption method involves a single key for both encryption and decryption processes?
a. Asymmetric encryption
b. Hashing
c. Symmetric encryption
d. Tokenization

Answer: c. Symmetric encryption. Explanation: Symmetric encryption uses a single key, known as a secret key, for both the encryption and decryption processes. This method is efficient for encrypting large amounts of data and is commonly used for securing data at rest.

In the context of data security, what is the primary purpose of hashing?
a. To encrypt data for secure transmission
b. To verify the integrity and authenticity of data
c. To provide a reversible method for data encryption
d. To assign permissions to data based on user roles

Answer: b. To verify the integrity and authenticity of data. Explanation: Hashing is used to ensure the integrity and authenticity of data by generating a fixed-size string of bytes (a hash value) from input data of any size. Hash functions are designed to be one-way, meaning the original data cannot be retrieved from the hash value, making them ideal for verifying data integrity without revealing the data itself.

What distinguishes asymmetric encryption from symmetric encryption?
a. Asymmetric encryption uses the same key for encryption and decryption.
b. Asymmetric encryption uses two different keys, a public key for encryption and a private key for decryption.
c. Asymmetric encryption is less secure than symmetric encryption.
d. Asymmetric encryption cannot be used for digital signatures.

Answer: b. Asymmetric encryption uses two different keys, a public key for encryption and a private key for decryption. Explanation: Asymmetric encryption, also known as public-key cryptography, involves a pair of keys: a public key, which can be shared publicly and is used for encryption, and a private key, which is kept secret and is used for decryption. This method allows for secure communication and data exchange even in the absence of a shared secret key.

How does Data Loss Prevention (DLP) software help protect sensitive information?
a. By physically securing the devices where data is stored
b. By detecting and preventing unauthorized access or transfer of sensitive data
c. By encrypting all data stored on a network
d. By creating backups of all data to prevent loss

Answer: b. By detecting and preventing unauthorized access or transfer of sensitive data. Explanation: DLP solutions monitor, detect, and block sensitive data while in use (data in motion), in motion (data in transit), and at rest, preventing unauthorized access and transfer. These solutions help enforce data security policies and compliance requirements by controlling data flows and identifying potential data breaches.

What role does Rights Management (often referred to as Digital Rights Management, DRM) play in data security?
a. It monitors network traffic to prevent data breaches.
b. It prevents the physical theft of digital devices.
c. It controls access and usage rights for digital content.
d. It encrypts emails to prevent interception.

Answer: c. It controls access and usage rights for digital content. Explanation: Rights Management, or DRM, is a technology used to control access to and usage rights of digital content. This includes preventing unauthorized copying, sharing, and viewing of digital media, ensuring that only authorized users can access and use the content in permitted ways.

In what scenario would you prefer to use symmetric encryption over asymmetric encryption?
a. When a high level of security is not required
b. For encrypting large volumes of data due to its faster processing times
c. For secure communication between parties without prior key exchange
d. For creating digital signatures

Answer: b. For encrypting large volumes of data due to its faster processing times. Explanation: Symmetric encryption is generally faster than asymmetric encryption and is thus preferred for encrypting large volumes of data, such as database encryption, file storage encryption, and secure internal communications, where both parties share a secret key securely.

What is a common use case for asymmetric encryption in data security?
a. Encrypting data stored on a mobile device
b. Securing online transactions and communications
c. Password hashing for user authentication
d. Data loss prevention within an internal network

Answer: b. Securing online transactions and communications. Explanation: Asymmetric encryption is commonly used to secure online transactions and communications, such as SSL/TLS for secure web browsing, email encryption, and secure file transfer protocols. It enables secure key exchange and communication between parties without the need to share a secret key in advance.

How can DLP solutions be integrated into an existing security infrastructure?
a. By replacing the existing firewall with DLP technology
b. By adding DLP capabilities to monitor, detect, and block sensitive data flows in the network, at endpoints, and in storage
c. By using DLP solely for email encryption

d. By focusing DLP implementation only on external threats

Answer: b. By adding DLP capabilities to monitor, detect, and block sensitive data flows in the network, at endpoints, and in storage. Explanation: DLP solutions can be integrated into an existing security infrastructure as an additional layer of protection, focusing on the internal and external movement and storage of sensitive data. This includes network DLP for data in transit, endpoint DLP for data in use, and storage DLP for data at rest.

What is a key advantage of implementing a DRM solution for digital content?
a. It makes content easily transferable between different platforms and devices.
b. It allows unlimited copying and distribution of digital content.
c. It ensures that only authorized users can access and use the content as intended.
d. It reduces the need for other types of data security controls.

Answer: c. It ensures that only authorized users can access and use the content as intended. Explanation: A key advantage of DRM is that it provides content creators and distributors with control over how their digital content is used, accessed, and distributed, ensuring that only authorized users can use the content in ways that comply with the specified usage rights, such as viewing, editing, copying, or sharing restrictions.

When choosing between hashing and encryption for securing data, what is a critical consideration?
a. Hashing is reversible, making it suitable for confidential communications.
b. Encryption, unlike hashing, does not verify data integrity.
c. Hashing is used for data integrity and authentication, not for confidentiality.
d. Encryption is primarily used for non-repudiation purposes.

Answer: c. Hashing is used for data integrity and authentication, not for confidentiality. Explanation: Hashing is a one-way function that is used to ensure data integrity and authenticity by generating a fixed-size hash value from input data. It is not used for encrypting data or keeping it confidential, as hash functions are designed to be irreversible, making it impossible to retrieve the original data from the hash value. Encryption, on the other hand, is used for securing data confidentiality and can be reversed with the correct key.

What is the primary goal of implementing the principle of least privilege in system design?
a) To ensure that all users have unrestricted access to system functionalities for convenience
b) To provide users with only the access necessary to perform their job functions
c) To create a single point of failure for easier management
d) To centralize all system controls under a single administrative account

Answer: b) To provide users with only the access necessary to perform their job functions. Explanation: The principle of least privilege aims to enhance security by restricting users' access rights to only what is strictly required to perform their job functions. This minimizes the potential damage from accidents, errors, or unauthorized use of privileged access.

How does separation of duties contribute to secure system design?
a) By ensuring that no single individual has control over all aspects of a system
b) By consolidating system responsibilities to a single department for efficiency
c) By requiring all users to share the same set of permissions for simplicity
d) By eliminating the need for audit trails since tasks are not divided

Answer: a) By ensuring that no single individual has control over all aspects of a system. Explanation: Separation of duties is a security concept that divides critical tasks and privileges among multiple users or systems to prevent fraud and errors. This ensures that no single individual has control over all aspects of any critical or sensitive process, thereby reducing the risk of misuse or abuse of power.

What defines fail-safe defaults in the context of secure design?
a) Systems are designed to grant all permissions by default until restrictions are applied
b) Systems default to a secure state, denying access unless explicitly granted

c) Systems are configured to fail open, maintaining accessibility under all conditions
d) Systems require manual intervention to activate security controls

Answer: b) Systems default to a secure state, denying access unless explicitly granted. Explanation: Fail-safe defaults refer to a design principle where, by default, systems are in a secure state and access is denied unless explicitly granted. This approach ensures that if a failure occurs or if access rights are not specifically defined, the system will default to a secure posture, minimizing potential vulnerabilities.

In the concept of defense in depth, what role does layered security play?
a) It eliminates the need for physical security measures
b) It relies on a single, impenetrable layer of security to protect assets
c) It employs multiple layers of security controls across the entire IT infrastructure
d) It focuses solely on perimeter security to defend against external threats

Answer: c) It employs multiple layers of security controls across the entire IT infrastructure. Explanation: Defense in depth is a strategy that employs multiple layers of security controls and measures across the entire IT infrastructure to protect data and systems. This approach ensures that if one layer is breached, additional layers provide continued protection, making it more difficult for threats to penetrate the system fully.

How can the principle of least privilege be effectively applied in a cloud computing environment?
a) By giving all users administrative privileges to avoid access issues
b) By using a single shared account for all cloud services to simplify management
c) By assigning permissions based on user roles and requiring justification for elevated access
d) By disabling logging and monitoring to improve performance

Answer: c) By assigning permissions based on user roles and requiring justification for elevated access. Explanation: In a cloud computing environment, the principle of least privilege can be effectively applied by assigning permissions based on user roles and responsibilities, ensuring that users have only the access necessary for their job functions. Elevated access should be granted sparingly and only with proper justification and approval, reducing the risk of unauthorized access or data breaches.

What design strategy can ensure system resilience and recovery capability in case of a failure?
a) Designing systems with a single critical component that, if failed, would cause system failure
b) Implementing redundancy and failover mechanisms for critical system components
c) Avoiding regular system backups to reduce data redundancy
d) Concentrating all data processing in a single location to simplify management

Answer: b) Implementing redundancy and failover mechanisms for critical system components. Explanation: Ensuring system resilience and recovery capability involves designing systems with redundancy and failover mechanisms for critical components. This means that if one component fails, another can take over without causing system downtime, thereby enhancing the system's overall reliability and availability.

How does the concept of secure by design influence the development of new software applications?
a) By prioritizing development speed and feature richness over security considerations
b) By integrating security considerations into every phase of the software development lifecycle
c) By adding security features only after the software development is complete
d) By outsourcing security to third-party services, absolving developers of security responsibilities

Answer: b) By integrating security considerations into every phase of the software development lifecycle. Explanation: Secure by design is a principle that integrates security considerations into every phase of the software development lifecycle, from planning and design to implementation and maintenance. This proactive approach ensures that security is a foundational aspect of the application, rather than an afterthought, leading to more secure and robust software.

What is a key benefit of implementing network segmentation in an organization's infrastructure design?
a) It allows for unrestricted communication between all network devices

b) It reduces the complexity of the network, making it easier to manage
c) It limits the spread of attacks within the network by isolating segments
d) It centralizes all network resources into a single segment for efficiency

Answer: c) It limits the spread of attacks within the network by isolating segments. Explanation: Network segmentation involves dividing a larger network into smaller, isolated segments or subnetworks. This design strategy limits the spread of attacks within the network, as it restricts attackers' lateral movement and access to sensitive information, thereby enhancing overall security.

In secure system design, what is the purpose of incorporating input validation mechanisms?
a) To ensure that all inputs are accepted without scrutiny for user convenience
b) To reject any input that does not meet the predefined criteria for format, type, and content
c) To automatically correct any input errors without notifying the user
d) To store all input directly in the database without preprocessing

Answer: b) To reject any input that does not meet the predefined criteria for format, type, and content. Explanation: Input validation mechanisms are incorporated into secure system design to ensure that only properly formatted and expected input is accepted by the system. This involves rejecting any input that does not meet specific criteria for format, type, and content, thereby preventing malicious input that could lead to vulnerabilities such as SQL injection or cross-site scripting (XSS) attacks.

What cryptographic algorithm is commonly used for secure symmetric key encryption and is known for its efficiency in various hardware and software environments?
a. RSA
b. AES
c. SHA-256
d. ECC

Answer: b. AES. Explanation: The Advanced Encryption Standard (AES) is widely used for symmetric key encryption due to its efficiency and strong security level. It is capable of encrypting data at rest and in transit, making it a versatile choice for securing sensitive information across different platforms.

For securing data in transit over the internet, which cryptographic protocol is commonly implemented?
a. HTTPS
b. S/MIME
c. PGP
d. WPA2

Answer: a. HTTPS. Explanation: HTTPS, which stands for Hypertext Transfer Protocol Secure, incorporates SSL/TLS protocols to provide a secure channel over the internet. It is widely used for secure communications between web browsers and servers, ensuring data in transit is encrypted and protected from interception.

In the context of Public Key Infrastructure (PKI), what role does a Certificate Authority (CA) play?
a. It generates private keys for users.
b. It issues and manages digital certificates, verifying the ownership of public keys.
c. It encrypts data using asymmetric cryptography.
d. It acts as a repository for storing encrypted data.

Answer: b. It issues and manages digital certificates, verifying the ownership of public keys. Explanation: A Certificate Authority (CA) is a trusted entity in PKI that issues digital certificates to validate the ownership of a public key. The digital certificate serves as a verification that the public key belongs to the individual, organization, device, or service identified in the certificate.

What is the primary function of digital signatures in the context of data security?
a. To encrypt data for confidentiality

b. To verify the integrity and authenticity of digital messages or documents
c. To generate symmetric encryption keys
d. To provide a secure channel for data transmission

Answer: b. To verify the integrity and authenticity of digital messages or documents. Explanation: Digital signatures use public key cryptography to verify the integrity and authenticity of digital messages or documents. The sender signs the document with their private key, and the recipient can verify the signature with the sender's public key, ensuring the data has not been tampered with and confirming the sender's identity.

Which cryptographic algorithm is best suited for establishing a secure session key during the initial handshake between two parties over an unsecured network?
a. AES
b. RSA
c. MD5
d. HMAC

Answer: b. RSA. Explanation: RSA, an asymmetric cryptographic algorithm, is well-suited for securely exchanging a session key over an unsecured network. Its ability to encrypt the session key with the recipient's public key ensures that only the holder of the corresponding private key can decrypt and use the session key for secure communication.

How does a hash function contribute to data integrity?
a. By encrypting the data to prevent unauthorized access
b. By generating a fixed-size output from input data, allowing for consistency checks
c. By generating a unique key for symmetric encryption
d. By creating a secure tunnel for data transmission

Answer: b. By generating a fixed-size output from input data, allowing for consistency checks. Explanation: A hash function processes input data to produce a fixed-size output (hash value), which is unique to the original data. Any alteration to the data results in a different hash value, allowing for effective integrity checks by comparing the computed hash value with the original one.

In the PKI ecosystem, what is the purpose of a digital certificate?
a. To serve as a digital form of identification for individuals or devices, linking public keys with identities
b. To encrypt email messages for confidentiality
c. To store private keys for user authentication
d. To act as a password for accessing secured networks

Answer: a. To serve as a digital form of identification for individuals or devices, linking public keys with identities. Explanation: A digital certificate in PKI binds a public key with the identity of the certificate's owner, such as an individual, organization, or device. This binding is established through the certificate authority's validation process and provides a way to verify the ownership of a public key for secure communications and transactions.

What advantage does asymmetric encryption offer over symmetric encryption?
a. It is faster and more efficient for encrypting large volumes of data.
b. It eliminates the need for the secure exchange of a secret key.
c. It uses shorter keys for the same level of security.
d. It is the only method capable of encrypting data at rest.

Answer: b. It eliminates the need for the secure exchange of a secret key. Explanation: Asymmetric encryption uses a pair of keys (public and private) for encryption and decryption, eliminating the need for the secure exchange of a secret key, which is a challenge in symmetric encryption. This facilitates secure communication between parties who have not previously exchanged secret keys.

What is the primary function of HVAC systems in maintaining data center security?
a) To provide a comfortable working environment for staff

b) To regulate temperature and humidity to protect equipment from overheating and moisture
c) To filter out cyber threats from the air
d) To circulate colored smoke as an indicator of system status

Answer: b) To regulate temperature and humidity to protect equipment from overheating and moisture. Explanation: HVAC (Heating, Ventilation, and Air Conditioning) systems play a critical role in data center security by regulating temperature and humidity. This ensures that sensitive equipment, particularly servers and storage devices, are kept within optimal operating conditions to prevent overheating, moisture damage, and subsequent hardware failures or data loss.

Which fire suppression system is most suitable for data centers and why?
a) Water sprinklers, for their widespread availability
b) Gas-based systems, because they minimize damage to electronic equipment
c) Foam suppressants, for their visual deterrent effect
d) Standard fire extinguishers, for ease of manual application

Answer: b) Gas-based systems, because they minimize damage to electronic equipment. Explanation: Gas-based fire suppression systems are most suitable for data centers as they extinguish fires without leaving residue or moisture that could harm sensitive electronic equipment. Unlike water sprinklers or foam suppressants, gas-based systems help to ensure the integrity and functionality of critical infrastructure during and after a fire incident.

What is the purpose of using mantraps in secure facilities?
a) To trap intruders physically until security personnel can respond
b) To control access to secure areas by allowing only one person to pass at a time
c) To provide a decorative entrance to the facility
d) To capture pests and rodents that may enter the facility

Answer: b) To control access to secure areas by allowing only one person to pass at a time. Explanation: Mantraps are physical security mechanisms designed to control access to secure areas by allowing only one person to pass through at a time. They typically consist of two sets of interlocking doors, where the second set of doors will not open until the first set has closed and proper authentication has been provided, thereby preventing tailgating and unauthorized access.

How do biometric access controls enhance physical security measures?
a) By using unique physical characteristics for individual identification, making it difficult to forge
b) By requiring a traditional key that can be easily duplicated
c) By using a standard password system accessible to all employees
d) By employing a color-coded badge system based on department

Answer: a) By using unique physical characteristics for individual identification, making it difficult to forge. Explanation: Biometric access controls enhance physical security by using unique individual characteristics, such as fingerprints, iris patterns, or facial recognition, for identification and authentication. This makes unauthorized access significantly more difficult compared to traditional keys or passwords, which can be shared, lost, or stolen.

In perimeter security, what is the main purpose of deploying surveillance cameras around a facility?
a) To broadcast the facility's interior to the public for transparency
b) To monitor and record activities for evidence and to deter unauthorized access
c) To provide live entertainment to employees during breaks
d) To track wildlife activity around the facility for environmental studies

Answer: b) To monitor and record activities for evidence and to deter unauthorized access. Explanation: The main purpose of deploying surveillance cameras around a facility's perimeter is to monitor and record activities continuously. This provides valuable evidence in the event of a security breach and acts as a deterrent to unauthorized individuals by increasing the likelihood of detection and intervention.

What role does fencing play in a comprehensive physical security strategy?

a) To provide an aesthetic boundary around the property
b) To serve as the primary and only line of defense against intrusions
c) To act as a physical barrier that delays and deters unauthorized access to the facility
d) To function as a guide for pedestrian traffic only, without any security implications

Answer: c) To act as a physical barrier that delays and deters unauthorized access to the facility. Explanation: Fencing plays a crucial role in a comprehensive physical security strategy by acting as a physical barrier that delays and deters unauthorized access. It serves as one of the first lines of defense by creating a boundary around the property, contributing to a layered security approach that includes additional controls beyond the perimeter.

Why is it important to have multiple layers of security, as in the principle of defense in depth?
a) To ensure that if one layer fails, others will provide additional protection
b) To create a more complex system that is harder for staff to understand
c) To increase the budget allocated to the security department
d) To comply with aesthetic standards for security equipment

Answer: a) To ensure that if one layer fails, others will provide additional protection. Explanation: The principle of defense in depth involves having multiple layers of security controls and measures in place so that if one layer fails or is bypassed, other layers will continue to provide protection. This redundancy increases the overall security posture by making it more difficult for an attacker to gain unauthorized access to assets.

What is the primary consideration when designing access controls for a secure facility?
a) Ensuring that they are visually pleasing and blend with the environment
b) Balancing security needs with operational efficiency and user convenience
c) Making them as complicated as possible to deter intruders
d) Using only the most expensive technology available

Answer: b) Balancing security needs with operational efficiency and user convenience. Explanation: When designing access controls for a secure facility, the primary consideration is to balance the need for security with operational efficiency and user convenience. Access controls should be robust enough to deter and prevent unauthorized access, but not so restrictive that they hinder the productivity of authorized users or create unnecessary bottlenecks.

In what way do environmental controls like fire suppression systems contribute to the overall security of a data center?
a) By enhancing the physical appearance of the data center
b) By providing a comfortable environment for visitors
c) By protecting hardware and data from damage due to environmental hazards
d) By increasing the data transfer speeds of the network

Answer: c) By protecting hardware and data from damage due to environmental hazards. Explanation: Environmental controls such as fire suppression systems are critical to the overall security of a data center as they protect the physical hardware and stored data from damage due to environmental hazards like fire. These systems are designed to quickly extinguish fires without harming sensitive equipment, thereby preserving the integrity and availability of data.

How does the implementation of security lighting contribute to the deterrence of unauthorized access after hours?
a) By providing a constant light source that allows intruders to see clearly
b) By illuminating potential hiding spots and entry points, reducing the cover of darkness for intruders
c) By using colored lighting to indicate security levels
d) By serving as a decorative feature to beautify the perimeter

Answer: b) By illuminating potential hiding spots and entry points, reducing the cover of darkness for intruders. Explanation: Security lighting is implemented as a deterrent to unauthorized access, especially after hours, by illuminating potential hiding spots and entry points around a facility's perimeter. This visibility reduces the concealment provided by darkness, making it more likely for intruders to be seen and reported, thereby enhancing the facility's overall security posture.

What function does Network Access Control (802.1x) serve in a secure network design?
a. It encrypts data packets to prevent eavesdropping.
b. It provides a framework for authenticating and authorizing devices before they access the network.
c. It increases the speed of the network by prioritizing traffic.
d. It serves as a physical barrier to unauthorized network access.

Answer: b. It provides a framework for authenticating and authorizing devices before they access the network. Explanation: Network Access Control (802.1x) acts as a gatekeeper to the network by enforcing authentication and authorization policies. Before devices are allowed access to network resources, they must be authenticated, typically using a username and password, certificates, or other methods. This ensures that only authorized devices can connect to the network, enhancing its security.

How do Intrusion Detection Systems (IDS) and Intrusion Prevention Systems (IPS) differ in their approach to network threats?
a. IDS detects and logs potential threats, while IPS takes active measures to block them.
b. IDS encrypts network traffic, whereas IPS decrypts it for inspection.
c. IDS is used exclusively for detecting malware, while IPS focuses on spam and phishing attacks.
d. IDS operates on the network perimeter, while IPS works on the endpoint devices.

Answer: a. IDS detects and logs potential threats, while IPS takes active measures to block them. Explanation: Intrusion Detection Systems (IDS) monitor network and system traffic for suspicious activity, logging potential threats for further investigation. In contrast, Intrusion Prevention Systems (IPS) not only detect threats but also take proactive steps to block or mitigate them, often in real-time, before they can cause harm.

What is the primary function of Security Information and Event Management (SIEM) systems in a secure network environment?
a. To serve as firewalls that control incoming and outgoing network traffic
b. To manage network devices and software installations
c. To collect, analyze, and report on security data from various sources in real-time
d. To encrypt data stored on network servers

Answer: c. To collect, analyze, and report on security data from various sources in real-time. Explanation: SIEM systems aggregate and analyze log and event data from various sources within the network, such as firewalls, IDS/IPS, and other security tools, to provide real-time analysis of security alerts. This enables organizations to detect, understand, and respond to security incidents more effectively.

In the context of secure network design, what advantage does segmenting the network provide?
a. It allows all users unrestricted access to the entire network for ease of use.
b. It reduces the overall cost of the network infrastructure by using fewer devices.
c. It limits the spread of attacks within the network by dividing it into smaller, controlled segments.
d. It simplifies network management by reducing the number of rules and policies.

Answer: c. It limits the spread of attacks within the network by dividing it into smaller, controlled segments. Explanation: Network segmentation divides the network into smaller, more manageable segments or subnetworks. This restricts the lateral movement of attackers within the network, as they would need to breach multiple segments to access sensitive information, thereby enhancing overall network security.

How does implementing a Demilitarized Zone (DMZ) enhance network security?
a. By acting as a dedicated space for file storage, thus freeing up internal network resources
b. By providing a buffer zone that separates the internal network from external traffic, hosting public-facing services
c. By increasing the bandwidth available for external communications
d. By serving as the sole authentication point for all users, both internal and external

Answer: b. By providing a buffer zone that separates the internal network from external traffic, hosting public-facing services. Explanation: A DMZ is a network segment that acts as a buffer zone between the public internet and an organization's internal network. It hosts public-facing services like web servers and email servers, minimizing the exposure of the internal network to potential threats and attacks from the internet.

What is the primary benefit of using a Virtual Private Network (VPN) for remote access in a secure network design?
a. It allows unlimited access to the internet without any security restrictions.
b. It provides a secure, encrypted connection over the internet for remote users accessing the internal network.
c. It enhances the speed of remote connections to internal network resources.
d. It eliminates the need for any form of user authentication.

Answer: b. It provides a secure, encrypted connection over the internet for remote users accessing the internal network.
Explanation: A VPN creates a secure and encrypted tunnel over the public internet, allowing remote users to access the organization's internal network resources securely. This encryption helps protect sensitive data and communications from interception or eavesdropping.

What feature distinguishes WPA3 from WPA2 in terms of wireless security?
a) WPA3 eliminates the need for passwords.
b) WPA3 introduces individualized data encryption to protect against eavesdropping on open networks.
c) WPA3 supports older hardware exclusively, enhancing backward compatibility.
d) WPA3 reduces the range of Wi-Fi signals for security purposes.

Answer: b) WPA3 introduces individualized data encryption to protect against eavesdropping on open networks.
Explanation: WPA3 enhances wireless security by introducing individualized data encryption, known as Wi-Fi Protected Access 3 - Enhanced Open, which provides users with data privacy on open public networks. This feature protects against eavesdropping by encrypting the connection between each device and the network, even if no password is used.

Which authentication method uses a certificate-based approach to validate client and server identities in a wireless network?
a) Pre-shared Key (PSK)
b) Extensible Authentication Protocol-Transport Layer Security (EAP-TLS)
c) Captive Portal
d) Wired Equivalent Privacy (WEP)

Answer: b) Extensible Authentication Protocol-Transport Layer Security (EAP-TLS). Explanation: EAP-TLS is a certificate-based authentication method that provides mutual authentication where both the client and the server validate each other's identities through certificates. This method is known for its high level of security, as it prevents unauthorized access by ensuring that both parties are trustworthy before establishing a connection.

How can a rogue access point pose a security threat in a wireless network?
a) By increasing the network's signal strength
b) By serving as a legitimate backup access point
c) By broadcasting a stronger signal to overshadow the legitimate network
d) By mimicking a legitimate access point to intercept and manipulate network traffic

Answer: d) By mimicking a legitimate access point to intercept and manipulate network traffic. Explanation: A rogue access point poses a significant security threat by impersonating a legitimate access point within the network. Unsuspecting users may connect to this malicious access point, allowing the attacker to intercept, manipulate, or redirect network traffic, potentially leading to data breaches and other security incidents.

What is the primary security concern associated with an "Evil Twin" attack in wireless networks?
a) It slows down the network speed by adding additional devices.
b) It encrypts all network traffic, making it unreadable.
c) It creates a duplicate of a legitimate access point to trick users into connecting.
d) It strengthens the network's encryption, locking out authorized users.

Answer: c) It creates a duplicate of a legitimate access point to trick users into connecting. Explanation: An "Evil Twin" attack involves setting up a malicious access point that appears identical to a legitimate one, often mirroring its SSID and other settings. Unsuspecting users connect to this deceptive twin, thinking it's a trusted network, which allows attackers to intercept sensitive data or launch further attacks against connected devices.

In the context of WPA2, what does the term PSK stand for, and what is its role?
a) Public Security Key, used for encrypting public Wi-Fi networks
b) Personal Safety Kit, a set of tools for network administrators
c) Pre-shared Key, a shared secret used for authentication in personal networks
d) Protocol Security Key, a dynamic key used for each session

Answer: c) Pre-shared Key, a shared secret used for authentication in personal networks. Explanation: PSK stands for Pre-shared Key, which is a mode of WPA2 used primarily in home and small office networks. It involves a shared secret or passphrase known to all authorized users of the network, which is used to authenticate devices trying to connect to the wireless network, ensuring that only those with the correct passphrase can gain access.

What role does the Extensible Authentication Protocol (EAP) play in wireless security?
a) It specifies packet formats for data transmission.
b) It provides a framework for wireless authentication methods.
c) It defines the encryption algorithms for wireless data.
d) It standardizes SSID broadcasting techniques.

Answer: b) It provides a framework for wireless authentication methods. Explanation: The Extensible Authentication Protocol (EAP) provides a general framework used in wireless networks and point-to-point connections to support multiple authentication methods. This flexibility allows networks to use various credentials and techniques, such as certificates, smart cards, or usernames and passwords, to authenticate users and devices.

How does implementing a wireless intrusion prevention system (WIPS) enhance network security?
a) By increasing the Wi-Fi signal strength across the network
b) By detecting and preventing unauthorized access and attacks on the wireless network
c) By automatically connecting users to the nearest access point
d) By encrypting traffic between wireless clients and access points with unique keys

Answer: b) By detecting and preventing unauthorized access and attacks on the wireless network. Explanation: A Wireless Intrusion Prevention System (WIPS) enhances network security by actively monitoring the wireless spectrum for unauthorized access points (rogue access points), wireless attacks, and other anomalies. When a threat is detected, WIPS can take preventive actions, such as alerting administrators, disconnecting unauthorized devices, or blocking malicious traffic, thereby protecting the network from potential breaches and attacks.

What advantage does WPA3 provide over WPA2 in terms of password-based authentication?
a) WPA3 eliminates the need for passwords altogether.
b) WPA3 introduces Simultaneous Authentication of Equals (SAE), which provides protection against offline dictionary attacks.
c) WPA3 allows for shorter, easier-to-remember passwords.
d) WPA3 requires passwords to be changed every 30 days automatically.

Answer: b) WPA3 introduces Simultaneous Authentication of Equals (SAE), which provides protection against offline dictionary attacks. Explanation: WPA3 enhances security over WPA2 by introducing Simultaneous Authentication of Equals (SAE), a new handshake process that replaces the Pre-shared Key (PSK) in WPA2. SAE provides better protection against offline dictionary attacks, where an attacker tries to guess the network password by trying numerous possibilities without interacting with the network.

In a wireless network, what is the primary security function of MAC address filtering?
a) To encrypt data packets based on the MAC address
b) To dynamically assign IP addresses to network devices

c) To restrict network access to devices with specific MAC addresses
d) To change the MAC address of devices periodically for anonymity

Answer: c) To restrict network access to devices with specific MAC addresses. Explanation: MAC (Media Access Control) address filtering is a security measure used in wireless networks to restrict network access to devices whose MAC addresses have been explicitly allowed. This creates a control list of devices that are permitted to connect to the network, adding a layer of security by preventing unauthorized devices from accessing the network even if they know the network's password.

What element of multi-factor authentication (MFA) involves something the user knows?
a) A fingerprint or retina scan
b) A mobile device or security token
c) A password or PIN
d) A physical key or smart card

Answer: c) A password or PIN. Explanation: In the context of multi-factor authentication, something the user knows typically refers to a knowledge factor, such as a password, PIN, or any other piece of information that the user must memorize and provide to verify their identity.

How does Single Sign-On (SSO) enhance user experience in an enterprise environment?
a) By requiring users to remember multiple passwords for different services
b) By allowing users to access multiple applications or services with one set of credentials
c) By increasing the time it takes to switch between different applications
d) By using a different password for each application to strengthen security

Answer: b) By allowing users to access multiple applications or services with one set of credentials. Explanation: Single Sign-On (SSO) enhances user experience by allowing individuals to access multiple applications or services within an organization using one set of credentials, eliminating the need to remember and enter different usernames and passwords for each service.

In the context of authentication, what does the "inherence" factor refer to?
a) Something the user possesses, such as a security token
b) Something the user knows, like a password
c) Something the user is, based on biometric characteristics
d) Something the user does, such as a specific way of typing

Answer: c) Something the user is, based on biometric characteristics. Explanation: The "inherence" factor in authentication refers to biometric characteristics that are unique to the individual, such as fingerprints, facial recognition, iris scans, or voice recognition. These inherent physical or behavioral traits are used to verify the user's identity.

What is a primary benefit of using federated identity management in a multi-organization ecosystem?
a) It requires users to maintain separate credentials for each organization's systems.
b) It consolidates user identities across different systems and organizations for simplified access control.
c) It restricts users to access resources only within their own organization.
d) It eliminates the need for encryption within the federated network.

Answer: b) It consolidates user identities across different systems and organizations for simplified access control. Explanation: Federated identity management allows different organizations to consolidate and manage user identities across various systems and applications. This enables users to access resources across organizational boundaries using a single set of credentials, simplifying access control and enhancing interoperability.

Which authentication mechanism involves something the user possesses?
a) A pattern the user draws on a screen
b) A hardware token or a smart card

c) A voice recognition system
d) A personal identification question

Answer: b) A hardware token or a smart card. Explanation: Something the user possesses, in the context of authentication, typically refers to a physical object that the user must have in their possession to gain access, such as a hardware token, smart card, mobile phone, or any other device used to verify identity.

How does multi-factor authentication (MFA) improve security over single-factor authentication methods?
a) By using only a password for all forms of authentication
b) By combining at least two different types of authentication factors
c) By decreasing the complexity of the authentication process
d) By implementing the same authentication factor multiple times

Answer: b) By combining at least two different types of authentication factors. Explanation: Multi-factor authentication improves security by requiring two or more distinct forms of evidence to verify a user's identity. This can include something the user knows (password), something the user has (security token), and something the user is (biometric verification), making unauthorized access significantly more difficult.

In a secure authentication system, what purpose does a man-in-the-middle (MITM) attack aim to serve?
a) To strengthen the encryption between the user and the authentication system
b) To intercept and possibly alter the communication between the user and the authentication system
c) To speed up the authentication process for the user
d) To provide an additional layer of security through obscurity

Answer: b) To intercept and possibly alter the communication between the user and the authentication system. Explanation: A man-in-the-middle (MITM) attack is a security threat where the attacker secretly intercepts and possibly alters the communication between two parties who believe they are directly communicating with each other. In the context of authentication, this can allow the attacker to steal credentials or manipulate the authentication process.

What does the use of a "possession" factor in authentication imply?
a) The user must remember a specific piece of information.
b) The user must demonstrate a unique physical or behavioral trait.
c) The user must verify their identity through a physical object they have.
d) The user must be located within a specific geographical location.

Answer: c) The user must verify their identity through a physical object they have. Explanation: A "possession" factor in authentication implies that the user must have a specific physical object in their possession to verify their identity. This could be a security token, a smart card, a mobile phone with an authentication app, or any device used to generate or receive a unique authentication code.

What is the significance of "contextual authentication" in enhancing security measures?
a) It relies solely on traditional passwords, disregarding any environmental or situational information.
b) It enhances security by considering the context of the access request, such as location, time, and device used.
c) It decreases security by adding unnecessary complexity to the authentication process.
d) It simplifies the authentication process by ignoring anomalous access patterns.

Answer: b) It enhances security by considering the context of the access request, such as location, time, and device used. Explanation: Contextual authentication adds an extra layer of security by taking into account the context in which an access request is made. This may include analyzing factors such as the user's location, the time of the request, the device being used, and any unusual access patterns, to determine the legitimacy of the request and adjust the authentication requirements accordingly.

What is the primary purpose of user account provisioning in identity and access management?
a. To ensure that all users have administrative privileges
b. To monitor and record user activities for compliance purposes

c. To grant users access to resources based on their roles and responsibilities
d. To restrict all users from accessing confidential information

Answer: c. To grant users access to resources based on their roles and responsibilities. Explanation: User account provisioning involves setting up user accounts and granting access to resources based on individuals' roles, responsibilities, and requirements within an organization. This ensures that users have the necessary permissions to perform their job functions while adhering to the principle of least privilege.

How does role-based access control (RBAC) enhance security within an organization?
a. By allowing unrestricted access to all users to simplify management
b. By assigning permissions to roles rather than individuals to streamline access management
c. By encrypting all data with a single key for ease of access
d. By requiring multifactor authentication for every access request

Answer: b. By assigning permissions to roles rather than individuals to streamline access management. Explanation: RBAC enhances security by assigning permissions to roles based on job functions rather than to individual users. This approach simplifies the management of user permissions, ensures consistent access control, and reduces the risk of unauthorized access.

What is a critical step in the identity and access lifecycle for ensuring that access rights remain appropriate over time?
a. Periodic access reviews and recertification
b. Constantly increasing access privileges
c. Disabling multifactor authentication
d. Sharing passwords among team members

Answer: a. Periodic access reviews and recertification. Explanation: Periodic access reviews and recertification are essential for ensuring that users' access rights remain appropriate and aligned with their current roles and responsibilities. This process involves regularly reviewing and validating user access to prevent privilege creep and ensure compliance with security policies.

In the context of access management, what does deprovisioning refer to?
a. Increasing the level of access based on user requests
b. Granting temporary access to guest users
c. Removing or restricting user access when it is no longer needed or when a user leaves the organization
d. Auditing and logging all access requests and activities

Answer: c. Removing or restricting user access when it is no longer needed or when a user leaves the organization. Explanation: Deprovisioning involves the removal or restriction of a user's access rights when they are no longer required, such as when an employee leaves the company or changes roles. This process is crucial for maintaining security and minimizing the risk of unauthorized access.

What benefit does implementing RBAC provide in terms of regulatory compliance?
a. It eliminates the need for compliance with any regulations
b. It automatically grants compliance with all global regulations
c. It helps in aligning access control policies with compliance requirements
d. It allows all users to participate in compliance audits

Answer: c. It helps in aligning access control policies with compliance requirements. Explanation: Implementing RBAC can help organizations align their access control policies with regulatory compliance requirements by ensuring that access to sensitive information and systems is restricted based on defined roles and responsibilities. This structured approach to access management can aid in meeting the access control criteria specified in various regulations.

How does automated user account provisioning benefit an organization's security posture?
a. By manually processing each account request to increase oversight
b. By reducing the time and potential errors associated with manual account setup processes

c. By increasing the complexity of the access request process
d. By delegating all access decisions to individual users

Answer: b. By reducing the time and potential errors associated with manual account setup processes. Explanation: Automated user account provisioning enhances an organization's security posture by streamlining the account setup process, reducing the time to grant necessary access, and minimizing the potential for errors that can occur with manual provisioning. This ensures timely and accurate assignment of access rights, enhancing overall security.

What is the significance of access reviews and recertification in maintaining effective access control?
a. They provide an ongoing record of all user activities for performance evaluations
b. They ensure that user access remains aligned with current roles, responsibilities, and security policies
c. They guarantee that all users have access to all systems at all times
d. They focus solely on increasing access levels to improve productivity

Answer: b. They ensure that user access remains aligned with current roles, responsibilities, and security policies. Explanation: Access reviews and recertification play a crucial role in maintaining effective access control by regularly verifying that user access rights are still appropriate based on their current job functions and responsibilities. This process helps in identifying and rectifying any discrepancies, such as unnecessary or excessive access rights, thereby upholding security policies and compliance.

What is the primary goal of implementing a user account modification process in identity management?
a. To ensure that all users have the same level of access
b. To update user access rights in response to role changes or evolving access requirements
c. To remove all access restrictions for ease of use
d. To centralize all user passwords in a single database

Answer: b. To update user access rights in response to role changes or evolving access requirements. Explanation: The user account modification process is essential for updating users' access rights to reflect changes in their roles, responsibilities, or access requirements. This ensures that access privileges remain relevant and appropriate, enhancing security and operational efficiency.

How do regular access reviews support the principle of least privilege?
a. By ensuring all users have maximum privileges for convenience
b. By periodically verifying and adjusting access rights to ensure users have only the necessary privileges
c. By allowing users to determine their own access needs
d. By permanently fixing user privileges to prevent changes

Answer: b. By periodically verifying and adjusting access rights to ensure users have only the necessary privileges. Explanation: Regular access reviews support the principle of least privilege by ensuring that users retain only the access rights necessary for their current job functions. This process involves assessing and adjusting privileges over time to prevent privilege accumulation and reduce the risk of unauthorized access or insider threats.

In what way does automating the identity and access lifecycle enhance security for an organization?
a. By completely removing human oversight from the access control process
b. By speeding up the provisioning and deprovisioning processes, reducing the window of opportunity for unauthorized access
c. By making all access decisions public to promote transparency
d. By relying solely on user discretion for access management

Answer: b. By speeding up the provisioning and deprovisioning processes, reducing the window of opportunity for unauthorized access. Explanation: Automating the identity and access lifecycle enhances security by ensuring that provisioning and deprovisioning processes are executed swiftly and accurately. This minimizes the risk of unauthorized access by ensuring that users have appropriate access rights as soon as they are needed and that these rights are promptly revoked when no longer necessary.

What is a critical factor to consider when granting vendor access to your organization's systems?
a) The geographical location of the vendor
b) The vendor's preference for communication tools
c) Ensuring access is based on the principle of least privilege
d) The color scheme of the vendor's branding

Answer: c) Ensuring access is based on the principle of least privilege. Explanation: When granting vendor access to an organization's systems, it is critical to apply the principle of least privilege. This means providing the vendor with only the access necessary to perform the required tasks, thereby minimizing potential risks to system security and data integrity.

How can organizations effectively evaluate a vendor's security posture?
a) By assessing the vendor's website design quality
b) Through comprehensive security assessments and audits of the vendor
c) Based on the number of employees in the vendor's company
d) Evaluating the vendor's office locations

Answer: b) Through comprehensive security assessments and audits of the vendor. Explanation: Evaluating a vendor's security posture effectively involves conducting comprehensive security assessments and audits. This may include reviewing the vendor's security policies, incident response plans, compliance certifications, and past security performance to ensure they meet the organization's security requirements.

What is a significant risk associated with inadequate vendor access management?
a) Increased collaboration efficiency
b) Enhanced system performance
c) Potential unauthorized access to sensitive data
d) More frequent system updates

Answer: c) Potential unauthorized access to sensitive data. Explanation: Inadequate vendor access management can lead to significant risks, such as potential unauthorized access to sensitive data. If vendors are given more access than necessary or if their access is not properly monitored and controlled, it can create vulnerabilities that might be exploited by malicious actors.

In the context of third-party IAM risks, what is the importance of continuous monitoring of vendor activities?
a) It helps in improving vendor relationship management
b) It ensures that vendor activities align with contractual obligations and security standards
c) It assists vendors in marketing their services
d) It reduces the need for internal security measures

Answer: b) It ensures that vendor activities align with contractual obligations and security standards. Explanation: Continuous monitoring of vendor activities is crucial to ensure that their actions remain in compliance with contractual obligations and adhere to the required security standards. This proactive approach helps in identifying and mitigating potential security risks associated with vendor access and operations.

What role does a Service Level Agreement (SLA) play in managing third-party IAM risks?
a) It outlines the recreational activities available to vendor employees
b) It specifies the security requirements and expectations from the vendor
c) It lists the vendor's clients other than the organization
d) It details the vendor's corporate social responsibility initiatives

Answer: b) It specifies the security requirements and expectations from the vendor. Explanation: A Service Level Agreement (SLA) is critical in managing third-party IAM risks as it explicitly specifies the security requirements, expectations, and responsibilities of the vendor. This legal document ensures that both parties have a clear understanding of the security measures that must be maintained, providing a basis for accountability and recourse in the event of non-compliance.

How should organizations manage the lifecycle of vendor access privileges?
a) By granting permanent access once a vendor is onboarded
b) Through regular reviews and adjustments of access rights as needed
c) By allowing vendors to determine their access levels
d) By using a single shared account for all vendors for simplicity

Answer: b) Through regular reviews and adjustments of access rights as needed. Explanation: Managing the lifecycle of vendor access privileges involves regular reviews and necessary adjustments to ensure that access rights remain appropriate to the vendor's role and responsibilities. This dynamic approach helps prevent access creep and reduces the risk of unauthorized access or data breaches.

What is a key consideration when integrating vendors into an organization's incident response plan?
a) Ensuring vendors have a detailed map of the organization's office
b) Defining clear roles, responsibilities, and communication protocols for vendors
c) Planning joint recreational activities to boost vendor morale
d) Focusing solely on incidents that impact the vendor directly

Answer: b) Defining clear roles, responsibilities, and communication protocols for vendors. Explanation: Integrating vendors into an organization's incident response plan requires defining clear roles, responsibilities, and communication protocols. This ensures that in the event of a security incident, both the organization and its vendors know exactly what actions to take, who to communicate with, and how to coordinate effectively to mitigate the impact.

What best practice should be followed when offboarding a vendor from an organization's systems?
a) Celebrating the vendor's departure with a party
b) Immediately revoking all access rights and credentials
c) Automatically transferring the vendor's access rights to a new vendor
d) Keeping the access active for an additional six months as a courtesy

Answer: b) Immediately revoking all access rights and credentials. Explanation: When offboarding a vendor, best practice dictates immediately revoking all their access rights and credentials to the organization's systems. This ensures that once the vendor's services are no longer required, they cannot access sensitive information or systems, thereby protecting against potential security risks.

In managing third-party IAM risks, what is the benefit of using federated identity management?
a) It allows unlimited access to all parties within the federation
b) It centralizes the management of identities across multiple systems and vendors, enhancing security and efficiency
c) It mandates that all vendors use the same passwords for simplicity
d) It eliminates the need for any form of authentication within the federation

Answer: b) It centralizes the management of identities across multiple systems and vendors, enhancing security and efficiency. Explanation: Federated identity management centralizes the management of user identities across multiple systems and vendors, allowing for more secure and efficient access control. By enabling users to access multiple applications with one set of credentials, it reduces password fatigue and streamlines the authentication process while maintaining high security standards.

What approach should be taken to assess a new vendor's security posture before granting them access to organizational systems?
a) Relying solely on the vendor's reputation
b) Conducting a thorough security assessment, including reviews of their policies, procedures, and compliance with relevant standards
c) Flipping a coin to decide on their trustworthiness
d) Basing the decision on the aesthetics of the vendor's website

Answer: b) Conducting a thorough security assessment, including reviews of their policies, procedures, and compliance with relevant standards. Explanation: Before granting a new vendor access to organizational systems, it is essential to conduct a thorough security assessment. This includes reviewing the vendor's security policies, procedures, incident response plans, and compliance with relevant standards and regulations. This due diligence ensures that the vendor adheres to the necessary security practices to protect the organization's data and systems.

In a Discretionary Access Control (DAC) system, who is primarily responsible for setting access policies for a resource?
a. The system administrator
b. The resource owner
c. A central security authority
d. All users equally

Answer: b. The resource owner. Explanation: In a Discretionary Access Control (DAC) system, the resource owner has the discretion to grant or restrict access to other users based on their own judgment. This model allows the owner of the resource, such as a file or a database record, to decide who can access it, making it flexible but potentially less secure if the owner does not strictly control access.

Which Access Control Model uses labels and clearances to make access decisions?
a. Role-Based Access Control (RBAC)
b. Mandatory Access Control (MAC)
c. Attribute-Based Access Control (ABAC)
d. Discretionary Access Control (DAC)

Answer: b. Mandatory Access Control (MAC). Explanation: Mandatory Access Control (MAC) is characterized by the use of labels (for resources) and clearances (for users) to make access decisions. In this model, access to resources is based on the information's classification and the user's clearance level, enforcing a strict policy that cannot be altered by users.

What is the main characteristic of Attribute-Based Access Control (ABAC)?
a. It controls access based on fixed roles assigned to users.
b. It allows access decisions to be based on a wide range of attributes, including user, resource, and environmental conditions.
c. It restricts access control management to the resource owner only.
d. It uses a centralized policy to enforce access decisions.

Answer: b. It allows access decisions to be based on a wide range of attributes, including user, resource, and environmental conditions. Explanation: Attribute-Based Access Control (ABAC) is distinguished by its ability to evaluate access based on a variety of attributes related to the user, the resource being accessed, and the current environmental context. This flexibility allows for highly dynamic and granular access control policies.

In which scenario is Role-Based Access Control (RBAC) most effective?
a. In environments where user roles frequently change
b. In high-security environments requiring data classification and clearance levels
c. In organizations with clearly defined job functions and responsibilities
d. In small teams where resource ownership is shared

Answer: c. In organizations with clearly defined job functions and responsibilities. Explanation: Role-Based Access Control (RBAC) is most effective in environments where organizations have clearly defined roles and responsibilities. This model simplifies access management by assigning permissions to roles rather than to individual users, making it easier to administer access rights, especially in larger organizations.

How does Mandatory Access Control (MAC) enhance security in high-security environments compared to Discretionary Access Control (DAC)?
a. By allowing users to freely share access with peers

b. By enforcing a centralized policy that does not allow individual users to alter permissions
c. By permitting resource owners to set their own access policies
d. By using role definitions to restrict access

Answer: b. By enforcing a centralized policy that does not allow individual users to alter permissions. Explanation: Mandatory Access Control (MAC) enhances security, particularly in high-security environments, by enforcing a centralized access control policy that does not permit individual users, including resource owners, to alter permissions. This ensures that access decisions are made based on predefined security classifications and user clearances, minimizing the risk of unauthorized access.

What advantage does ABAC offer over RBAC in complex and dynamic IT environments?
a. ABAC restricts access control decisions to predefined roles, simplifying management.
b. ABAC provides flexibility by considering context, such as time of access, in addition to user roles.
c. ABAC relies on resource owners to define access policies, enhancing user autonomy.
d. ABAC requires fewer resources for implementation and management.

Answer: b. ABAC provides flexibility by considering context, such as time of access, in addition to user roles. Explanation: Attribute-Based Access Control (ABAC) offers significant advantages in complex and dynamic IT environments by providing the flexibility to consider a multitude of attributes, including user characteristics, resource attributes, and contextual information such as time or location, in making access decisions. This allows for more nuanced and adaptable access control policies.

How does the principle of "least privilege" apply within the RBAC model?
a. By granting users the maximum permissions available within their role
b. By ensuring users are assigned to roles that provide only the access necessary to perform their duties
c. By automatically assigning every new user to the most restrictive role
d. By allowing users to self-assign roles based on their judgment

Answer: b. By ensuring users are assigned to roles that provide only the access necessary to perform their duties. Explanation: The principle of "least privilege" in the Role-Based Access Control (RBAC) model ensures that users are assigned only to those roles that provide the minimum level of access necessary to perform their job functions. This minimizes the potential for unauthorized access and reduces the risk of damage from user errors or malicious actions.

What is a primary security concern with the Discretionary Access Control (DAC) model?
a. It is too rigid for most modern computing environments.
b. It does not allow for the transfer of access rights.
c. It potentially allows for the propagation of excessive permissions due to its reliance on user discretion.
d. It requires a complex attribute management system.

Answer: c. It potentially allows for the propagation of excessive permissions due to its reliance on user discretion. Explanation: A primary security concern with the Discretionary Access Control (DAC) model is that it can lead to the propagation of excessive permissions. Since the model relies on the discretion of individual users, particularly resource owners, to manage access, it can result in overly permissive access rights if not carefully managed, increasing security risks.

In the context of ABAC, how can environmental conditions influence access control decisions?
a. Environmental conditions are irrelevant in the ABAC model.
b. They can restrict access based on factors like time of day or the accessing device's security posture.
c. They mandate that all access decisions are made during predefined environmental conditions only.
d. They ensure that all users have the same access level regardless of environmental conditions.

Answer: b. They can restrict access based on factors like time of day or the accessing device's security posture. Explanation: In Attribute-Based Access Control (ABAC), environmental conditions play a crucial role in access decisions. Factors such as the time of day, the location from which access is requested, or the security posture of the accessing device can influence whether access is granted or denied, allowing for dynamic and context-aware security policies.

Made in the USA
Middletown, DE
07 May 2024

54023561R00157